CHEMISTRY

Hans van Kessel
Bellerose Composite High School

Frank Jenkins
Ross Sheppard High School

Oliver Lantz
Harry Ainlay High School

Dick Tompkins
Old Scona Academic High School

Michael V. Falk
Harry Ainlay High School

Contributing Authors

Michael Dzwiniel
Harry Ainlay High School

George H. Klimiuk
(Retired)
McNally Composite High School

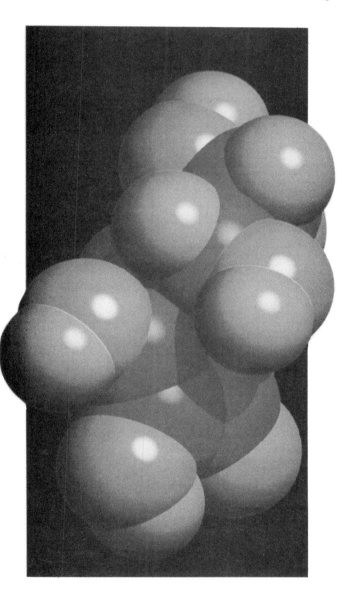

*Teacher's Resource
Masters*

Nelson Canada

I(T)P ™
International Thomson Publishing
The trademark ITP is used under license

© Nelson Canada,
A Division of Thomson Canada Limited, 1994

Published in 1994 by
Nelson Canada,
A Division of Thomson Canada Limited
1120 Birchmount Road
Scarborough, Ontario M1K 5G4

All investigations in the textbook and additional investigations in the Teacher's Resource Masters have been designed to be as safe as possible, and have been reviewed by professionals specifically for that purpose. As well, appropriate warnings concerning potential safety hazards are included where applicable to particular investigations. However, responsibility for safety remains with the student, the classroom teacher, the school principal, and the school board.

SAFETY REVIEW: **Margaret Redway**/Consultant, Fraser Science Awareness Inc.

ISBN 0-17-603975-9

Canadian Cataloguing in Publication Data

Main entry under title:
Nelson chemistry.   Teacher's resource masters
Supplement to: Jenkins, Frank, 1944 –   Nelson chemistry
ISBN   0-17-603975-9

1. Chemistry – Problems, exercises, etc.
I. Van Kessel, Hans.   II. Jenkins, Frank, 1944 –
Nelson chemistry.

QD33.J42 1994   540   C93-094526-3

Printed and Bound in Canada
1 2 3 4 5 6 7 8 9 WC 9 8 7 6 5 4

This book is printed on acid-free paper. The choice of paper reflects Nelson Canada's goal of using, within the publishing process, the available resources, technology, and suppliers that are as environment friendly as possible.

COVER:
Daryl Benson/Masterfile

Dr. E. Keller/Kristallographisches Institut der Universität Freiburg, Germany (inset photo)

CREDITS
Executive Editor: **Lynn Fisher**
Project Editor/Co-ordinator: **Winnie Siu**
Supervising Production Editor: **Cecilia Chan**

Art Director: **Bruce Bond**
Design: **Suzanne Peden**
Illustrations: **VISU*TronX***

THE *NELSON CHEMISTRY* PROGRAM
Nelson Chemistry
Nelson Chemistry, Teacher's Edition
Nelson Chemistry, Solutions Manual
Nelson Chemistry, Teacher's Resource Masters
Nelson Chemistry, Computer Test Bank

TABLE OF CONTENTS

ACKNOWLEDGMENTS

Nelson Chemistry represents the culmination of a curriculum project that was initiated and nurtured by dedicated classroom teachers over a long period of time. A dissatisfaction with existing chemistry materials in 1973 led to the development of the ALCHEM materials during the 1970s by Frank Jenkins, Dean Hunt, Dale Jackson, Dick Tompkins, Tom Mowat, Eugene Kuzub, Michael Dzwiniel, George Klimiuk, Myron Baziuk, Hans van Kessel, Oliver Lantz, and Mike Falk. The Edmonton Public School Board provided some initial funding but the project would not have survived and thrived were it not for the efforts of the chemistry teachers and students at Queen Elizabeth Composite High School, and also Gerry Mikytyshyn and his industrial arts students. Included in this extraordinary mix of people were Don Witwicky of Edmonton Public Schools, the professors in the University of Alberta Departments of Chemistry and Secondary Education, and chemical engineers in many Alberta industries.

Ten years after the start of ALCHEM, some of the same teacher-authors — Frank Jenkins, Hans van Kessel, Michael Dzwiniel, Dick Tompkins, Oliver Lantz, Mike Falk, and George Klimiuk — started the development of the STSC Chemistry project. Concepts and classroom strategies for chemistry education were broadened to include the nature of science, technology, STS issues, and communication as explicit curriculum emphases. Dr. Heidi Kass at the University of Alberta and Sharon Thomas at the Calgary Public School Board were both very helpful as we were evolving the STSC philosophy and materials. The students, pilot teachers, support staff, and administrators at several Edmonton-area high schools — Paul Kane, Queen Elizabeth, McNally, Strathcona, and, in particular, Harry Ainlay — greatly assisted the development and evaluation of this new direction in chemical education. Other pilot teachers and their students at schools like Forest Lawn and Central Memorial High Schools in Calgary also made significant contributions. The authors are very appreciative of the support by thousands of students and many teachers at the following schools and colleges who used the STSC Chemistry materials — Alberta Vocational Centres, Beaumont, Bellerose, Bonnie Doon, Bowness, Camrose, Central Memorial, Concordia College, County Central, Crescent Heights, Crowsnest, Delia, Ecole Mallaig, Fairview College, Father Lacombe, Forest Lawn, Frank Maddock, Harry Ainlay, Harry Collinge, Immanuel Christian, Jasper Place, J. Percy Page, J.T. Foster, Louis St. Laurent, Matthew Halton, McNally, Mistassiniy, Morrin, Oilfields, Old Scona, Paul Kane, Peace River, Provost, P.W. Kaesar, Queen Elizabeth (Calgary and Edmonton), Ross Sheppard, Savanna, Sir John Franklin, Spirit River, Spruce Grove, St. Joseph, Strathcona Christian Academy, Strathcona Composite, Sundre, William Aberhart, and others.

Karitann Publishers was started by Karen Jenkins, Rita Dzwiniel, and Anne van Kessel to publish the STSC Chemistry materials, and Karen has reliably distributed the materials to all participating schools for ten years. Liz Makaryshyn and Kevin Stevenson provided expert desktop publishing and artwork, respectively, while Harry Nichol at Quality Color Press has printed the textbooks for many years. Timely financial support was generously provided by the Government of Canada through the Canadian Studies Directorate and the Public Awareness Program for Science and Technology, and by the Alberta Foundation for the Literary Arts.

The excellent personnel at Nelson Canada, particularly Bill Allan and Lynn Fisher, believed in our philosophy and our project, and have done an outstanding job in the production of our materials. Thank you, as well, to all of the editors at Trifolium Press and to the reviewers for their contributions to the student materials. Most important to us is the unbelievable support from our families who have endured our "after-hours and holiday" work schedules for twenty years. Thank you, Karen, Anne, Marlene, Gerry, and Helen. Their contribution to our work and to chemistry education in general is greater than they realize.

Frank Jenkins, Hans van Kessel, Dick Tompkins, Mike Falk, Oliver Lantz

TRANSPARENCY MASTERS

Blackline masters are provided in this section of *Nelson Chemistry Teacher's Resource Masters* to:

- **increase the efficiency of presentation** by grouping examples and diagrams provided in the textbook, *Nelson Chemistry*. When you need to speed up a presentation, especially near the end of a class period or near exam time, transparencies can be very useful.

- **increase the achievement level of visual learners** by providing written examples and diagrams to support your oral presentation.

- **help organize and even extend ideas** by going beyond the textbook with more diagrams, flow charts, and summaries. You might also include concept maps prepared by students in this category.

We suggest that you prepare a transparency of each blackline master and keep the transparencies in a three-ring binder. Then make a note in your *Nelson Chemistry Teacher's Edition* where you have a related transparency available. Each transparency is numbered and has a page reference to the textbook. Here are some suggestions for the use of the transparencies and the use of the overhead projector.

- Always refer to the textbook page that the transparency accompanies.

- Use a colored (permanent ink) felt/flow pen or highlighter to write on the black and white transparency to aid your presentation.

- Use colored arrows, boxes, or circles to help draw students' attention to the important parts of the transparency.

- Stand beside the screen rather than the overhead projector (OHP) so that you do not block the students' view of the screen. Students can look at you and the screen simultaneously and you can easily point to the screen to emphasize and reinforce your oral presentation. Or, try to move back and forth between the OHP and the screen to help maintain eye contact with students during the use of the transparency.

- Keep sufficient light in the classroom so that students can still refer to *Nelson Chemistry* while viewing the transparency.

- Use your OHP also with transparencies that you prepare from pages in the Extra Practice section of this book, in the *Solutions Manual*, or in the *Test Bank* to help you increase the efficiency and effectiveness of your presentation.

- Use your OHP for viewing chemical reactions in petri dishes, in transparent spot plates, or in test tubes on a horizontal stage (e.g., Alyea) adaptor.

- Use your OHP with an LCD panel to view computer simulations or computer analysis of evidence gathered by students in the laboratory.

- Use your OHP with an LCD panel and electronic **probes attached** to a computer to collect and display evidence gathered during a demonstration, e.g., pH curves, temperature changes, and voltages of a cell.

View this section as providing the basis for a series of transparencies that would increase your range of instructional strategies to help you respond to the wide range of learning styles of your students.

MASTER LIST OF TRANSPARENCIES

1 EXPERIMENTAL WORK AND REPORTS

(page 18)

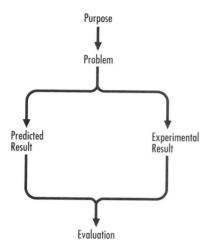

Many investigations involve comparing a predicted result with an experimental result.

(page 19 and page 522)

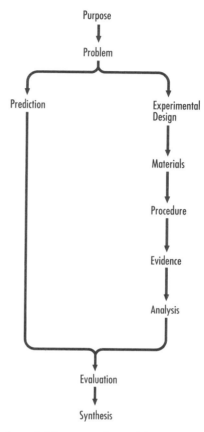

A model for scientific problem solving and reporting, based on the idea of concept-testing developed by Karl Popper.

(page 522)

> ✔ **Problem**
> ✔ **Prediction**
> ✔ **Design**
> ✔ **Materials**
> ✔ **Procedure**
> ✔ **Evidence**
> ✔ **Analysis**
> ✔ **Evaluation**
> ✔ **Synthesis**

Students are required to complete the sections of the report with check marks.

9

2 CLASSIFYING KNOWLEDGE

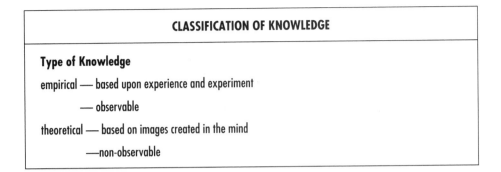

CLASSIFICATION OF KNOWLEDGE

Type of Knowledge

empirical — based upon experience and experiment

— observable

theoretical — based on images created in the mind

—non-observable

Parallel and Interacting Streams of Work

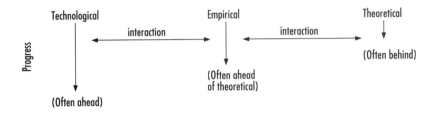

Classifying Scientific Knowledge

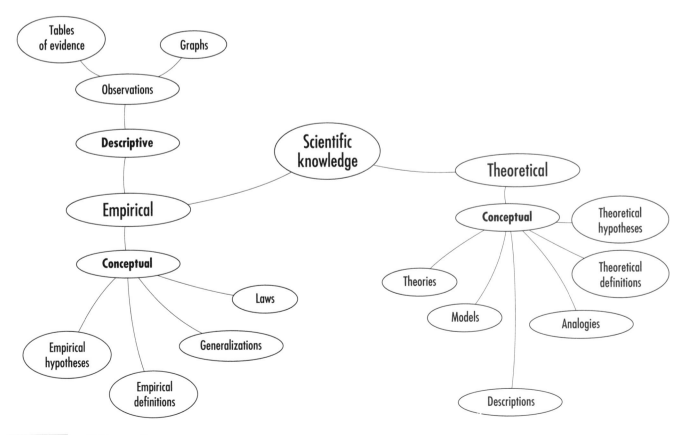

3   TABLES AND GRAPHS

(page 30)

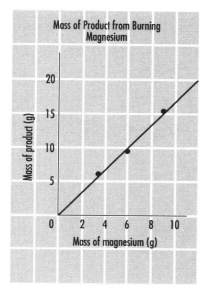

Figure 1.3
The relationship between the mass of magnesium that reacted and the mass of product obtained.

(page 30)

Table 1.2

MASS OF MAGNESIUM BURNED AND MASS OF ASH PRODUCED		
Trial	Mass of Magnesium (g)	Mass of Product (g)
1	3.6	6.0
2	6.0	9.9
3	9.1	15.1

(page 31)

Table 1.4

VARIABLES	
Type of Variables	**Definition**
manipulated variable (also called the independent variable)	the property that is systematically changed during an experiment
responding variable (also called the dependent variable)	the property that is measured as changes are made to the manipulated variable
controlled variable	a property that is kept constant throughout an experiment

(page 548)

Title for Table	
Manipulated variable (unit)	Responding variable (unit)
(data without units)	

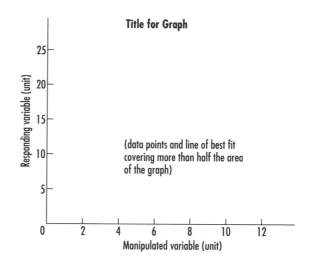

11

4   WORKPLACE HAZARDOUS MATERIAL INFORMATION SYSTEM SYMBOLS

(page 35)

WHMIS

Class A:
Compressed gas

Class B:
Flammable and combustible material

Class C:
Oxidizing material

Class D:
Poisonous and infectious material
1. Materials causing immediate and serious toxic effect

Class D:
2. Materials causing other toxic effects

Class D:
3. Biohazardous infectious material

Class E:
Corrosive material

Class F:
Dangerously reactive material

5 LABORATORY EQUIPMENT

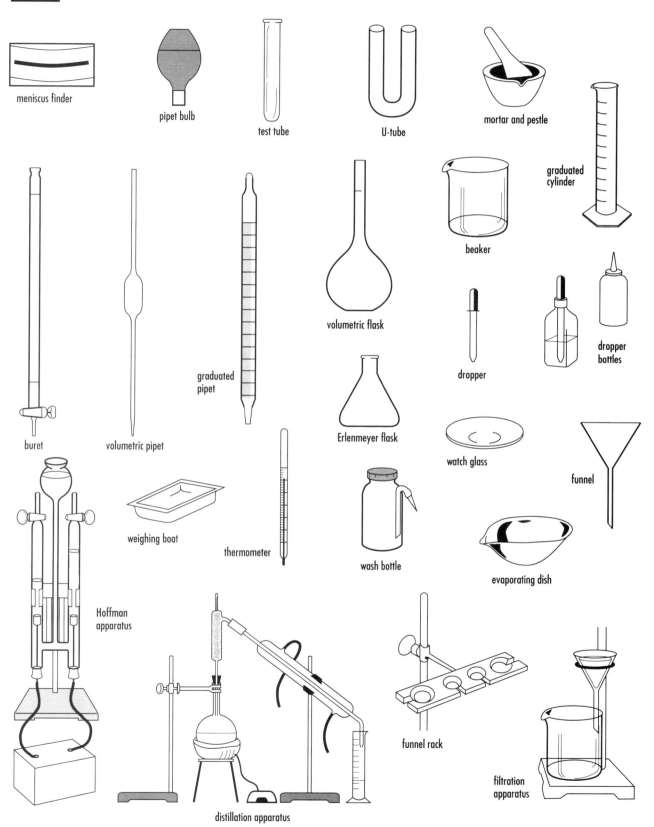

meniscus finder

pipet bulb

test tube

U-tube

mortar and pestle

graduated cylinder

beaker

buret

volumetric pipet

graduated pipet

volumetric flask

Erlenmeyer flask

dropper

dropper bottles

watch glass

funnel

Hoffman apparatus

weighing boat

thermometer

wash bottle

evaporating dish

funnel rack

filtration apparatus

distillation apparatus

6 CLASSIFYING MATTER

(page 36)

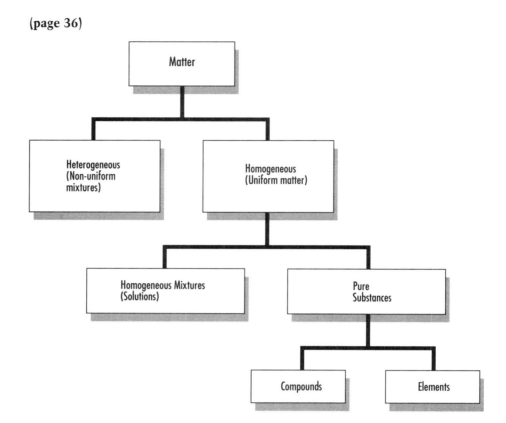

Figure 1.8
A classification of matter.

(page 36)

Table 1.5

DEFINITIONS OF ELEMENTS AND COMPOUNDS			
Substance	**Empirical Definition**	**Theoretical Definition**	**Examples**
element	substance that cannot be broken down chemically into simpler units by heat or electricity	substance composed of only one kind of atom	Mg (magnesium), O_2 (oxygen), C (carbon)
compound	substance that can be decomposed chemically by heat or electricity	substance composed of two or more kinds of atoms	H_2O (water), NaCl (table salt), $C_{12}H_{22}O_{11}$ (sugar)

7 LABORATORY BURNER

(See page 529 for the procedure of using a laboratory burner.)

(page 38)

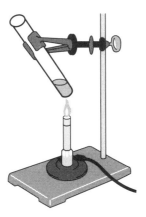

Figure 1.10
Heating substances. (a) An Erlenmeyer flask is used to funnel vapors. (b) A test tube is used when heating small quantities of a chemical. (c) A crucible is required when a substance must be heated strongly.

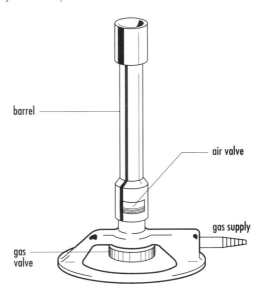

barrel

air valve

gas supply

gas valve

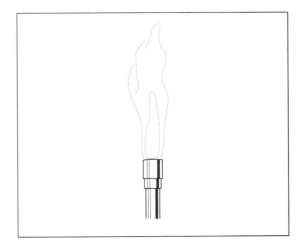

To light the burner, turn the gas on and the air off. A yellow flame is obtained. A yellow flame is a relatively cool flame and is easier to obtain than a blue flame when lighting a burner. A yellow flame is not used for heating objects because it contains a lot of black soot.

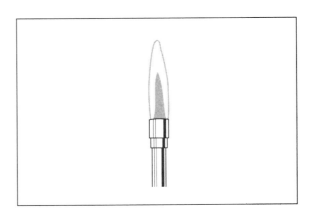

To get a hot flame, increase the air supply. A blue flame is obtained. A pale, almost invisible flame is much hotter than a yellow flame. The hottest point is at the tip of the inner blue cone.

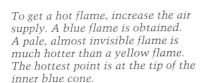

15

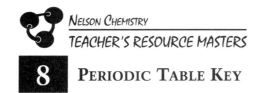

8 PERIODIC TABLE KEY

(page 53)

(theoretical)	**Key**	(empirical)

	26	55.85	atomic molar mass (g/mol)
atomic number	26	55.85	atomic molar mass (g/mol)
electronegativity	1.8	1535	melting point (°C)
common ion charge	3+	2750	boiling point (°C)
other ion charge	2+	7.87	density (g/cm³)

Fe

iron

element symbol
element name

density (g/cm^3)
density of gases (g/L)
gases in red
liquids in green
synthetic in blue

Figure 2.10
This key, which also appears on this book's inside front cover, helps you to determine the meaning of the numbers in the periodic table.

(page 83)

Figure 3.7
This information from the periodic table indicates that the most common ion formed from Fe atoms is Fe^{3+}. Some metals have more than two possible ion charges, but only the most common two are listed in the periodic table.

(theoretical)	**Key**	(empirical)
atomic number	26	55.85	atomic molar mass (g/mol)
electronegativity	1.8	1535	melting point (°C)
common ion charge	3+	2750	boiling point (°C)
other ion charge	2+	7.87	density (g/cm³)

Fe

iron

element symbol
element name

density (g/cm^3)
density of gases (g/L)
gases in red
liquids in green
synthetic in blue

9 CLASSIFYING ELEMENTS

(page 53)

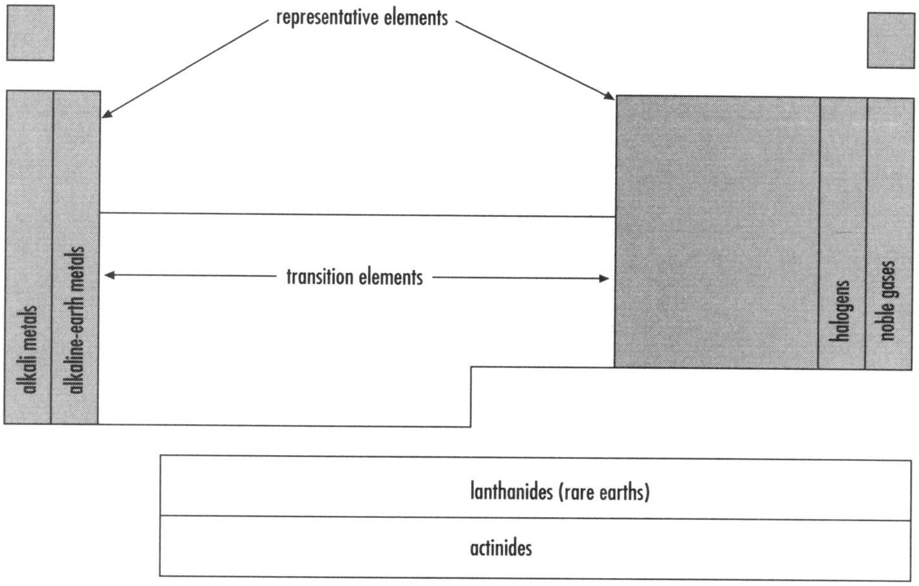

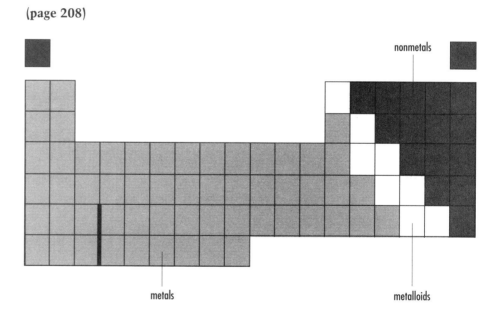

Figure 2.11
Commonly used names, identifying different sections of the periodic table. The representative elements are shown in green.

(page 208)

Figure 8.8
Metalloids occupy the region near the staircase line of the periodic table. This class of elements represents a revision of the classification of elements as either metals or nonmetals.

17

10 (A) ATOMIC THEORIES AND MODELS

Theory	**Model**	**Analogy**

1. Dalton's Theory
   - John Dalton
   - 1803
   - indivisible atoms

(featureless sphere)

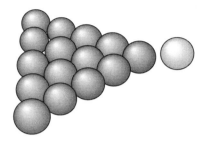

(billiard balls)

Figure 2.15(a) (page 58)

2. Thomson's Theory
   - J. J. Thomson
   - 1897+
   - positive sphere with imbedded electrons
   - net charge of zero

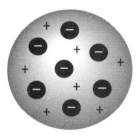

(uniform charge distribution)

(raisin bun)

Figure 2.15(b) (page 58)

3. Nagaoka's Theory
   - H. Nagaoka
   - 1904
   - positive sphere with a ring of electrons

(flat ring of electrons)

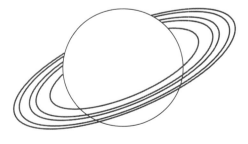

(Saturn)

10 (B) ATOMIC THEORIES AND MODELS

Theory	Model	Analogy

4. Rutherford's Theory
   - Ernest Rutherford
   - 1914
   - small positive nucleus surrounded by electrons

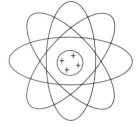

(nuclear model)

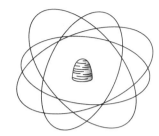

(beehive)

5. Bohr's Theory
   - Niels Bohr
   - 1921
   - magic numbers of electrons in quantized energy levels

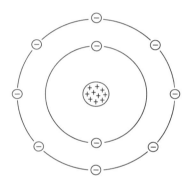

(nuclear model plus orbiting electrons)

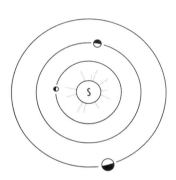

(planets orbiting around the sun)

6. Quantum Mechanics Theory
   - Erwin Schrödinger
   - 1926
   - electrons in probability orbitals (wave patterns)

Figure 2.14 (page 62)

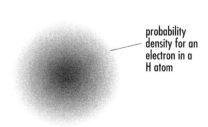

probability density for an electron in a H atom

(probability pattern of electrons)

(rotating fan blades)

11 RUTHERFORD SCATTERING EXPERIMENT

Thomson's Theory
Figure 2.15(a) (Page 58)

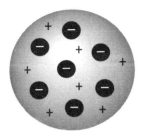

(page 58)

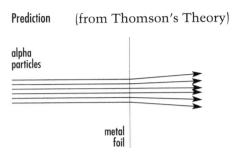

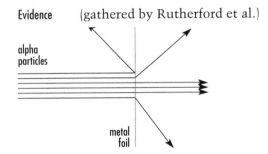

Figure 2.16
*Rutherford's experimental
observations were dramatically
different from what he had
expected.*

Synthesis of New Theory (Rutherford)

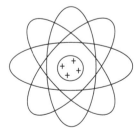

12 TESTING THE BOHR MODEL OF THE ATOM

**Evidence
(page 60)**

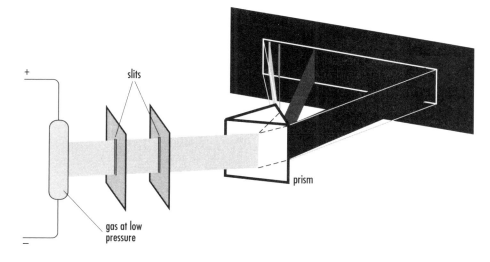

Figure 2.20
When electricity is passed through a gaseous element at low pressure, the gas emits light of only certain wavelengths, which can be seen if the light is passed through a prism. Every gas produces a unique pattern of colored lines, called a line spectrum.

**Explanation
(page 60)**

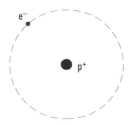

Figure 2.21
The Bohr model of a hydrogen atom in its lowest energy state includes the nucleus (one proton) and a single electron in the first orbit.

(page 61)

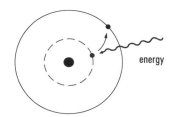

(a) An electron gains a quantum of energy.

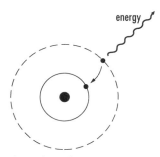

(b) An electron loses a quantum of energy.

Figure 2.22
*(a) Energy is absorbed, causing electrons to rise to a higher energy orbit.
(b) Energy is released as electrons fall to a lower energy orbit.*

13 CLASSIFYING COMPOUNDS

(page 73)

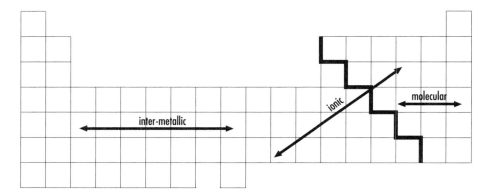

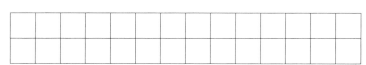

Figure 3.1
From two classes of elements there can be three classes of compounds. These classes of compounds are called ionic, molecular, and inter-metallic.

Intermetallic — metal + metal

Ionic — metal + nonmetal

Molecular — nonmetal + nonmetal

14 CLASSIFYING CHEMICAL REACTIONS

(page 101)

Table 4.4

CHEMICAL REACTIONS	
Reaction Type	**Generalization**
formation	elements $\rightarrow$ compound
simple decomposition	compound $\rightarrow$ elements
complete combustion	substance + oxygen $\rightarrow$ most common oxides
single replacement	element + compound $\rightarrow$ element + compound
double replacement	compound + compound $\rightarrow$ compound + compound

(pages 102-105)

1. **Formation Reactions**

   $2\ Mg_{(s)}\ +\ O_{2(g)}\ \rightarrow\ 2\ MgO_{(s)}$

2. **Simple Decomposition Reactions**

   $2\ H_2O_{(l)}\ \rightarrow\ 2\ H_{2(g)}\ +\ O_{2(g)}$

3. **Complete Combustion Reactions**

   $2\ C_4H_{10(g)}\ +\ 13\ O_{2(g)}\ \rightarrow\ 8\ CO_{2(g)}\ +\ 10\ H_2O_{(g)}$

4. **Single Replacement Reactions**

   $Cu_{(s)}\ +\ 2\ AgNO_{3(aq)}\ \rightarrow\ 2\ Ag_{(s)}\ +\ 10\ Cu(NO_3)_{2(aq)}$

   metal + compound $\rightarrow$ metal + compound

   $Cl_{2(g)}\ +\ 2NaI_{(aq)}\ \rightarrow\ I_{2(s)}\ +\ 2\ NaCl_{(aq)}$

   nonmetal + compound $\rightarrow$ nonmetal + compound

5. **Double Replacement Reactions**

   Precipitation
   $CaCl_{2(aq)}\ +\ Na_2CO_{3(aq)}\ \rightarrow\ CaCO_{3(s)}\ +\ 2\ NaCl_{(aq)}$

   Neutralization
   $HCl_{(aq)}\ +\ KOH_{(aq)}\ \rightarrow\ HOH_{(l)}\ +\ KCl_{(aq)}$

   acid + base $\rightarrow$ water + ionic compound
   (a salt)

15 CLASSIFYING MATTER AND SOLUBILITY

(page 113)

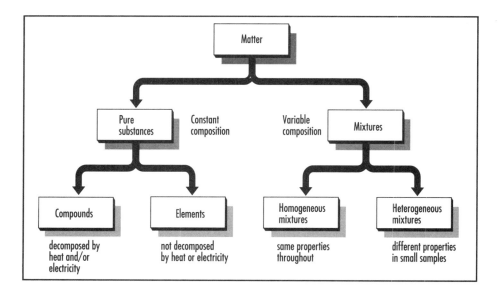

Figure 5.1
Matter is classified according to physical and chemical properties. This classification helps chemists to organize and communicate large quantities of knowledge about substances.

(inside back cover)

SOLUBILITY OF IONIC COMPOUNDS AT SATP – GENERALIZATIONS							
Anion	**Cl^-, Br^-, I^-**	**S^{2-}**	**OH^-**	**SO_4^{2-}**	**CO_3^{2-}, PO_4^{3-}, SO_3^{2-}**	**CH_3COO^-**	**NO_3^-**
High Solubility (aq) ≥ 0.1 mol/L (at SATP)	most	Group 1, NH_4^+ Group 2	Group 1, NH_4^+ Sr^{2+}, Ba^{2+}, Tl^+	most	Group 1, NH_4^+	most	all
Low Solubility (s) < 0.1 mol/L (at SATP)	Ag^+, Pb^{2+}, Tl^+, Hg_2^{2+} (Hg^+), Cu^+	most	most	Ag^+, Pb^{2+}, Ca^{2+}, Ba^{2+}, Sr^{2+}, Ra^{2+}	most	Ag^+	none
All Group 1 compounds, including acids, and all ammonium compounds, are assumed to have high solubility in water.							

24

CHAPTER 5   PAGE 113

16 DIAGNOSTIC TESTS

(page 537)

SOME STANDARD DIAGNOSTIC TESTS	
Substance Tested	**Diagnostic Test**
water	If cobalt(II) chloride paper is exposed to a liquid or vapor, and the paper turns from blue to pink, then water is likely present.
oxygen	If a glowing splint is inserted into the test tube, and the splint glows brighter or relights, then oxygen gas is likely present.
hydrogen	If a flame is inserted into the test tube, and a squeal or pop is heard, then hydrogen is likely present.
carbon dioxide	If the unknown gas is bubbled into a limewater solution, and the limewater turns cloudy, then carbon dioxide is likely present.
halogens	If a few millilitres of a chlorinated hydrocarbon solvent is added, with shaking, to a solution in a test tube, and the color of the solvent appears to be • light yellow-green, then chlorine is likely present. • orange, then bromine is likely present. • purple, then iodine is likely present.
acid	If strips of blue and red litmus paper are dipped into the solution, and the blue litmus turns red, then an acid is present.
base	If strips of blue and red litmus paper are dipped into the solution, and the red litmus turns blue, then a base is present.
neutral solution	If strips of blue and red litmus paper are dipped into the solution, and neither litmus changes color, then only neutral substances are likely present.
neutral ionic solution	If a neutral solution is tested for conductivity with a multimeter, and the solution conducts a current, then a neutral ionic substance is likely present.
neutral molecular solution	If a neutral solution is tested for conductivity with a multimeter, and the solution does not conduct a current, then a neutral molecular substance is likely present.
	There are thousands of diagnostic tests. You can create some of these, using data from the periodic table (inside front cover of this book); and from the data tables in Appendix F, pages 550 to 553, and on the inside back cover.

(inside back cover)

ION COLORS			
Ion	**Flame Color**	**Ion**	**Solution Color**
Li^+	bright red	Groups 1, 2, 17	colorless
Na^+	yellow	Cr^{2+}	blue
K^+	violet	Cr^{3+}	green
		Co^{2+}	pink
Ca^{2+}	yellow-red	Cu^+	green
Sr^{2+}	bright red	Cu^{2+}	blue
Ba^{2+}	yellow-green	Fe^{2+}	pale green
		Fe^{3+}	yellow-brown
Cu^{2+}	blue (halides)	Mn^{2+}	pale pink
	green (others)	Ni^{2+}	green
		CrO_4^{2-}	yellow
Pb^{2+}	light blue-grey	$Cr_2O_7^{2-}$	orange
Zn^{2+}	whitish green	MnO_4^-	purple

(Solubility can also be used to construct diagnostic tests.)

17 CLASSES OF SUBSTANCES IN SOLUTION

(page 117)

Empirical and Theoretical Definitions of Solutions

Table 5.1

ACIDS, BASES, AND NEUTRAL SUBSTANCES		
Type of Substance	**Empirical Definition**	**Theoretical Definition**
acids	• turn blue litmus red and are electrolytes • neutralize bases	• some hydrogen compounds ionize to produce $H^+_{(aq)}$ ions • $H^+_{(aq)}$ ions react with $OH^-_{(aq)}$ ions to produce water
bases	• turn red litmus blue and are electrolytes • neutralize acids	• ionic hydroxides dissociate to produce $OH^-_{(aq)}$ ions • $OH^-_{(aq)}$ ions react with $H^+_{(aq)}$ ions to produce water
neutral substances	• do not affect litmus • some are electrolytes • some are non-electrolytes	• no $H^+_{(aq)}$ or $OH^-_{(aq)}$ ions are formed • some are ions in solution • some are molecules in solution

(page 118)

Entities in a Water Environment

Table 5.2

MAJOR ENTITIES PRESENT IN A WATER ENVIRONMENT			
Type of Substance	**Solubility in Water**	**Typical Pure Substance**	**Major Entities Present when Substance is Placed in Water**
ionic compounds	high	$NaCl_{(s)}$	$Na^+_{(aq)}$, $Cl^-_{(aq)}$, $H_2O_{(l)}$
	low	$CaCO_{3(s)}$	$CaCO_{3(s)}$, $H_2O_{(l)}$
bases	high	$NaOH_{(s)}$	$Na^+_{(aq)}$, $OH^-_{(aq)}$, $H_2O_{(l)}$
	low	$Ca(OH)_{2(s)}$	$Ca(OH)_{2(s)}$, $H_2O_{(l)}$
molecular substances	high	$C_{12}H_{22}O_{11(s)}$	$C_{12}H_{22}O_{11(aq)}$, $H_2O_{(l)}$
	low	$C_8H_{18(l)}$	$C_8H_{18(l)}$, $H_2O_{(l)}$
strong acids	high	$HCl_{(g)}$	$H^+_{(aq)}$, $Cl^-_{(aq)}$, $H_2O_{(l)}$
weak acids	high	$CH_3COOH_{(l)}$	$CH_3COOH_{(aq)}$, $H_2O_{(l)}$
elements	low	$Cu_{(s)}$	$Cu_{(s)}$, $H_2O_{(l)}$
	low	$N_{2(g)}$	$N_{2(g)}$, $H_2O_{(l)}$

18 QUALITATIVE ANALYSIS

(page 122)

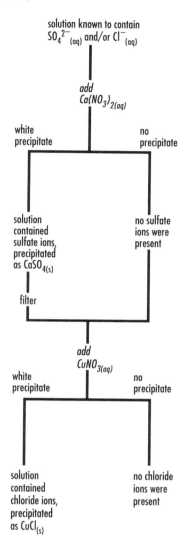

Figure 5.8
Analysis of a solution for sulfate and chloride ions.

19 FILTRATION

(See page 534, Figure C15 and page 535, Figure C17 for photos indicating how to use the filtration apparatus.)

The tip of the funnel should touch the inside wall of the collecting beaker.

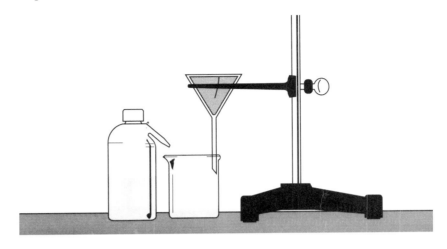

(page 534)

To prepare a filter paper, fold it in half twice and then remove the outside corner as shown.

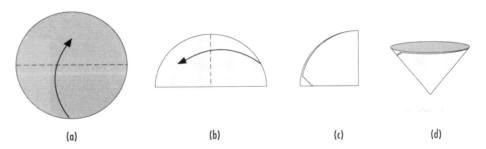

Figure C16
To prepare a filter paper, fold it in half twice and then remove an outside corner as shown.

(a)   (b)   (c)   (d)

The separation technique of pouring off clear liquid is called decanting. Pouring along the stirring rod prevents drops of liquid from going down the outside of the beaker when you stop pouring.

20 READING BALANCES

(See page 531, Figure C7 for photos indicating how to use a mechanical balance.)

Mechanical Balances

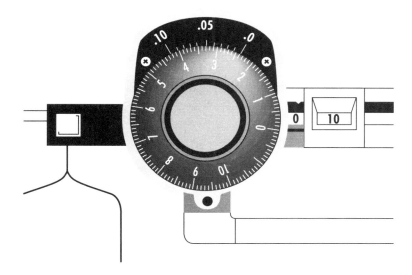

The dial reading on this balance with a vernier scale is 2.34 g. To read the hundredth of a gram, look below the zero on the vernier and then look for the line on the vernier that lines up best with a line on the dial.

(See page 530, Figure C6 for a photo of an electronic balance.)

Electronic Balances

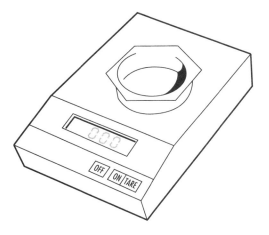

In many school laboratories, electronic balances that measure masses to within 0.01 g or 0.0001 g have replaced mechanical balances. These balances are efficient and produce reliable and reproducible measurements of mass. Push the TARE button to zero the balance.

21 pH SCALE

(page 131)

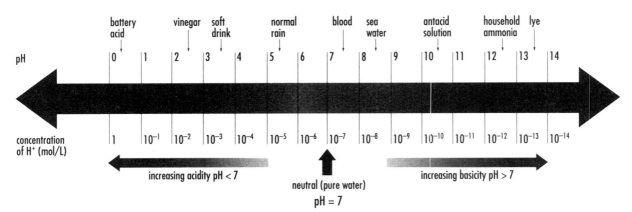

Figure 5.11
The pH scale can communicate a broad range of hydrogen ion concentrations, in a wide variety of substances.

22 PIPETTING

(See pages 532–533 for photos indicating how to use a volumetric pipet.)

When you use a pipet, you should take note of the following:

1. A volumetric pipet delivers the volume printed on the label if the temperature is near room temperature.

2. Squeeze the bulb in the palm of your hand and prepare to place your finger (not thumb) on the top of the pipet.

3. Completely release your fingers from the bulb. Your thumb placed across the top of the bulb maintains a good seal. Setting the pipet tip on the bottom slows the rise or fall of the liquid.

4. To allow the liquid to drop slowly to the calibration line, it is necessary for your finger and the pipet top to be dry. Rotate the pipet between your thumb and fingers. Also keep the tip on the bottom to slow down the flow. You may use a dispensing bulb if it is available.

5. A volumetric pipet drains best by gravity when held vertically. When the tip is then placed against the inside wall of the container at about a 45° angle, a small volume is expected to remain in the tip.

© NELSON CANADA,
A DIVISION OF THOMSON CANADA LIMITED, 1994

31

23 DYNAMIC EQUILIBRIUM

(page 138)

1. Both test tubes contain a saturated solution of iodine and excess iodine crystals.

2. Radioactive iodine crystals are added to one sample, and a saturated solution of radioactive iodine is added to the other sample.

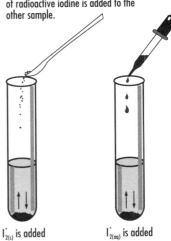

$I^*_{2(s)}$ is added $I^*_{2(aq)}$ is added

3. After a few minutes, the radioactivity is dispersed throughout the mixtures.

Figure 5.16
Radioactive iodine (I_2^*), added to a saturated solution of normal iodine (I_2), is eventually distributed throughout the mixture.

$I^*_{2(s)} \rightleftharpoons I^*_{2(aq)}$ $I^*_{2(aq)} \rightleftharpoons I^*_{2(s)}$

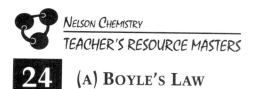

24   (A) BOYLE'S LAW

(page 145)

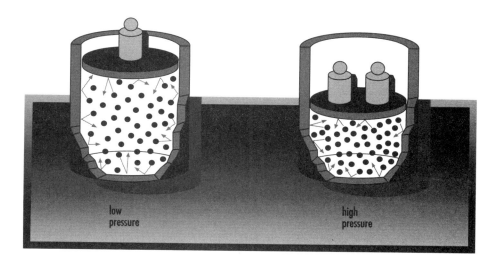

Figure 6.2
*As the pressure on a gas increases,
the volume of the gas decreases.*

If the initial volume (v_1) and the intial pressure (p_1) of a given amount of gas are changed to different values (v_2 and p_2), then

$$p_1v_1 = k$$

and $$p_2v_2 = k$$

therefore, $$p_1v_1 = p_2v_2 \qquad \text{(Boyle's Law)}$$

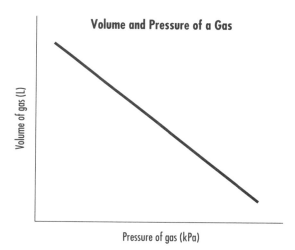

© NELSON CANADA,
A DIVISION OF THOMSON CANADA LIMITED, 1994

24 (B) CHARLES' LAW

(page 147)

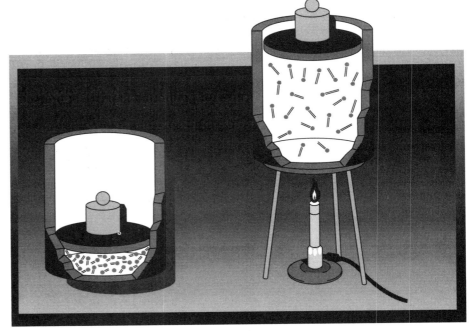

Figure 6.5
The volume of a gas in a container with a movable piston increases as the temperature of the gas increases.

(page 147)

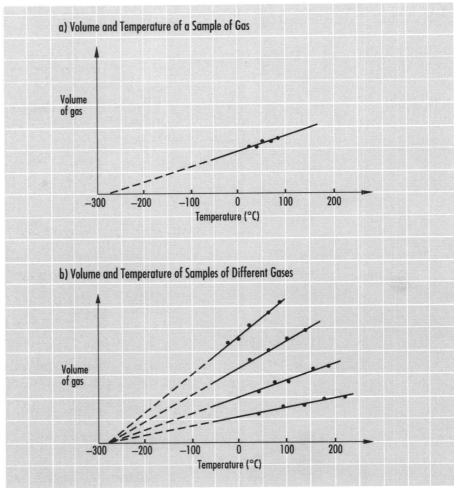

a) Volume and Temperature of a Sample of Gas

Volume of gas

Temperature (°C)
−300   −200   −100   0   100   200

b) Volume and Temperature of Samples of Different Gases

Volume of gas

Temperature (°C)
−300   −200   −100   0   100   200

Figure 6.6
When the graphs of several careful volume-temperature experiments are extrapolated, all the lines meet at absolute zero, −273°C.

25 GRAVIMETRIC STOICHIOMETRY

(pages 170–171)

EXAMPLE _____

Suppose that you decomposed 1.00 g of malachite. What mass of copper(II) oxide would be formed?

$$Cu(OH)_2 \cdot CuCO_{3(s)} \rightarrow 2\,CuO_{(s)} + CO_{2(g)} + H_2O_{(g)}$$

1.00 g m
221.13 g/mol   79.55 g/mol

$$n_{Cu(OH)_2 \cdot CuCO_3} = 1.00\ g \times \frac{1\ mol}{221.13\ g} = 0.004\ 52\ mol$$

$$n_{CuO} = 0.004\ 52\ mol \times \frac{2}{1} = 0.009\ 04\ mol$$

Always round off to the least number of significant digits when communicating a numerical result on paper. Never round off in your calculator.

$$m_{CuO} = 0.009\ 04\ mol \times \frac{79.55\ g}{1\ mol} = 0.719\ g$$

According to the stoichiometric method, the mass of copper(II) oxide produced by the decomposition of 1.00 g of malachite is 0.719 g.

or $m_{CuO} = 1.00\ g \times \dfrac{1\ mol}{221.13\ g} \times \dfrac{2}{1} \times \dfrac{79.55\ g}{1\ mol} = 0.719\ g$

EXAMPLE _____

Iron is the most widely used metal in North America (Figure 7.5). It may be produced by the reaction of iron(III) oxide, from iron ore, with carbon monoxide to produce iron metal and carbon dioxide. What mass of iron(III) oxide is required to produce 1000 g of iron?

$$Fe_2O_{3(s)} + 3\,CO_{(g)} \rightarrow 2\,Fe_{(s)} + 3\,CO_{2(g)}$$

m 1000 g
159.70 g/mol   55.85 g/mol

$$n_{Fe} = 1000\ g \times \frac{1\ mol}{55.85\ g} = 17.91\ mol$$

$$n_{Fe_2O_3} = 17.91\ mol \times \frac{1}{2} = 8.953\ mol$$

$$m_{Fe_2O_3} = 8.953\ mol \times \frac{159.70\ g}{1\ mol} = 1430\ g\ or\ 1.430\ kg$$

Based on the method of stoichiometry, the mass of iron(III) oxide that is required to produce 1000 g of iron is 1.430 kg.

or $m_{Fe_2O_3} = 1000 \times \dfrac{1\ mol}{55.85\ g} \times \dfrac{1}{2} \times \dfrac{159.70\ g}{1\ mol} = 1430\ g$

26 GAS VOLUME STOICHIOMETRY

EXAMPLE ———————————————————————————

Hydrogen gas is produced when sodium metal is added to water. What mass of sodium is necessary to produce 20.0 L of hydrogen at SATP?

$$2\,Na_{(s)} \;+\; 2\,HOH_{(l)} \;\;\rightarrow\;\; H_{2(g)} \;+\; 2\,NaOH_{(aq)}$$

$$\begin{array}{ll} m & 20.0\ L \\ 22.99\ g/mol & 24.8\ L/mol \end{array}$$

$$n_{H_2} \;=\; 20.0\ L \;\times\; \frac{1\ mol}{24.8\ L} \;=\; 0.806\ mol$$

$$n_{Na} \;=\; 0.806\ mol \;\times\; \frac{2}{1} \;=\; 1.61\ mol$$

$$m_{Na} \;=\; 1.61\ mol \;\times\; \frac{22.99\ g}{1\ mol} \;=\; 37.1\ g$$

According to the method of stoichiometry, the mass of sodium required to produce 20.0 L of hydrogen gas is 37.1 g.

or $\;\; m_{Na} \;=\; 20.0\ L \times \dfrac{1\ mol}{24.8\ L} \;\times\; \dfrac{2}{1} \;\times\; \dfrac{22.99\ g}{1\ mol} \;=\; 37.1\ g$

(page 177)

EXAMPLE ———————————————————————————

In an industrial application known as the Haber process (page 443), ammonia to be used as fertilizer results from the reaction of nitrogen and hydrogen. What volume of ammonia at 450 kPa pressure and 80°C can be obtained from the complete reaction of 7.5 kg of hydroge?

$$N_{2(g)} \;+\; 3\,H_{2(g)} \;\;\rightarrow\;\; 2\,NH_{3(g)}$$

$$\begin{array}{ll} & 7.5\ kg & v \\ & 2.02\ g/mol & 450\ kPa,\ 80°C \end{array}$$

$$n_{H_2} \;=\; 7.5\ kg \;\times\; \frac{1\ mol}{2.02\ g} \;=\; 3.7\ kmol$$

$$n_{NH_3} \;=\; 3.7\ kmol \;\times\; \frac{2}{3} \;=\; 2.5\ kmol$$

$$v_{NH_3} \;=\; \frac{nRT}{p} \;=\; \frac{2.5\ kmol \;\times\; \dfrac{8.31\ kPa\cdot L}{mol\cdot K} \;\times\; 353\ K}{450\ kPa}$$

$$=\; 16\ kL$$

According to the stoichiometric method, the volume of ammonia produced from 7.5 kg of hydrogen is 16 kL.

or $\;\; v_{NH_3} \;=\; 7.5\ kg \;\times\; \dfrac{1\ mol}{2.02\ g} \;\times\; \dfrac{2}{3} \;\times\; \dfrac{8.31\ kPa\cdot L}{1\ mol\cdot K} \;\times\; \dfrac{353\ K}{450\ kPa} \;=\; 16\ kL$

27 SOLUTION (VOLUMETRIC) STOICHIOMETRY

(page 180)

EXAMPLE ——————————————————————

Solutions of ammonia and phosphoric acid are used to produce ammonium hydrogen phosphate fertilizer (Figure 7.12). What volume of 14.8 mol/L $NH_{3(aq)}$ is needed for the ammonia to react completely with 1.00 ML of 12.9 mol/L $H_3PO_{4(aq)}$ to produce fertilizer?

$$2\,NH_{3(aq)} \quad + \quad H_3PO_{4(aq)} \quad \rightarrow \quad (NH_4)_2HPO_{4(aq)}$$
$$v \qquad\qquad\qquad 1.00\,ML$$
$$14.8\,mol/L \qquad 12.9\,mol/L$$

$$n_{H_3PO_4} = 1.00\,ML \times \frac{12.9\,mol}{1\,L} = 12.9\,Mmol$$

$$n_{NH_3} = 12.9\,Mmol \times \frac{2}{1} = 25.8\,Mmol$$

$$v_{NH_3} = 25.8\,Mmol \times \frac{1\,L}{14.8\,mol} = 1.74\,ML$$

According to the stoichiometric method, the required volume of ammonia solution is 1.74 ML.

or $\quad v_{NH_3} = 1.00\,ML \times \dfrac{12.9\,mol}{1\,L} \times \dfrac{2}{1} \times \dfrac{1\,L}{14.8\,mol} = 1.74\,ML$

(page 181)

EXAMPLE ——————————————————————

As part of a chemical analysis, a technician determines the concentration of a sulfuric acid solution. In the experiment, a 10.00 mL sample of sulfuric acid reacts completely with 15.9 mL of 0.150 mol/L potassium hydroxide. Calculate the concentration of the sulfuric acid.

$$H_2SO_{4(aq)} \quad + \quad 2\,KOH_{(aq)} \quad \rightarrow \quad 2\,HOH_{(1)} + K_2SO_{4(aq)}$$
$$10.00\,mL \qquad 15.9\,mL$$
$$C \qquad\qquad\quad 0.150\,mol/L$$

$$n_{KOH} = 15.9\,mL \times \frac{0.150\,mol}{1\,L} = 2.39\,mmol$$

$$n_{H_2SO_4} = 2.39\,mmol \times \frac{1}{2} = 1.19\,mmol$$

$$C_{H_2SO_4} = \frac{1.19\,mmol}{10.00\,mL} = 0.119\,mol/L$$

According to the evidence gathered and the stoichiometric method, the molar concentration of the sulfuric acid is 0.119 mol/L.

or $\quad C_{H_2SO_4} = 15.9\,mL \times \dfrac{0.150\,mol}{1\,L} \times \dfrac{1}{2} \times \dfrac{1}{10.00\,mL} = 0.119\,mol/L$

28 STOICHIOMETRY

(page 181)

SUMMARY: GRAVIMETRIC, GAS, AND SOLUTION STOICHIOMETRY

Step 1: Write a balanced chemical equation and list the measurements and conversion factors for the given and required substances.

Step 2: Convert the given measurement to an amount in moles by using the appropriate conversion factor.

Step 3: Calculate the amount of the required substance by using the mole ratio from the balanced equation.

Step 4: Convert the calculated amount to the final required quantity by using the appropriate conversion factor or the ideal gas law.

EXERCISE

23. Ammonium sulfate fertilizer is manufactured by having sulfuric acid react with ammonia. In a laboratory study of this process, 50.0 mL of sulfuric acid reacts with 24.4 mL of a 2.20 mol/L ammonia solution to produce the ammonium sulfate solution. From this evidence, calculate the concentration of the sulfuric acid at this stage in the process.

(page 68 of *Nelson Chemistry Solutions Manual*)

$$H_2SO_{4(aq)} + 2NH_{3(aq)} \rightarrow (NH_4)_2SO_{4(aq)}$$

50.0 mL   24.4 mL

C 2.20 mol/L

$$n_{NH_3} = 24.4 \text{ mL} \times \frac{2.20 \text{ mol}}{1 \text{ L}} = 53.7 \text{ mmol}$$

$$n_{H_2SO_4} = 53.7 \text{ mmol} \times \frac{1}{2} = 26.8 \text{ mmol}$$

$$C_{H_2SO_4} = \frac{26.8 \text{ mmol}}{50.0 \text{ mL}} = 0.537 \text{ mol/L}$$

The concentration of sulfuric acid is 0.537 mol/L according to the method of stoichiometry.

or $C_{H_2SO_4} = 24.4 \text{ mL} \times \dfrac{2.20 \text{ mol}}{1 \text{ L}} \times \dfrac{1}{2} \times \dfrac{1}{50.0 \text{ mL}} = 0.537 \text{ mol/L}$

38

CHAPTER 7   PAGE 181

29 THE SOLVAY PROCESS: A CASE STUDY

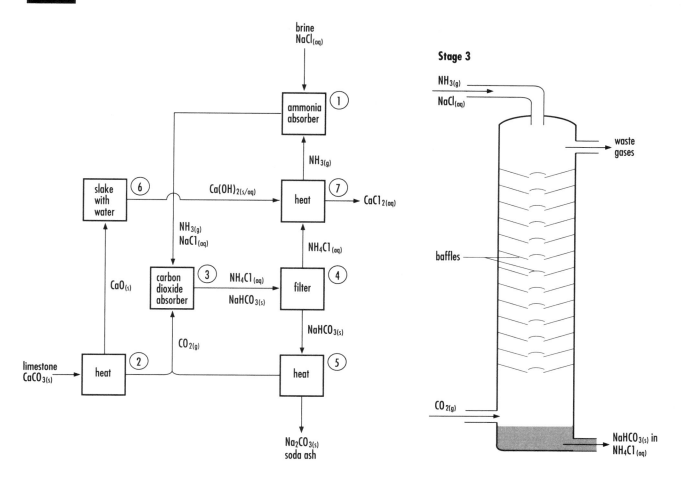

Stage Reactions

Stage Reactions

2 $\quad CaCO_{3(s)} \rightarrow CaO_{(s)} + CO_{2(g)}$

1, 3 $\quad 2\ CO_{2(g)} + 2\ NH_{3(g)} + 2\ H_2O_{(l)} \rightarrow 2\ NH_4HCO_{3(aq)}$

3, 4 $\quad 2\ NH_4HCO_{3(aq)} + 2\ NaCl_{(aq)} \rightarrow 2\ NH_4Cl_{(aq)} + 2\ NaHCO_{3(s)}$ (intermediate)

5 $\quad 2\ NaHCO_{3(s)} \rightarrow Na_2CO_{3(s)} + H_2O_{(g)} + CO_{2(g)}$

6 $\quad CaO_{(s)} + H_2O_{(l)} \rightarrow Ca(OH)_{2(s)}$

7 $\quad Ca(OH)_{2(s)} + 2\ NH_4Cl_{(aq)} \rightarrow 2\ NH_{3(g)} + CaCl_{2(aq)} + 2\ H_2O_{(l)}$

Net $\quad CaCO_{3(s)} + 2\ NaCl_{(aq)} \rightarrow Na_2CO_{3(aq)} + CaCl_{2(aq)}$
$\qquad\qquad$ raw materials $\rightarrow$ primary product $+$ by-product

30 SOLUTION PREPARATION FROM SOLID

(See Investigation 7.5, page 186 and page 535 for the procedure.)

1. **Calculation (pages 185–186)**

$$n_{Na_2CO_3} = 250.0 \text{ mL} \times \frac{0.100 \text{ mol}}{1 \text{L}} = 25.0 \text{ mmol}$$

$$m_{Na_2CO_3} = 25.0 \text{ mmol} \times \frac{105.99 \text{ g}}{1 \text{ mol}}$$

2. **Mass measurement**

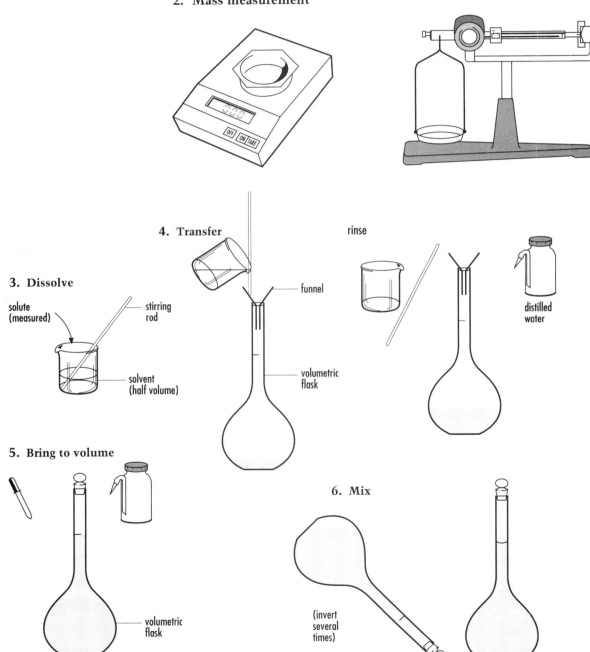

3. **Dissolve**

solute (measured)
stirring rod
solvent (half volume)

4. **Transfer**

rinse
funnel
volumetric flask
distilled water

5. **Bring to volume**

volumetric flask

6. **Mix**

(invert several times)

31 SOLUTION PREPARATION BY DILUTION

(See Investigation 7.6, page 188)

1. Calculate the volume of the concentrated reagent required.

$$n_i = n_f$$
$$v_i C_i = v_f C_f$$
$$v_i \times 0.500 \text{ mol/L} = 100.0 \text{ mL} \times 0.05000 \text{ mol/L}$$
$$v_i = 10.0 \text{ mL}$$

2.

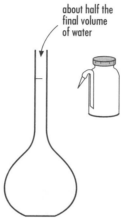

about half the final volume of water

Add approximately one-half of the final volume of pure water to the volumetric flask.

3.

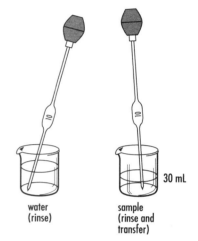

water (rinse)

sample (rinse and transfer)

30 mL

Measure the required volume of stock solution using a pipet. (Refer to "Using a pipet" on page 532.)

4.

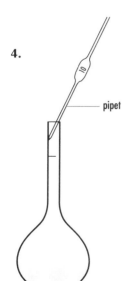

pipet

Transfer the stock solution slowly into the volumetric flask — while mixing if heat is generated.

5.

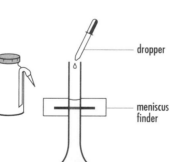

dropper

meniscus finder

volumetric flask

Add pure water using a wash bottle and then a medicine dropper and a meniscus finder until the bottom of the meniscus is on the calibration line.

6.

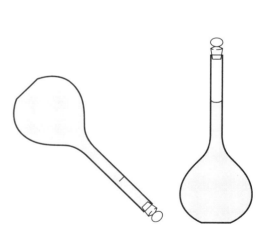

Stopper and mix the solution by slowly inverting the flask several times.

32 VOLUMETRIC EQUIPMENT — TITRATION

(See page 536 for the description of how a buret is used in titration.)

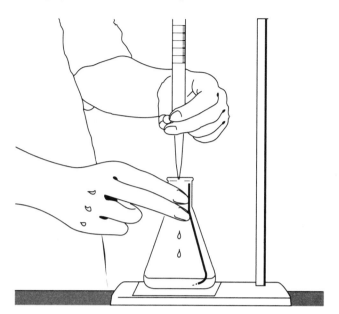

Near the endpoint, continuous gentle swirling of the solution is particularly important.

(page 191)

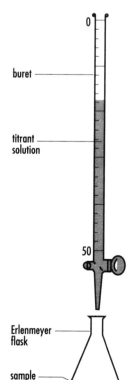

buret

titrant solution

50

Erlenmeyer flask

sample solution

Figure 7.21
An initial reading of volume is made on the buret before any titrant is added to the sample solution. Then titrant is added until the reaction is complete; that is, when a drop of titrant changes the color of the sample. The final buret reading is then taken. The difference in buret readings is the volume of titrant added.

33 SOLUTION PREPARATION AND TITRATION

Preparation

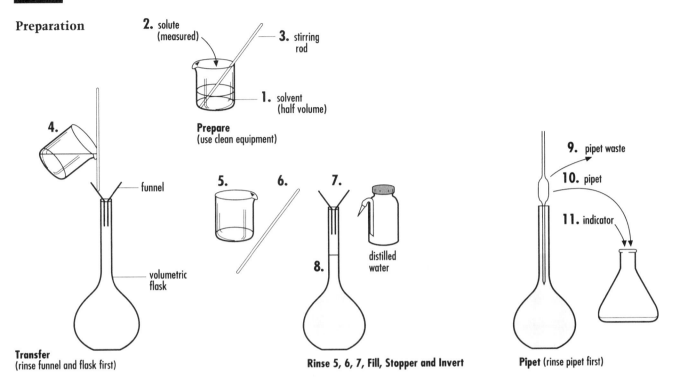

2. solute (measured)

3. stirring rod

1. solvent (half volume)

Prepare (use clean equipment)

4.

funnel

volumetric flask

Transfer (rinse funnel and flask first)

5.   **6.**   **7.**

distilled water

8.

Rinse 5, 6, 7, Fill, Stopper and Invert

9. pipet waste

10. pipet

11. indicator

Pipet (rinse pipet first)

Titration

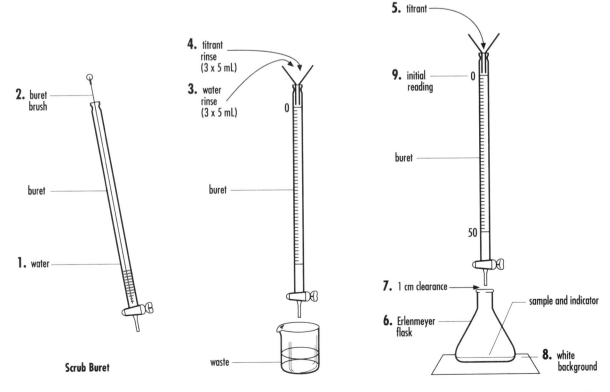

2. buret brush

buret

1. water

Scrub Buret

4. titrant rinse (3 x 5 mL)

3. water rinse (3 x 5 mL)

0

buret

waste

5. titrant

9. initial reading

0

buret

50

7. 1 cm clearance

6. Erlenmeyer flask

sample and indicator

8. white background

43

34 BONDING IN ELEMENTS

TRENDS IN EMPIRICAL PROPERTIES OF NON-TRANSITION ELEMENTS								
	Group Number							
Property	1	2	13	14	15	16	17	18
Appearance	shiny, silver		shiny, grey					dull, colored
Physical state	solids		solids			solids, liquids, gases		
Mechanical strength	flexible		brittle					brittle
Electrical conductivity	high		slight					none
Structure	continuous compact, close packed			continuous open, tetrahedral		individual particles various structures		
Heat of vaporization	relatively high				highest			very low

←————— metals —————→ ←—— metalloids ——→ ←— nonmetals —→

Metallic

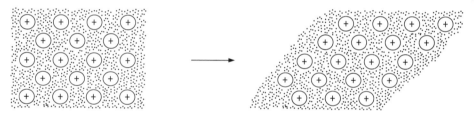

Electron Sea Model of a metal showing a deforming of the crystal shape to explain the flexible character of metals.

Network Covalent
(page 209)

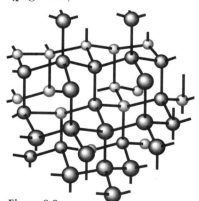

Figure 8.9
Metalloids like silicon have a diamond-like structure. The structure of diamond, shown here, is theoretically described as a network of covalently bonded atoms forming a macromolecule the size of the crystal itself.

Molecular

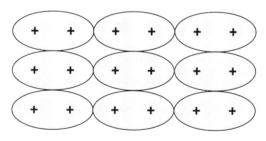

London (dispersion) forces explain the boiling points of molecular elements. The protons in the nucleus of one molecule attract the electrons in an adjacent molecule.

35 EMPIRICAL FORMULAS

(page 215)

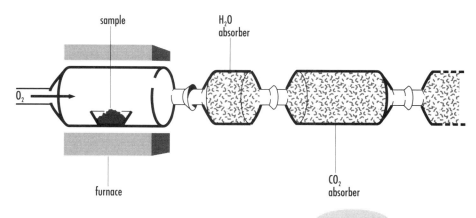

Figure 8.15
A substance burned in a combustion analyzer produces oxides that are captured by absorbers in chemical traps. The initial and final masses of each trap indicate the masses of the oxides produced. These masses are then used in the calculation of the percentage composition of the substance burned.

(page 217)

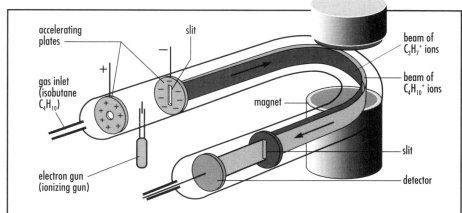

Figure 8.16
A mass spectrometer is used to determine the masses of ionized particles by measuring the amount of deflection in the path of the particles as they pass through a magnetic field.

(page 217–218)

EXAMPLE

Complete the Analysis of the investigation report.

Problem

What is the molecular formula of the fluid in a butane lighter?

Evidence

From combustion analysis: percent by mass of carbon = 82.5%
percent by mass of hydrogen = 17.5%

From mass spectrometry: molar mass = 58 g/mol

Analysis

Assume one mole (58 g) of the compound is analyzed.

$$m_C = \frac{82.5}{100} \times 58 \text{ g} = 48 \text{ g}, \quad n_C = 48 \text{ g} \times \frac{1 \text{ mol}}{12.01 \text{ g}} = 4.0 \text{ mol}$$

$$m_H = \frac{17.5}{100} \times 58 \text{ g} = 10 \text{ g}, \quad n_H = 10 \text{ g} \times \frac{1 \text{ mol}}{1.01 \text{ g}} = 10 \text{ mol}$$

The mole ratio of carbon atoms to hydrogen atoms in the compound analyzed is 4:10.

According to the evidence gathered in this investigation, the empirical molecular formula of the fluid in a butane lighter is C_4H_{10}.

36 CLASSES OF COMPOUNDS AND ACCOMPANYING BONDING

(page 73)

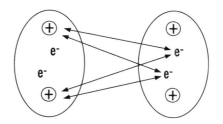

Figure 3.1
From two classes of elements there can be three classes of compounds. These classes of compounds are called ionic, molecular, and inter-metallic.

Ionic (page 78)

Figure 3.4
To agree with the explanation of the empirical formula for sodium chloride, the model of a sodium chloride crystal must represent both a 1:1 ratio of ions and the shape of the salt crystal.

Intermolecular Forces
- London forces
- dipole-dipole forces
- hydrogen bonding

Molecular: London

London (dispersion) forces help to explain the boiling points of all molecular substances.

Molecular: Dipole-Dipole (page 227)

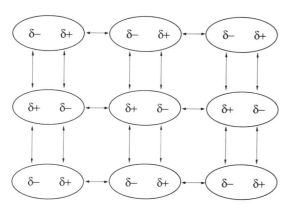

Figure 8.26
A polar molecule is simultaneously attracted to all the other polar molecules around it.

Molecular: Hydrogen (page 229)

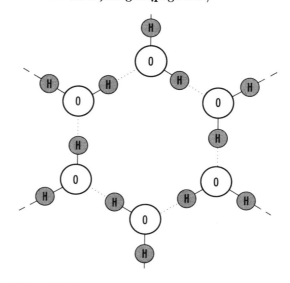

Figure 8.28
In ice, hydrogen bonds between the molecules result in a regular hexagonal crystal structure.

37 POLAR MOLECULES

(page 228)

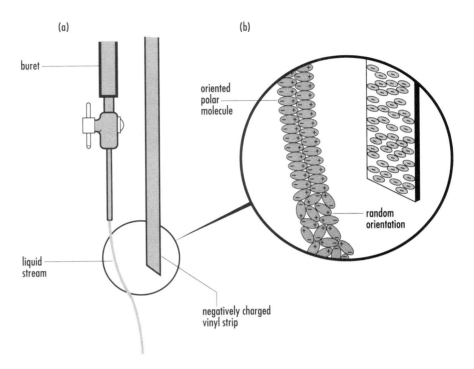

(a)

(b)

Figure 8.27
(a) Testing a liquid with a charged strip provides evidence for the existence of polar molecules in a substance.
(b) Polar molecules in a liquid become oriented so that their positive poles are closer to a negatively charged material. Near a positively charged material they become oriented in the opposite direction. Polar molecules are thus attracted by either kind of charge.

(page 227)

H — Cl

δ^+ δ^-

Figure 8.25
In an HCl molecule, the electrons are pulled more strongly to the chlorine end, resulting in a polar covalent bond. In diatomic molecules such as this, a polar bond causes the molecule itself to be polar.

Rules for Predicting Polarity of Molecules

CLASSIFICATION OF POLAR SUBSTANCES	
AB	binary molecular compounds such as $HCl_{(g)}$ and $CO_{(g)}$
N_xA_y	molecular compounds of nitrogen such as $NH_{3(g)}$
O_xA_y	molecular compounds of oxygen such as $H_2O_{2(l)}$ and $OCl_{2(g)}$ where oxygen is a central atom
$C_xA_yB_z$	molecular compounds of carbon with two different nonmetals such as $CHCl_{3(l)}$ and $C_2H_5OH_{(l)}$

Empirically, substances not affected by electric charges are called non-polar. Theoretically, these substances are believed to be composed of non-polar molecules.

CLASSIFICATION OF NON-POLAR SUBSTANCES	
A_x	molecular elements such as $Cl_{2(g)}$ and $P_{4(s)}$
C_xA_y	molecular compounds of carbon and one nonmetal such as $CO_{2(g)}$, $CCl_{4(g)}$, and $CH_{4(g)}$ and other hydrocarbons

38 FAMILIES OF ORGANIC COMPOUNDS

(page 240)

Table 9.1

FAMILIES OF ORGANIC COMPOUNDS		
Family Name	**General Formula**	**Example**
alkanes	$-\overset{\mid}{\underset{\mid}{C}}-\overset{\mid}{\underset{\mid}{C}}-$	propane, $CH_3-CH_2-CH_3$
alkenes	$-\overset{\mid}{C}=\overset{\mid}{C}-$	propene (propylene), $CH_2=CH-CH_3$
alkynes	$-C\equiv C-$	propyne, $CH\equiv C-CH_3$
aromatics	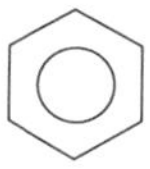 ⬡	toluene, 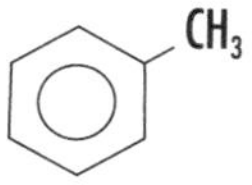 ⬡$-CH_3$
organic halides	$R-X$	chloropropane, $CH_3-CH_2-CH_2-Cl$
alcohols	$R-OH$	propanol, $CH_3-CH_2-CH_2-OH$
carboxylic acids	$R(H)-\overset{O}{\overset{\|}{C}}-OH$	propanoic acid, $CH_3-CH_2-\overset{O}{\overset{\|}{C}}-OH$
aldehydes	$R(H)-\overset{O}{\overset{\|}{C}}-H$	propanal, $CH_3-CH_2-\overset{O}{\overset{\|}{C}}-H$
ketones	$R_1-\overset{O}{\overset{\|}{C}}-R_2$	propanone (acetone), $CH_3-\overset{O}{\overset{\|}{C}}-CH_3$
esters	$R_1(H)-\overset{O}{\overset{\|}{C}}-O-R_2$	methyl ethanoate (methyl acetate), $CH_3-\overset{O}{\overset{\|}{C}}-O-CH_3$
amines	$R_1-\overset{R_2(H)}{\overset{\mid}{N}}-R_3(H)$	propylamine, $CH_3-CH_2-CH_2-\overset{H}{\overset{\mid}{N}}-H$
amides	$R_1(H)-\overset{O}{\overset{\|}{C}}-\overset{R_2(H)}{\overset{\mid}{N}}-R_3(H)$	propanamide, $CH_3-CH_2-\overset{O}{\overset{\|}{C}}-\overset{H}{\overset{\mid}{N}}-H$

39 FRACTIONAL DISTILLATION

(page 242)

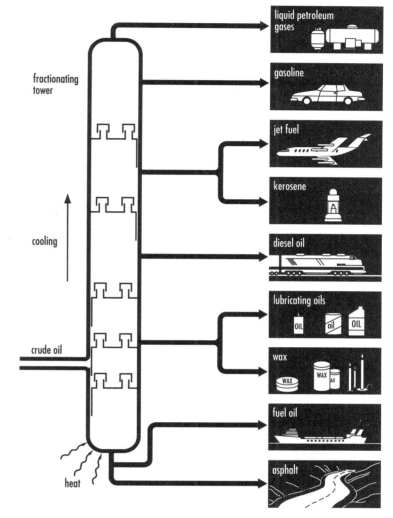

Figure 9.8
A fractional distillation tower contains trays positioned at various levels. Heated crude oil enters near the bottom of the tower. The bottom of the tower is kept hot, and the temperature gradually decreases towards the top of the tower. The lower the boiling point of a fraction, the higher the tray on which it condenses.

Laboratory Distillation Apparatus

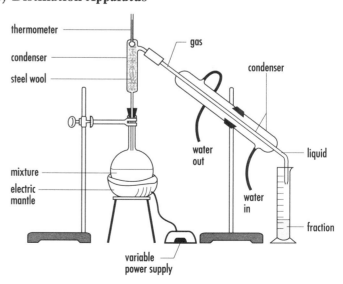

40 PERSPECTIVES ON THE USE OF FOSSIL FUELS

(page 256)

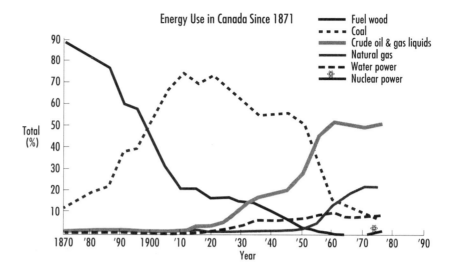

Energy Use in Canada Since 1871

— Fuel wood
···· Coal
— Crude oil & gas liquids
— Natural gas
-- Water power
✸ Nuclear power

(page 256)

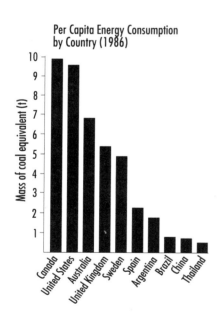

Per Capita Energy Consumption by Country (1986)

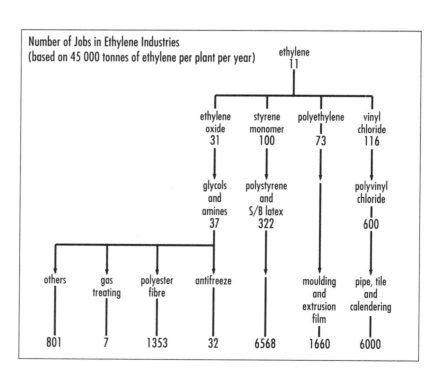

Number of Jobs in Ethylene Industries
(based on 45 000 tonnes of ethylene per plant per year)

41 CLASSES OF ORGANIC REACTIONS

SUMMARY: ORGANIC REACTIONS OF HYDROCARBONS

Cracking

large molecules $\xrightarrow{\text{catalyst}}$ smaller molecules

$$C_{17}H_{36(l)} \xrightarrow{\text{heat}} C_9H_{20(l)} + C_8H_{16(l)}$$

Reforming

small molecule(s) $\xrightarrow{\text{catalyst}}$ larger molecule with more branches

$$C_5H_{12(l)} + C_5H_{12(l)} \xrightarrow{\text{heat}} C_{10}H_{22(l)} + H_{2(g)}$$

Complete Combustion

compound + $O_{2(g)} \rightarrow$ most common oxides

$$2C_8H_{18(l)} + 25\,O_{2(g)} \rightarrow 16\,CO_{2(g)} + 18\,H_2O_{(g)} + \text{energy}$$

Addition (hydrogenation)

alkene or alkyne + $H_{2(g)} \rightarrow$ alkane

$$CH_2{=}CH{-}CH_2{-}CH_3 + H{-}H \rightarrow CH_3{-}CH_2{-}CH_2{-}CH_3$$

SUMMARY: ORGANIC REACTIONS OF HYDROCARBON DERIVATIVES

Addition

alkene/alkyne + small molecules $\rightarrow$
                              hydrocarbon or hydrocarbon derivative

$$CH_2{=}CH{-}CH_3 + H{-}Cl \rightarrow$$
$$CH_2Cl{-}CH_2{-}CH_3 + CH_3{-}CHCl{-}CH_3$$

Substitution

alkane/aromatic + halogen $\xrightarrow{\text{light}}$ organic halide + hydrogen halide

$$CH_3{-}CH_2{-}CH_3 + Br{-}Br \rightarrow$$
$$CH_2Br{-}CH_2{-}CH_3 + CH_3{-}CHBr{-}CH_3 + H{-}Br$$

Elimination

alkyl halide + $OH^- \rightarrow$ alkene + water + halide ion

and   alcohol $\xrightarrow{\text{acid}}$ alkene + water

$$CH_3{-}CH_2{-}CH_2Br + OH^- \rightarrow CH_2{=}CH{-}CH_3 + H{-}O{-}H + Br^-$$

Condensation (esterification)

carboxylic acid + alcohol $\rightarrow$ ester + water

$$CH_3{-}COOH + HO{-}CH_3 \rightarrow CH_3{-}COO{-}CH_3 + H{-}O{-}H$$

42 HEATING CURVES

(page 291)

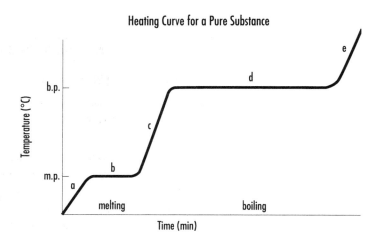

Heating Curve for a Pure Substance

Figure 10.8
*As a pure substance is heated, the following changes can occur: **a** – an increase in the temperature of the solid (q); **b** – a solid to liquid phase change (ΔH); **c** – an increase in the temperature of the liquid (q); **d** – a liquid to gas phase change (ΔH); and **e** – an increase in temperature of the gas (q).*

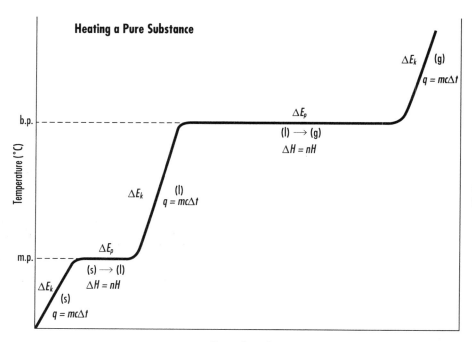

Heating a Pure Substance

43 PHASE CHANGES

Energy Changes
(page 292)

Table 10.2

COMPARISON OF ENERGY CHANGES		
Energy Change	**Empirical Evidence**	**Theoretical Explanation**
flow of heat, q	a temperature change with no change in state or chemicals	ΔE_k as a result of an increase or decrease in the speed of the particles
phase change, ΔH	exothermic or endothermic change forming a new state of matter	ΔE_P as a result of changes in the intermolecular bonds among particles
chemical reaction, ΔH	exothermic or endothermic change forming new chemical substances	ΔE_P as a result of changes in the ionic or covalent bonds among ions or atoms
nuclear reaction, ΔH	exothermic or endothermic change forming new elements or subatomic particles	ΔE_P as a result of changes in the nuclear bonds among nuclear particles (nucleons)

(page 293)

Figure 10.11
A solid absorbs energy as it changes to a liquid. A liquid in turn absorbs additional energy to change to a gas. If the direction of the phase changes is reversed, energy is released.

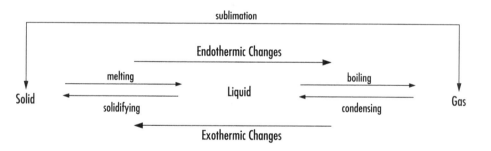

(page 293)

Table 10.3

MOLAR ENTHALPIES OF PHASE CHANGES FOR SELECTED SUBSTANCES			
Chemical Name	**Formula**	**Molar Enthalpy of Fusion (kJ/mol)**	**Molar Enthalpy of Vaporization (kJ/mol)**
sodium	Na	2.6	101
chlorine	Cl_2	6.40	20.4
sodium chloride	NaCl	28	171
water	H_2O	6.03	40.8
ammonia	NH_3	–	1.37
freon-12	CCl_2F_2	–	34.99
ethylene glycol	$C_2H_4(OH)_2$	–	58.8
sodium sulfate-10-water	$Na_2SO_4 \cdot 10H_2O$	78.0*	–

* This value represents molar enthalpy of solution (see page 301).

44 EXOTHERMIC AND ENDOTHERMIC CHANGE

(page 318)

Figure 11.5
This figure shows potential energy diagrams for (a) exothermic and (b) endothermic chemical changes. A potential energy diagram represents a balanced chemical equation with the reactants and products positioned at different values on the vertical energy scale. The horizontal axis represents the progress of the reaction.

Figure 11.6
The standard molar enthalpy of formation for magnesium oxide is obtained from the data table in Appendix F (page 551). Since this formation is observed to be exothermic, the reactants must have a higher potential energy than the product.

Figure 11.7
The standard molar enthalpy of the decomposition of water is obtained by reversing the sign of the standard molar enthalpy of formtion of water. Since this reaction is endothermic, the reactant (water) must have a lower potential energy than the products (hydrogen and oxygen).

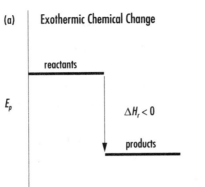

(a) Exothermic Chemical Change

E_p

reactants

$\Delta H_r < 0$

products

Reaction progress

(b) Endothermic Chemical Change

E_p

products

$\Delta H_r > 0$

reactants

Reaction progress

The Formation of Magnesium Oxide

E_p (kJ)

$Mg_{(s)} + \frac{1}{2} O_{2(g)}$

$\Delta H_f^\circ = -601.6$ kJ

$MgO_{(s)}$

Reaction progress

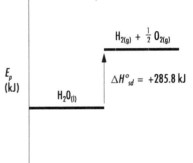

The Simple Decomposition of Liquid Water

E_p (kJ)

$H_{2(g)} + \frac{1}{2} O_{2(g)}$

$\Delta H_{sd}^\circ = +285.8$ kJ

$H_2O_{(l)}$

Reaction progress

SUMMARY: FOUR WAYS OF COMMUNICATING ENERGY CHANGES

	Exothermic Changes	Endothermic Changes
1. Molar Enthalpy	$H < 0$	$H > 0$
2. Enthalpy Change	reactants $\rightarrow$ products; $\Delta H < 0$	reactants $\rightarrow$ products; $\Delta H > 0$
3. Term in a Balanced Equation	reactants $\rightarrow$ products + energy	reactants + energy $\rightarrow$ products
4. Potential Energy Diagram	E_p (reactants) > E_p (products)	E_p (reactants) < E_p (products)

Examples:

1. $H_f^\circ = -601.6$ kJ/mol
   MgO

   $H_{sd}^\circ = +285.8$ kJ/mol
   H_2O

2. $Mg_{(s)} + \frac{1}{2} O_{2(g)} \rightarrow MgO_{(s)}$ $\Delta H = -601.6$ kJ

   $H_2O_{(l)} \rightarrow H_{2(g)} + \frac{1}{2} O_{2(g)}$ $\Delta H = +285.8$ kJ

3. $Mg_{(s)} + \frac{1}{2} O_{2(g)} \rightarrow MgO_{(s)} + 601.6$ kJ

   $H_2O_{(l)} + 285.8$ kJ $\rightarrow H_{2(g)} + \frac{1}{2} O_{2(g)}$

4. (as above)

45   HESS'S LAW

(pages 322–323)

SUMMARY: ENTHALPY OF REACTION AND HESS'S LAW

To determine an enthalpy change of a reaction by using Hess's law, follow these steps:

1. Write the net reaction equation, if it is not given.
2. Manipulate the given equations so they will add to yield the net equation.
3. Cancel and add the remaining reactants and products.
4. Add the component enthalpy changes to obtain the net enthalpy change.
5. Determine the molar enthalpy, if required.

EXAMPLE

Problem

What is the standard molar enthalpy of formation of butane?

Experimental Design

Since the formation of butane cannot be determined calorimetrically, Hess's law is chosen as the method to obtain the value of the standard molar enthalpy of formation.

Evidence

The following values were determined by calorimetry.

$$(1)\quad C_4H_{10(g)} + \tfrac{13}{2}O_{2(g)} \rightarrow 4\,CO_{2(g)} + 5\,H_2O_{(g)} \qquad \Delta H_c^\circ = -2657.4\text{ kJ}$$

$$(2)\quad C_{(s)} + O_{2(g)} \rightarrow CO_{2(g)} \qquad \Delta H_f^\circ = -393.5\text{ kJ}$$

$$(3)\quad 2\,H_{2(g)} + O_{2(g)} \rightarrow 2\,H_2O_{(g)} \qquad \Delta H_f^\circ = -483.6\text{ kJ}$$

Analysis

$$4\,CO_{2(g)} + 5\,H_2O_{(g)} \rightarrow C_4H_{10(g)} + \tfrac{13}{2}O_{2(g)} \qquad \Delta H^\circ = +2657.4\text{ kJ}$$

$$4\,C_{(s)} + 4\,O_{2(g)} \rightarrow 4\,CO_{2(g)} \qquad \Delta H^\circ = -1574.0\text{ kJ}$$

$$5\,H_{2(g)} + \tfrac{5}{2}O_{(g)} \rightarrow 5\,H_2O_{(g)} \qquad \Delta H^\circ = -1209.0\text{ kJ}$$

Net $\quad 4\,C_{(s)} + 5\,H_{2(g)} \rightarrow C_4H_{10(g)} \qquad \Delta H^\circ = -125.6\text{ kJ}$

$$H_f^\circ{}_{C_4H_{10}} = \frac{\Delta H_f^\circ}{n} = \frac{-125.6\text{ kJ}}{1\text{ mol}} = -125.6\text{ kJ/mol}$$

According to the evidence gathered and Hess's law, the standard molar enthalpy of formation of butane is –125.6 kJ/mol.

46 MOLAR ENTHALPIES OF FORMATION

(page 328)

SUMMARY: USING ENTHALPIES OF FORMATION TO PREDICT ΔH_r

According to Hess's law, the net enthalpy change for a chemical reaction is equal to the sum of the enthalpies of formation of the products minus the sum of the enthalpies of formation of the reactants.

$$\Delta H_r = \Sigma n H_{fp} - \Sigma n H_{fr}$$

(page 328)

EXAMPLE ———————————————————————————

What is the standard molar enthalpy for the slaking of lime?

$$CaO_{(s)} + H_2O_{(l)} \rightarrow Ca(OH)_{2(s)} \qquad\qquad \Delta H°_r = ?$$

$$
\begin{aligned}
\Delta H°_r &= \underset{Ca(OH)_2}{\Sigma n H°_{fp}} - \underset{CaO + H_2O}{\Sigma n H°_{fr}} \\[4pt]
&= \underset{Ca(OH)_2}{n H°_f} - (\underset{CaO}{n H°_f} + \underset{H_2O}{n H°_f}) \\[4pt]
&= -986.1 \text{ kJ} - (-920.7 \text{ kJ}) \\[4pt]
&= -65.4 \text{ kJ}
\end{aligned}
$$

According to Hess's law and empirically determined molar enthalpies of formation, the standard enthalpy change for the slaking of lime is reported as follows.

$$CaO_{(s)} + H_2O_{(l)} \rightarrow Ca(OH)_{2(s)} \qquad\qquad \Delta H°_r = -65.4 \text{ kJ}$$

Therefore, the $H°_r$ for $Ca(OH)_2$ in this reaction is –65.4 kJ/mol.

(page 329)

EXAMPLE ———————————————————————————

What is the standard molar enthalpy of combustion of methane fuel?

$$CH_{4(g)} + 2\,O_{2(g)} \rightarrow CO_{2(g)} + 2\,H_2O_{(g)}$$

$$
\begin{aligned}
\Delta H°_c &= \Sigma n H°_{fp} - \Sigma n H°_{fr} \\[4pt]
&= \left(1 \text{ mol} \times -393.5\,\frac{kJ}{mol} + 2 \text{ mol} \times -241.8\,\frac{kJ}{mol}\right) \\[4pt]
&\quad - \left(1 \text{ mol} \times -74.4\,\frac{kJ}{mol} + 2 \text{ mol} \times 0\,\frac{kJ}{mol}\right) \\[4pt]
&= -877.1 \text{ kJ} - (-74.4 \text{ kJ}) \\[4pt]
&= -802.7 \text{ kJ}
\end{aligned}
$$

$$
\begin{aligned}
\underset{CH_4}{H°_c} &= \frac{\Delta H°_c}{n} \\[4pt]
&= \frac{-802.7 \text{ kJ}}{1 \text{ mol}} = -802.7 \text{ kJ/mol}
\end{aligned}
$$

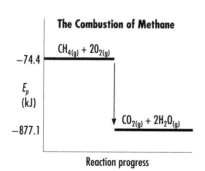

The Combustion of Methane

$CH_{4(g)} + 2O_{2(g)}$

–74.4

E_p (kJ)

$CO_{2(g)} + 2H_2O_{(g)}$

–877.1

Reaction progress

47 NUCLEAR FISSION AND FUSION

**Fission
(page 338)**

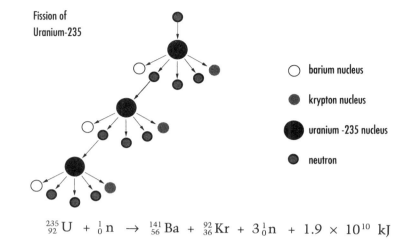

Fission of
Uranium-235

○ barium nucleus

● krypton nucleus

● uranium -235 nucleus

● neutron

(page 332)

$$^{235}_{92}U + ^1_0n \rightarrow ^{141}_{56}Ba + ^{92}_{36}Kr + 3^1_0n + 1.9 \times 10^{10} \text{ kJ}$$

**Fusion
(page 331)**

(See Figure 11.18)
Direct solar radiation provides the energy required for green plants to produce food and oxygen daily. Indirectly, solar energy is also the source of energy from winds, water, and fossil fuels. According to current theory, fossil fuels are the remains of plants and animals that originally depended on sunlight for energy. Fossil fuels are therefore considered a stored form

Figure 11.19
A potential energy diagram of a nuclear fusion reaction that is used in the research and development of nuclear fusion reactors.

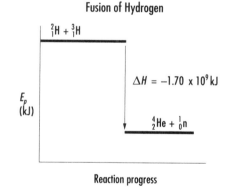

Fusion of Hydrogen

$^2_1H + ^3_1H$

E_p
(kJ)

$\Delta H = -1.70 \times 10^9 \text{ kJ}$

$^4_2He + ^1_0n$

Reaction progress

For convenience when comparing enthalpy changes, scientific notation is combined with the SI prefix, k, in kJ.

$$^2_1H + ^3_1H \rightarrow ^4_2He + ^1_0n \qquad\qquad \Delta H = -1.70 \times 10^9 \text{ kJ}$$

$$^2_1H + ^3_1H \rightarrow ^4_2He + ^1_0n + 1.70 \times 10^9 \text{ kJ}$$

48 REDUCTION AND OXIDATION

(page 347)

The Development of Metallurgical Processes

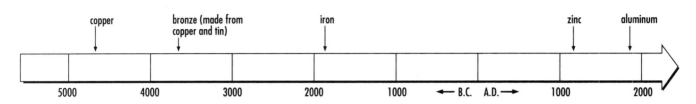

Figure 12.1
The technology of metallurgy has a long history, preceding by thousands of years the scientific understanding of the processes.

(page 347)

Reduction

ore + reducing agent $\rightarrow$ metal + . . .

$$Fe_2O_{3(s)} + 3\,CO_{(g)} \rightarrow 2\,Fe_{(s)} + 3\,CO_{2(g)}$$
$$SnO_{2(s)} + C_{(s)} \rightarrow Sn_{(s)} + CO_{2(g)}$$
$$CuS_{(s)} + H_{2(g)} \rightarrow Cu_{(s)} + H_2S_{(g)}$$

(page 348)

Oxidation

metal + oxidizing agent $\rightarrow$ ionic compound

$$2\,Mg_{(s)} + O_{2(g)} \rightarrow 2\,MgO_{(s)}$$
$$2\,Al_{(s)} + 3\,Cl_{2(g)} \rightarrow 2\,AlCl_{3(s)}$$
$$Cu_{(s)} + Br_{2(g)} \rightarrow CuBr_{2(s)}$$

49 Redox Table Building

Design 1: *Evidence*

INVESTIGATION 12.2 REACTIONS OF METALS AND METAL IONS				
	$Cu_{(s)}$	$Pb_{(s)}$	$Ag_{(s)}$	$Zn_{(s)}$
$Cu^{2+}_{(aq)}$	—	√	—	√
$Pb^{2+}_{(aq)}$	—	—	—	√
$Ag^{+}_{(aq)}$	√	√	—	√
$Zn^{2+}_{(aq)}$	—	—	—	—

Analysis

(page 354)

Table 12.1

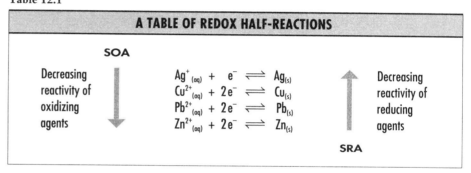

A TABLE OF REDOX HALF-REACTIONS

SOA

Decreasing reactivity of oxidizing agents

$$Ag^{+}_{(aq)} + e^{-} \rightleftharpoons Ag_{(s)}$$
$$Cu^{2+}_{(aq)} + 2\,e^{-} \rightleftharpoons Cu_{(s)}$$
$$Pb^{2+}_{(aq)} + 2\,e^{-} \rightleftharpoons Pb_{(s)}$$
$$Zn^{2+}_{(aq)} + 2\,e^{-} \rightleftharpoons Zn_{(s)}$$

Decreasing reactivity of reducing agents

SRA

(page 357)

OA   **RA**

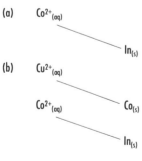

(a) $Co^{2+}_{(aq)}$ — $In_{(s)}$

(b) $Cu^{2+}_{(aq)}$ — $Co_{(s)}$

$Co^{2+}_{(aq)}$ — $In_{(s)}$

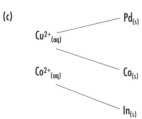

(c) $Cu^{2+}_{(aq)}$ — $Pd_{(s)}$

$Co^{2+}_{(aq)}$ — $Co_{(s)}$

— $In_{(s)}$

Figure 12.9
The relative position of a pair of oxidizing and reducing agents indicates whether a reaction will be spontaneous or not.

Figure 12.8
The redox spontaneity rule.

(page 355)

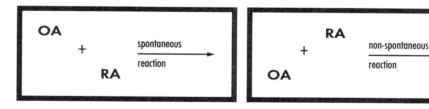

Design 2: *Evidence*
(page 357)

$$3\,Co^{2+}_{(aq)} + 2\,In_{(s)} \rightarrow 2\,In^{3+}_{(aq)} + 3\,Co_{(s)}$$
$$Cu^{2+}_{(aq)} + Co_{(s)} \rightarrow Co^{2+}_{(aq)} + Cu_{(s)}$$
$$Cu^{2+}_{(aq)} + Pd_{(s)} \rightarrow \text{no evidence of reaction}$$

Analysis
(page 357)

SOA $Pd^{2+}_{(aq)} + 2\,e^{-} \rightleftharpoons Pd_{(s)}$

$Cu^{2+}_{(aq)} + 2\,e^{-} \rightleftharpoons Cu_{(s)}$

$Co^{2+}_{(aq)} + 2\,e^{-} \rightleftharpoons Co_{(s)}$

$In^{3+}_{(aq)} + 3\,e^{-} \rightleftharpoons In_{(s)}$ SRA

50 PREDICTING REDOX REACTIONS

(page 361)

Table 12.2

HINTS FOR LISTING AND LABELLING ENTITIES
• Aqueous solutions contain $H_2O_{(l)}$ molecules.
• Acidic solutions contain $H^+_{(aq)}$ ions.
• Basic solutions contain $OH^-_{(aq)}$ ions.
• Some oxidizing or reducing agents are combinations; for example, the combination of $MnO_4^-_{(aq)}$ and $H^+_{(aq)}$.
• $H_2O_{(l)}$, $Fe^{2+}_{(aq)}$, and $Sn^{2+}_{(aq)}$ may act as either oxidizing or reducing agents.

EXAMPLE

$$\overset{OA}{K^+_{(aq)}} \quad \overset{OA}{MnO_4^-_{(aq)}} \overset{OA}{H^+_{(aq)}} \quad \underset{RA}{\overset{OA}{Fe^{2+}_{(aq)}}} \quad \overset{OA}{SO_4^{2-}_{(aq)}} \underset{RA}{\overset{OA}{H_2O_{(l)}}}$$

$$MnO_4^-_{(aq)} + 8\,H^+_{(aq)} + 5\,e^- \rightarrow Mn^{2+}_{(aq)} + 4\,H_2O_{(l)}$$

$$5\,[\,Fe^{2+}_{(aq)} \rightarrow Fe^{3+}_{(aq)} + e^-\,]$$

$$MnO_4^-_{(aq)} + 8\,H^+_{(aq)} + 5\,Fe^{2+}_{(aq)} \overset{spont.}{\rightarrow} 5\,Fe^{3+}_{(aq)} + Mn^{2+}_{(aq)} + 4\,H_2O_{(l)}$$

(page 364)

Example

In a chemical industry, could copper pipe be used to transport a hydrochloric acid solution? To answer this question,

(a) predict the redox reaction and its spontaneity, and

(b) describe two diagnostic tests that could be done to test your prediction.

(a)
$$\underset{SRA}{\overset{}{Cu_{(s)}}} \quad \underset{}{\overset{SOA}{H^+_{(aq)}}} \quad \underset{RA}{\overset{}{Cl^-_{(aq)}}} \underset{RA}{} \quad \underset{RA}{\overset{OA}{H_2O_{(l)}}}$$

$$2\,H^+_{(aq)} + 2\,e^- \rightarrow H_{2(g)}$$

$$Cu_{(s)} \rightarrow Cu^{2+}_{(aq)} + 2\,e^-$$

$$2\,H^+_{(aq)} + Cu_{(s)} \xrightarrow{\text{non-spont.}} H_{2(g)} + Cu^{2+}_{(aq)}$$

Since the reaction is non-spontaneous, it should be possible to use a copper pipe to carry hydrochloric acid.

(b) *If the mixture is observed, and no gas is produced, then* it is likely that no hydrogen gas was produced (Figure 12.13). If the color of the solution is observed, and the color did not change to blue, then copper probably did not react to produce copper(II) ions. (If the solution is tested for pH before and after adding the copper, and the pH did not increase, then hydrogen ions probably did not react.)

51 REDOX STOICHIOMETRY

(page 367)

Evidence

TITRATION OF TIN(II) SOLUTION				
(volume of $KMnO_{4(aq)}$ required to react with 10.00 mL of acidic 0.0500 mol/L tin(II) chloride)				
Trial	**1**	**2**	**3**	**4**
Final buret reading (mL)	18.4	35.3	17.3	34.1
Initial buret reading (mL)	1.0	18.4	0.6	17.3
Volume of $KMnO_{4(aq)}$ (mL)	17.4	16.9	16.7	16.8
Endpoint color	dark pink	light pink	light pink	light pink

Analysis

The first endpoint was overshot and was not used in the average for the analysis. At the endpoint, an average of 16.8 mL of permanganate solution was used.

$$
\begin{array}{cccccc}
\textbf{OA} & \overset{\textbf{SOA}}{\diagup}\overset{\textbf{OA}}{} & & \textbf{OA} & & \textbf{OA} \\
K^+_{(aq)} & MnO_4^-{}_{(aq)}\ \ H^+_{(aq)} & Sn^{2+}{}_{(aq)} & Cl^-_{(aq)} & H_2O_{(1)} \\
& \textbf{SRA} & \textbf{RA} & \overset{}{\diagup}\ \textbf{RA} \\
& & & \textbf{RA}
\end{array}
$$

$$2\,[MnO_4^-{}_{(aq)} + 8\,H^+_{(aq)} + 5\,e^- \rightarrow Mn^{2+}{}_{(aq)} + 4\,H_2O_{(1)}]$$

$$5\,[Sn^{2+}{}_{(aq)} \rightarrow Sn^{4+}{}_{(aq)} + 2\,e^-]$$

$$2\,MnO_4^-{}_{(aq)} + 16\,H^+_{(aq)} + 5\,Sn^{2+}{}_{(aq)} \rightarrow 2\,Mn^{2+}{}_{(aq)} + 8\,H_2O_{(1)} + 5\,Sn^{4+}{}_{(aq)}$$

$$
\begin{array}{cc}
16.8 \text{ mL} & 10.00 \text{ mL} \\
C & 0.0500 \text{ mol/L}
\end{array}
$$

$$n_{Sn^{2+}} = 10.00 \text{ mL} \times 0.0500\,\frac{mol}{L} = 0.500 \text{ mmol}$$

$$n_{MnO_4^-} = 0.500 \text{ mmol} \times \frac{2}{5} = 0.200 \text{ mmol}$$

The *standardized* potassium permanganate solution can now be used as a strong oxidizing agent in further titrations. A laboratory technician might standardize the solution in the morning, and then re-standardize at noon and at the end of the day, to increase the certainty of the results.

$$C_{MnO_4^-} = \frac{0.200 \text{ mmol}}{16.8 \text{ mL}} = 0.0119 \text{ mol/L}$$

According to the evidence gathered and the stoichiometric analysis, the molar concentration of the potassium permanganate solution is 0.0119 mol/L or 11.9 mmol/L.

© NELSON CANADA,
A DIVISION OF THOMSON CANADA LIMITED, 1994

52 OXIDATION NUMBERS AND BALANCING REDOX EQUATIONS

(page 373)

The oxidation numbers in Table 12.4 apply to most compounds. However, there are some exceptions:

- Hydrogen has an oxidation number of −1 when bonded to a less electronegative element; for example, hydrides such as $LiH_{(s)}$.
- Oxygen has an oxidation number of −1 in peroxides such as H_2O_2; its oxidation number is +2 in the compound OF_2.

Oxidation Numbers

Table 12.4

COMMON OXIDATION NUMBERS		
Atom or Ion	**Oxidation Number**	**Examples**
all atoms in elements	0	Na is 0, Cl in Cl_2 is 0
hydrogen in compounds	+1	H in HCl is +1
oxygen in compounds	−2	O in H_2O is −2
all monatomic ions	charge on the ion	Na^+ is +1, S^{2-} is −2

$$x + 4(+1) = 0$$
$$x = -4$$
$$\overset{x\ +1}{CH_4} \quad or \quad \overset{-4\ +1}{CH_4}$$

$$x + 4(-2) = -1$$
$$x = +7$$
$$\overset{x\quad -2}{MnO_4^-} \quad or \quad \overset{+7\ -2}{MnO_4^-}$$

(page 374)

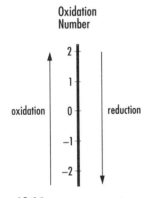

Oxidation Number

oxidation ← → reduction

Figure 12.16
In a redox reaction, both oxidation and reduction occur.

Oxidation and Reduction
(page 375)

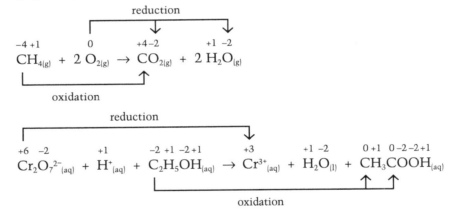

$$\overset{-4\ +1}{CH_{4(g)}} + 2\ \overset{0}{O_{2(g)}} \rightarrow \overset{+4\ -2}{CO_{2(g)}} + 2\ \overset{+1\ -2}{H_2O_{(g)}}$$

reduction

oxidation

$$\overset{+6\ -2}{Cr_2O_7^{2-}}_{(aq)} + \overset{+1}{H^+}_{(aq)} + \overset{-2\ +1\ -2\ +1}{C_2H_5OH}_{(aq)} \rightarrow \overset{+3}{Cr^{3+}}_{(aq)} + \overset{+1\ -2}{H_2O}_{(l)} + \overset{0\ +1\ 0\ -2\ -2\ +1}{CH_3COOH}_{(aq)}$$

reduction

oxidation

Balancing Redox Equations

$$3Ag_2S_{(s)} \qquad + \qquad 2Al_{(s)} \qquad \rightarrow \qquad Al_2S_{3(s)} + 6Ag_{(s)}$$

$1\ e^-/Ag$	$3\ e^-/Al$
$2\ e^-/Ag_2S$	$3\ e^-/Al$
$(\times 3)$	$(\times 2)$

(page 378)

EXAMPLE

Balance the chemical equation for the oxidation of ethanol by dichromate ions in a breathalyzer (page 380).

$$\overset{+6\ -2}{2\ Cr_2O_7^{2-}}_{(aq)} + \overset{+1}{16\ H^+}_{(aq)} + \overset{-2\ +1\ -2\ +1}{3\ C_2H_5OH}_{(aq)} \rightarrow \overset{+3}{4\ Cr^{3+}}_{(aq)} + 11\ H_2O_{(l)} +$$

$$\overset{0\ +1\ 0\ -2\ -2\ +1}{3\ CH_3COOH}_{(aq)}$$

$\underset{+6}{} $	$\underset{-2}{}$	$\underset{+3}{}$
$3e^-/Cr$	$2e^-/C$	
$6e^-/Cr_2O_7^{2-}$	$4e^-/C_2H_5OH$	$\underset{0\quad 0}{}$

53 BLOOD ALCOHOL ANALYSIS

Design 1: Breathalyzers — An Indirect Method

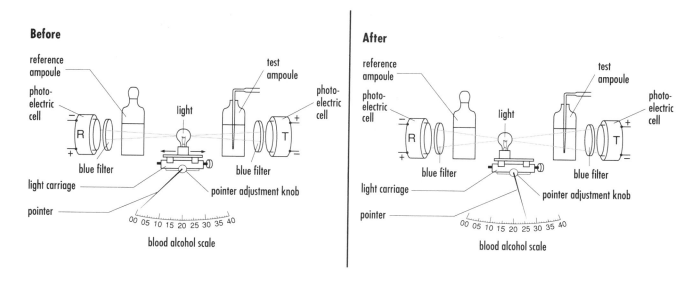

The reference and test ampoules contain acidified potassium dichromate. Alcohol-containing breath blown into the test ampoule decreases the color of the solution, allowing more light through the ampoule and into the photoelectric cell. The larger electric current from the photoelectric cell is recorded on the electric meter as a blood alcohol content.

Design 2: Back Titration of Blood — A Direct Method

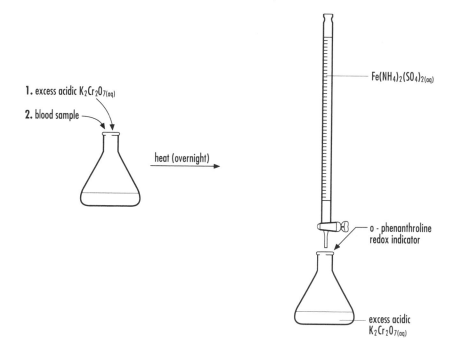

54 (A) CONSUMER CELLS

Dry Cells (primary)
(page 391)

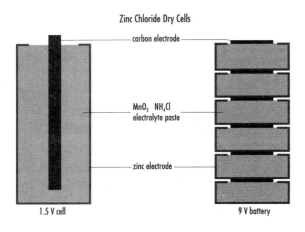

Figure 13.8
Six 1.5 V dry cells in series make a 9 V battery.

Lead-Acid Car Battery (secondary)
(page 392)

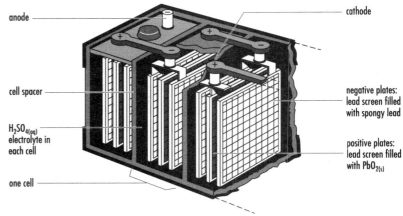

Figure 13.9
The anodes of a lead-acid car battery are composed of spongy lead and the cathodes are composed of lead(IV) oxide on a metal screen. The large electrode surface area is designed to deliver the current required to start a car engine.

(page 392)

Table 13.2

PRIMARY, SECONDARY, AND FUEL CELLS			
Type	**Name of Cell**	**Cell Reactions**	**Characteristics and Uses**
primary cells	dry cell (1.5 V)	$2 MnO_{2(s)} + 2 NH_4^+{}_{(aq)} + 2 e^- \rightarrow Mn_2O_{3(s)} + 2 NH_{3(aq)} + H_2O_{(1)}$ $Zn_{(s)} \rightarrow Zn^{2+}{}_{(aq)} + 2 e^-$	• inexpensive, portable, many sizes • flashlights, radios, many other consumer items
	alkaline dry cell (1.5 V)	$2 MnO_{2(s)} + H_2O_{(1)} + 2 e^- \rightarrow Mn_2O_{3(s)} + 2 OH^-{}_{(aq)}$ $Zn_{(s)} + 2 OH^-{}_{(aq)} \rightarrow ZnO_{(s)} + H_2O_{(1)} + 2 e^-$	• longer shelf life; higher currents for longer periods compared with dry cell • same uses as dry cell
	mercury cell (1.35 V)	$HgO_{(s)} + H_2O_{(1)} + 2 e^- \rightarrow Hg_{(1)} + 2 OH^-$ $Zn_{(s)} + 2 OH^-{}_{(aq)} \rightarrow ZnO_{(s)} + H_2O_{(1)} + 2 e^-$	• small cell; constant voltage during its active life • hearing aids, watches
secondary cells	Ni-Cad cell (1.25 V)	$2 NiO(OH)_{(s)} + 2 H_2O_{(1)} + 2 e^- \rightarrow 2 Ni(OH)_{2(s)} + 2 OH^-$ $Cd_{(s)} + 2 OH^-{}_{(aq)} \rightarrow Cd(OH)_{2(s)} + 2 e^-$	• can be completely sealed; lightweight but expensive • all normal dry cell uses, as well as power tools, shavers
	lead-acid cell (2.0 V)	$PbO_{2(s)} + 4 H^+{}_{(aq)} + SO_4^{2-}{}_{(aq)} + 2 e^- \rightarrow PbSO_{4(s)} + 2 H_2O_{(1)}$ $Pb_{(s)} + SO_4^{2-}{}_{(aq)} \rightarrow PbSO_{4(s)} + 2 e^-$	• very large currents; reliable for many recharges • all vehicles
fuel cells	aluminum-air cell (2 V)	$3 O_{2(g)} + 6 H_2O_{(1)} + 12 e^- \rightarrow 12 OH^-{}_{(aq)}$ $4 Al_{(s)} \rightarrow 4 Al^{3+}{}_{(aq)} + 12 e^-$	• very high energy density; made from readily available aluminum alloys • designed for electric cars

54 (B) CONSUMER CELLS

Mercury Cell (primary)

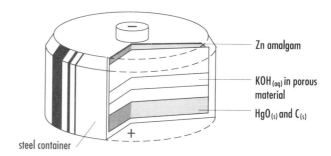

Zn amalgam

KOH$_{(aq)}$ in porous material

HgO$_{(s)}$ and C$_{(s)}$

steel container

cathode $\quad HgO_{(s)} + H_2O_{(l)} + 2e^- \rightarrow Hg_{(l)} + 2OH^-_{(aq)}$ $\qquad E_r^\circ = +0.10$ V

anode $\quad Zn_{(s)} + 2OH^-_{(aq)} \rightarrow ZnO_{(s)} + H_2O_{(l)} + 2e^-$ $\qquad E_r^\circ = -1.25$ V

Ni-Cad Cell (secondary)

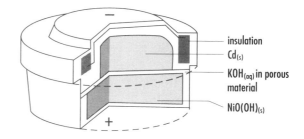

insulation

Cd$_{(s)}$

KOH$_{(aq)}$ in porous material

NiO(OH)$_{(s)}$

cathode $\quad 2NiO(OH)_{(s)} + 2H_2O_{(l)} + 2e^- \rightarrow 2Ni(OH)_{2(s)} + 2OH^-_{(aq)}$

$$E_r^\circ = +0.49 \text{ V}$$

anode $\quad Cd_{(s)} + 2OH^-_{(aq)} \rightarrow Cd(OH)_{2(s)} + 2e^-$ $\qquad E_r^\circ = -0.76$ V

(page 393)

Molicel® (secondary)

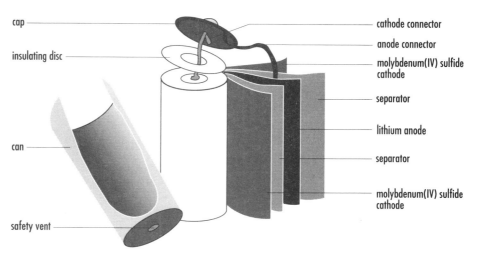

cap

insulating disc

can

safety vent

cathode connector

anode connector

molybdenum(IV) sulfide cathode

separator

lithium anode

separator

molybdenum(IV) sulfide cathode

Figure 13.10
Invented and manufactured in British Columbia, the Molicel® is a high-energy, rechargeable cell. It has long layers of lithium anode and molybdenum(IV) sulfide cathode in a unique, jelly-roll design. The Molicel® has three times the energy density and five times the power capability of the lead-acid cell in a car battery.

54 (C) BOOSTING CAR BATTERIES

A battery may explode when being boosted or when being used for boosting. Hydrogen gas generated by a side reaction — the electrolysis of water — may be ignited by a spark.

Charging: $2 H_2O_{(l)}$ + electrical energy $\rightarrow 2 H_{2(g)} + O_{2(g)}$

Explosion: $2 H_{2(g)} + O_{2(g)}$ + spark $\rightarrow 2 H_2O_{(g)}$ + energy

The hydrogen explosion can burst the battery case and expel the sulfuric acid and electrolyte in a dangerous manner.

Safety Procedure

1. Wear eye protection (goggles or glasses).

2. Assuming a negative–to–ground electrical system, connect positive (red) to positive and then negative (black) to ground.

3. Make the last connection and the first disconnection — usually negative (black) — away from one of the batteries and away from any hydrogen being produced. These operations produce a spark that might ignite any hydrogen present.

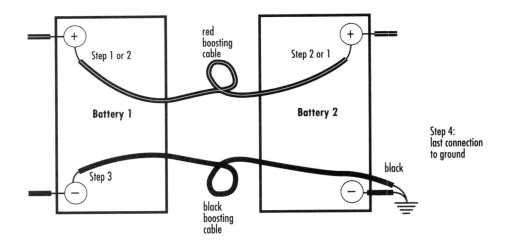

54 (D) FUEL CELLS

Aluminum-Oxygen Fuel Cell

An aluminum-oxygen fuel cell has been developed for possible use in electric cars. Oxygen is obtained from the air and the aluminum anodes can be replaced as required. The criterion of replacement qualifies the aluminum-oxygen cell to be called a **fuel cell**.

Typical Efficiencies

Gasoline lawnmower 12%
Automobile engine 17.23%
Electric power plant .... 30 – 40%
Fuel cells .... 40 – 70%

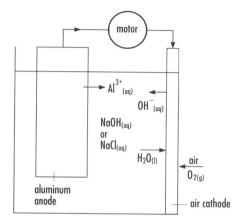

$$3\ O_{2(g)} + 6\ H_2O_{(l)} + 12\ e^- \rightarrow 12\ OH^-_{(aq)}$$
$$4\ Al_{(s)} \rightarrow 4\ Al^{3+}_{(aq)} + 12\ e^-$$
$$4\ Al_{(s)} + 3\ O_{2(g)} + 6\ H_2O_{(l)} \rightarrow 4\ Al(OH)_{3(s)} \qquad \Delta E° = 2\ V$$

Hydrogen-Oxygen Fuel Cell

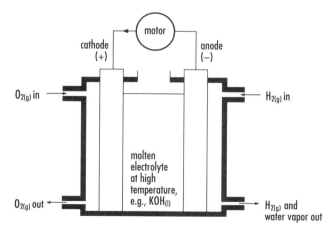

$$2\ H_{2(g)} + O_{2(g)} \rightarrow 2\ H_2O_{(g)}$$

© NELSON CANADA,
A DIVISION OF THOMSON CANADA LIMITED, 1994

67

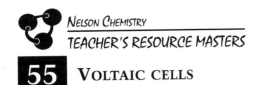
55 VOLTAIC CELLS

Write the half-reaction and net-reaction equations along with $\Delta E°$'s for the voltaic cells. Label the electrodes and electrolytes, anode and cathode, positive and negative electrodes, and electron and ion motion.

Cell 1

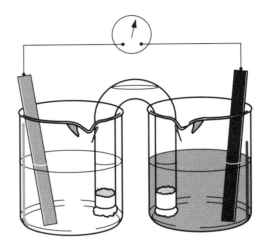

Cell 2

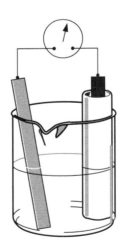

56 REFERENCE HALF-CELL

(page 401)

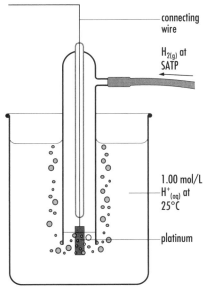

$Pt_{(s)} \mid H_{2(g)}, H^+_{(aq)} \qquad E^\circ_r = 0.00 \text{ V}$

Figure 13.17
The standard hydrogen half-cell is used internationally as the reference half-cell in electrochemical research.

(page 403)

E° (V)

$+0.34$ — $Cu^{2+}_{(aq)} + 2e^- \rightleftharpoons Cu_{(s)}$

0.00 — $2H^+_{(aq)} + 2e^- \rightleftharpoons H_{2(g)}$ $\Big\}$ 0.34 V

$\Big\}$ 1.66 V

-1.66 — $Al^{3+}_{(aq)} + 3e^- \rightleftharpoons Al_{(s)}$

Figure 13.20
Measurements of standard cell potentials show that the reduction potential of $Cu^{2+}_{(aq)}$ is 0.34 V greater than that of $H^+_{(aq)}$, which is 1.66 V greater than that of $Al^{3+}_{(aq)}$. If a standard copper-aluminum cell were measured, we would expect copper to be the cathode, with a reduction potential 2.00 V above that of the aluminum anode.

(page 402)

EXAMPLE

The reduction potential of the standard copper half-cell is 0.34 V and the reduction potential of the standard zinc half-cell is –0.76 V.

(a) Determine a revised set of half-cell reduction potentials for the half-reactions if the reference half-cell is changed and one of the half-cells were arbitrarily assigned a value of 0.00 V.

(b) What is the difference between the reduction potentials of the copper and zinc half-cells in each case?

$$\Delta E^\circ = E^\circ_{r\,(cathode)} - E^\circ_{r\,(anode)}$$

(a)

Reduction Half-Reaction	Reference Half-Cell		
	Hydrogen E°_r (V)	Copper E°_r (V)	Zinc E°_r (V)
$Cu^{2+}_{(aq)} + 2e^- \rightarrow Cu_{(s)}$	+0.34	0.00	+1.10
$2H^+_{(aq)} + 2e^- \rightarrow H_{2(g)}$	0.00	–0.34	+0.76
$Zn^{2+}_{(aq)} + 2e^- \rightarrow Zn_{(s)}$	–0.76	–1.10	0.00

(b) In all cases, $\Delta E^\circ = 1.10$ V for a copper-zinc cell.

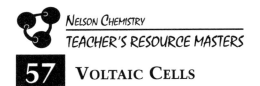

57 VOLTAIC CELLS

(page 402)

Hydrogen-Copper Cell

$$Pt_{(s)} \mid H_{2(g)}, H^+_{(aq)} \parallel Cu^{2+}_{(aq)} \mid Cu_{(s)} \qquad\qquad \Delta E^\circ = 0.34 \text{ V}$$

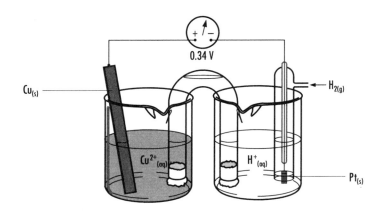

Figure 13.18
A copper-hydrogen standard cell.

cathode	$Cu^{2+}_{(aq)} + 2\,e^- \rightarrow Cu_{(s}$	$E^\circ_r = 0.34$ V
anode	$H_{2(g)} \rightarrow 2\,H^+_{(aq)} + 2\,e^-$	$E^\circ_r = 0.00$ V
net	$Cu^{2+}_{(aq)} + H_{2(g)} \rightarrow Cu_{(s)} + 2\,H^+_{(aq)}$	$\Delta E^\circ = 0.34$ V

(page 403)

Aluminum-Hydrogen Cell

$$Al_{(s)} \mid Al^{3+}_{(aq)} \parallel H^+_{(aq)}, H_{2(g)} \mid Pt_{(s)} \qquad\qquad \Delta E^\circ = 1.66 \text{ V}$$

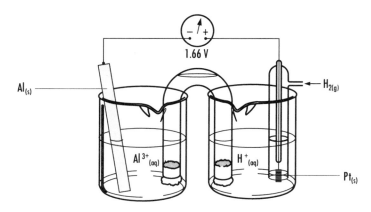

Figure 13.19
An aluminum-hydrogen standard cell.

cathode	$3\,[\, 2\,H^+_{(aq)} + 2\,e^- \rightarrow H_{2(g)} \,]$	$E^\circ_r = 0.00$ V
anode	$2\,[\, Al_{(s)} \rightarrow Al^{3+}_{(aq)} + 3\,e^- \,]$	$E^\circ_r = -1.66$ V
net	$6\,H^+_{(aq)} + 2\,Al_{(s)} \rightarrow 3\,H_{2(g)} + 2\,Al^{3+}_{(aq)}$	$\Delta E^\circ = 1.66$ V

58 READING A VOLTMETER

(page 532)

Voltage Measurements of Batteries

1. Set the dial to the appropriate value on the direct current volts (DCV) scale; for example, 3 V.

2. The black lead (labelled negative or COM) is normally connected to the anode and the red lead (positive) is connected to the cathode of a voltaic cell.

3. Make a firm contact between each metal probe and an electrode of the cell. (Press firmly with the pointed probe or use leads with an alligator clip.)

4. On analog meters (those with a needle), be sure to read the scale corresponding to the meter value you set in step 1.

5. If the needle attempts to move to the left off the scale or a digital meter registers a negative number, then switch the connections to the cell.

59 VOLTAIC AND ELECTROLYTIC CELLS

(page 408)

Table 13.4

COMPARISON OF VOLTAIC AND ELECTROLYTIC CELLS		
	Voltaic Cell	**Electrolytic Cell**
Spontaneity	spontaneous reaction	non-spontaneous reaction
Standard cell potential, $\Delta E°$	positive	negative
Cathode	• positive electrode • strongest oxidizing agent undergoes a reduction	• negative electrode • strongest oxidizing agent undergoes a reduction
Anode	• negative electrode • strongest reducing agent undergoes an oxidation	• positive electrode • strongest reducing agent undergoes an oxidation
Electron movement	anode → cathode	anode → cathode
Ion movement	anions → anode cations → cathode	anions → anode cations → cathode

cations to cathode
electrons to cathode
SOAC/GERC
* cathode positive
* $\Delta E° = +$ value
* spontaneous
* voltaic cell

"reactants" ⟶
⟵ "products" + **electrical energy**

* electrolytic cell
* non-spontaneous
* $\Delta E° = -$ value
* cathode negative
SOAC/GERC
electrons to cathode
cations to cathode

* differences
SOAC/GERC — Strongest Oxidizing Agent at the Cathode
Gains Electrons and is Reduced at the Cathode

60 ELECTROLYSIS IN TECHNOLOGY

(page 409)
See Figure 13.23.

Aqueous Electrolysis (production of iodine)

$$\begin{array}{ccc} \textbf{OA} & & \textbf{SOA} \\ K^+_{(aq)}, & I^-_{(aq)}, & H_2O_{(l)} \\ \textbf{SRA} & \textbf{RA} \end{array}$$

cathode $\quad 2\,H_2O_{(l)} + 2\,e^- \rightarrow H_{2(g)} + 2\,OH^-_{(aq)}$
gas bubbles   blue litmus

anode $\quad 2\,I^-_{(aq)} \rightarrow I_{2(s)} + 2\,e^-$
yellow-brown
(purple in chlorinated hydrocarbons)

net $\quad 2\,H_2O_{(l)} + 2\,I^-_{(aq)} \rightarrow H_{2(g)} + 2\,OH^-_{(aq)} + I_{2(s)}$

(page 413)

Molten Electrolysis (production of lithium)

$$\begin{array}{cc} & \textbf{SOA} \\ Li^+_{(l)}, & Cl^-_{(l)} \qquad t > 605°C \\ & \textbf{SRA} \end{array}$$

cathode $\quad 2\,Li^+_{(l)} + 2\,e^- \rightarrow 2\,Li_{(l)}$
anode $\quad\quad\quad\;\; 2\,Cl^-_{(l)} \rightarrow Cl_{2(g)} + 2\,e^-$

net $\quad 2\,Li^+_{(l)} + 2\,Cl^-_{(l)} \rightarrow 2\,Li_{(l)} + Cl_{2(g)}$

(page 414)
See Figure 13.24 and Figure 13.25.

Molten-Ionic Solvent (production of aluminum)

$$\begin{array}{cc} & \textbf{SOA} \\ Al^{3+}_{(cryolite)}, & O^{2-}_{(cryolite)} \\ & \textbf{SRA} \end{array}$$

cathode $\quad 4\,[\,Al^{3+}_{(cryolite)} + 3\,e^- \rightarrow Al_{(l)}\,]$
anode $\quad\quad\;\; 3\,[\,2\,O^{2-}_{(cryolite)} \rightarrow O_{2(g)} + 4\,e^-\,]$

net $\quad 4\,Al^{3+}_{(cryolite)} + 6\,O^{2-}_{(cryolite)} \rightarrow 4\,Al_{(l)} + 3\,O_{2(g)}$

The overall effect is a decomposition reaction.

$$2\,Al_2O_{3(s)} \rightarrow 4\,Al_{(s)} + 3\,O_{2(g)}$$

(page 415)
See Figure 13.26 and
Figure 13.27.

Refining of Metals (purifying copper)

cathode	reduction of copper	$Cu^{2+}_{(aq)} + 2\,e^- \rightarrow Cu_{(s)}$
anode	oxidation of copper	$Cu_{(s)} \rightarrow Cu^{2+}_{(aq)} + 2\,e^-$
	oxidation of zinc	$Zn_{(s)} \rightarrow Zn^{2+}_{(aq)} + 2\,e^-$
	oxidation of iron	$Fe_{(s)} \rightarrow Fe^{2+}_{(aq)} + 2\,e^-$

61 CELL STOICHIOMETRY

(page 418)

EXAMPLE ——————————————————————————

Convert a current of 1.74 A for 30.0 min into an amount of electrons. Recall that 1 A = 1 C/s (Table 13.1, page 389).

$$n_{e^-} = \frac{1.74 \text{ C}}{\text{s}} \times 30.0 \text{ min} \times \frac{60 \text{ s}}{1 \text{ min}} \times \frac{1 \text{ mol}}{9.65 \times 10^4 \text{ C}} = 0.0325 \text{ mol}$$

or

$$n_{e^-} = \frac{It}{F} = \frac{1.74 \text{ C/s} \times 30.0 \text{ min} \times 60 \text{ s/min}}{9.65 \times 10^4 \text{ C/mol}} = 0.0325 \text{ mol}$$

(page 418)

EXAMPLE ——————————————————————————

What is the mass of copper deposited at the cathode of a copper-refining electrolytic cell operated at 12.0 A for 40.0 min (Figure 13.27, page 415)?

$$Cu^{2+}_{(aq)} \quad + \quad 2 \text{ e}^- \quad \rightarrow \quad Cu_{(s)}$$

40.0 min m

12.0 A   63.55 g/mol

9.65×10^4 C/mol

$$n_{e^-} = \frac{12.0 \text{ C}}{\text{s}} \times 40.0 \text{ min} \times \frac{60 \text{ s}}{1 \text{ min}} \times \frac{1 \text{ mol}}{9.65 \times 10^4 \text{ C}} = 0.298 \text{ mol}$$

$$n_{Cu} = 0.298 \text{ mol} \times \frac{1}{2} = 0.149 \text{ mol}$$

$$m_{Cu} = 0.149 \text{ mol} \times \frac{63.55 \text{ g}}{1 \text{ mol}} = 9.48 \text{ g}$$

According to the stoichiometric method and the laws of electrolysis, the mass of copper deposited is 9.48 g.

or

$$m_{Cu} = \frac{12.0 \text{ C}}{\text{s}} \times 40.0 \text{ min} \times \frac{60 \text{ s}}{1 \text{ min}} \times \frac{1 \text{ mol}}{9.65 \times 10^4 \text{ C}} \times \frac{1}{2} \times \frac{63.55 \text{ g}}{1 \text{ mol}}$$

62 K_w CALCULATIONS

(page 444)

EXAMPLE

A 0.15 mol/L solution of hydrochloric acid at 25°C is found to have a hydrogen ion concentration of 0.15 mol/L. Calculate the concentration of the hydroxide ions.

$$HCl_{(aq)} \rightarrow H^+_{(aq)} + Cl^-_{(aq)}$$

$$[H^+_{(aq)}] = [HCl_{(aq)}] = 0.15 \text{ mol/L}$$

$$[OH^-_{(aq)}] = \frac{K_w}{[H^+_{(aq)}]}$$

$$= \frac{1.0 \times 10^{-14} \text{ (mol/L)}^2}{0.15 \text{ mol/L}}$$

$$= 6.7 \times 10^{-14} \text{ mol/L}$$

EXAMPLE

Calculate the hydrogen ion concentration in a 0.25 mol/L solution of barium hydroxide.

$$Ba(OH)_{2(s)} \rightarrow Ba^{2+}_{(aq)} + 2\,OH^-_{(aq)}$$

$$[OH^-_{(aq)}] = 2 \times [Ba(OH)_{2(aq)}] = 2 \times 0.25 \text{ mol/L} = 0.50 \text{ mol/L}$$

$$[H^+_{(aq)}] = \frac{K_w}{[OH^-_{(aq)}]}$$

$$= \frac{1.0 \times 10^{-14} \text{ (mol/L)}^2}{0.50 \text{ mol/L}}$$

$$= 2.0 \times 10^{-14} \text{ mol/L}$$

EXAMPLE

Determine the hydrogen ion and hydroxide ion concentrations in 500 mL of an aqueous solution containing 2.6 g of dissolved sodium hydroxide.

$$n_{NaOH} = 2.6 \text{ g} \times \frac{1 \text{ mol}}{40.00 \text{ g}} = 0.065 \text{ mol}$$

$$[NaOH_{(aq)}] = \frac{0.065 \text{ mol}}{0.500 \text{ L}} = 0.13 \text{ mol/L}$$

$$NaOH_{(s)} \rightarrow Na^+_{(aq)} + OH^-_{(aq)}$$

$$[OH^-_{(aq)}] = [NaOH_{(aq)}] = 0.13 \text{ mol/L}$$

$$[H^+_{(aq)}] = \frac{K_w}{[OH^-_{(aq)}]}$$

$$= \frac{1.0 \times 10^{-14} \text{ (mol/L)}^2}{0.13 \text{ mol/L}}$$

$$= 7.7 \times 10^{-14} \text{ mol/L}$$

63 pH CALCULATIONS

(page 131)

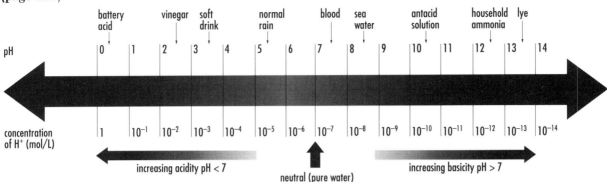

Figure 5.11
The pH scale can communicate a broad range of hydrogen ion concentrations, in a wide variety of substances.

(page 446)

$$pH = -\log[H^+_{(aq)}]$$

EXAMPLE

Communicate a hydrogen ion concentration of 4.7×10^{-11} mol/L as a pH value.

$$
\begin{aligned}
pH &= -\log[H^+_{(aq)}] \\
&= -\log(4.7 \times 10^{-11}) \quad \text{(two significant digits)} \\
&= 10.33 \quad \text{(two digits following the decimal point)}
\end{aligned}
$$

On many calculators, $-\log(4.7 \times 10^{-11})$ may be entered by pushing the following sequence of keys.

| 4 | • | 7 | EXP |
| 1 | 1 | +/– | log | +/– |

$$[H^+_{(aq)}] = 10^{-pH}$$

(page 446)

EXAMPLE

Communicate a pH of 10.33 as a hydrogen ion concentration.

$$
\begin{aligned}
[H^+_{(aq)}] &= 10^{-pH} \\
&= 10^{-10.33} \text{ mol/L} \quad \text{(two digits following the decimal point)} \\
&= 4.7 \times 10^{-11} \text{ mol/L} \quad \text{(two significant digits)}
\end{aligned}
$$

On many calculators, $10^{-10.33}$ may be entered by pushing the following sequence of keys.

1	0	•	3	3
+/–	either	INV	log
or	2nd	log

(page 447)

According to the rules of logarithms,

$$\log(ab) = \log(a) + \log(b)$$

Using the equilibrium law for the ionization of water,

$$
\begin{aligned}
[H^+_{(aq)}][OH^-_{(aq)}] &= K_w \\
\log[H^+_{(aq)}] + \log[OH^-_{(aq)}] &= \log(K_w) \\
(-pH) + (-pOH) &= -14.00 \\
pH + pOH &= 14.00
\end{aligned}
$$

SUMMARY: pH AND pOH

$$K_w = [H^+_{(aq)}][OH^-_{(aq)}]$$

$$pH = -\log[H^+_{(aq)}] \qquad pOH = -\log[OH^-_{(aq)}]$$

$$[H^+_{(aq)}] = 10^{-pH} \qquad [OH^-_{(aq)}] = 10^{-pOH}$$

$$pH + pOH = 14.00 \text{ (at SATP)}$$

64 pH METER

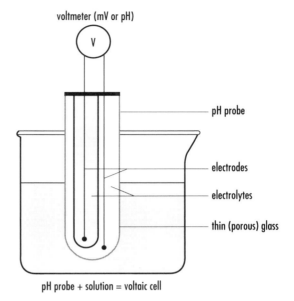

voltmeter (mV or pH)

pH probe

electrodes

electrolytes

thin (porous) glass

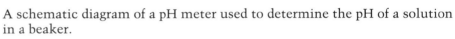

pH probe + solution = voltaic cell

A schematic diagram of a pH meter used to determine the pH of a solution in a beaker.

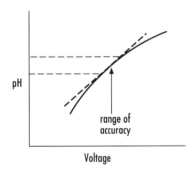

pH

range of accuracy

Voltage

The relationship of voltage generated by a pH meter and pH.

Figure 14.13
Arnold Beckman invented the pH meter in 1935, 26 years after Sören Sörenson had developed the concept of pH for communicating hydrogen ion concentration.

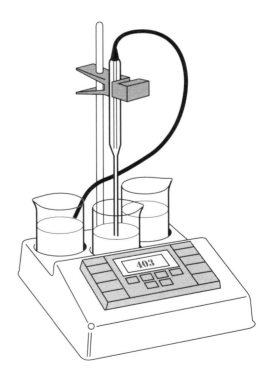

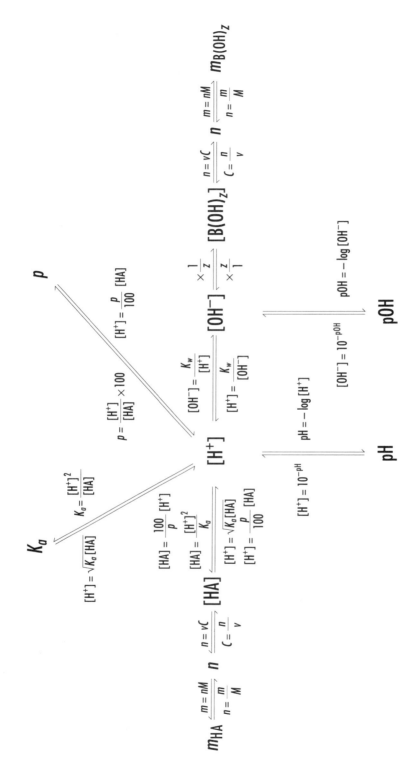

66 (A) ACIDS AND BASES

(page 453)

Suppose you measured the pH of a 0.25 mol/L carbonic acid solution to be 3.48. What is the K_a for carbonic acid?

$$[H^+_{(aq)}] = 10^{-pH}$$
$$= 10^{-3.48} \text{ mol/L}$$
$$= 3.3 \times 10^{-4} \text{ mol/L}$$

$$H_2CO_{3(aq)} \rightleftharpoons H^+_{(aq)} + HCO_3^-_{(aq)}$$

$$K_a = \frac{[H^+_{(aq)}][HCO_3^-_{(aq)}]}{[H_2CO_{3(aq)}]}$$

$$= \frac{(3.3 \times 10^{-4} \text{ mol/L})^2}{0.25 \text{ mol/L}}$$

$$= 4.4 \times 10^{-7} \text{ mol/L}$$

Although units are often omitted from equilibrium constants, they are usually included with acid ionization constants (K_a). Units for all K_a values are mol/L.

(page 453)

Strong Bases

$$NaOH_{(aq)} \rightarrow Na^+_{(aq)} + OH^-_{(aq)}$$

$$Ba(OH)_{2(aq)} \rightarrow Ba^{2+}_{(aq)} + 2\,OH^-_{(aq)}$$

(pages 453–454)

Lab Exercise 14E Qualitative Analysis

Complete the Analysis of the investigation report.

Problem

Which of the unknown solutions provided is $HBr_{(aq)}$, $CH_3COOH_{(aq)}$, $NaCl_{(aq)}$, $C_{12}H_{22}O_{11(aq)}$, $Ba(OH)_{2(aq)}$, and $KOH_{(aq)}$?

Experimental Design

The solutions, which have been prepared with equal concentrations, are each tested with a conductivity apparatus and with both red and blue litmus.

Evidence

LITMUS AND CONDUCTIVITY TESTS ON UNKNOWN SOLUTIONS			
Solution	**Red Litmus**	**Blue Litmus**	**Conductivity**
1	blue	no change	very high
2	no change	red	low
3	no change	no change	none
4	no change	red	high
5	no change	no change	high
6	blue	no change	high

66 (B) ACIDS AND BASES ANALYSIS FLOWCHART

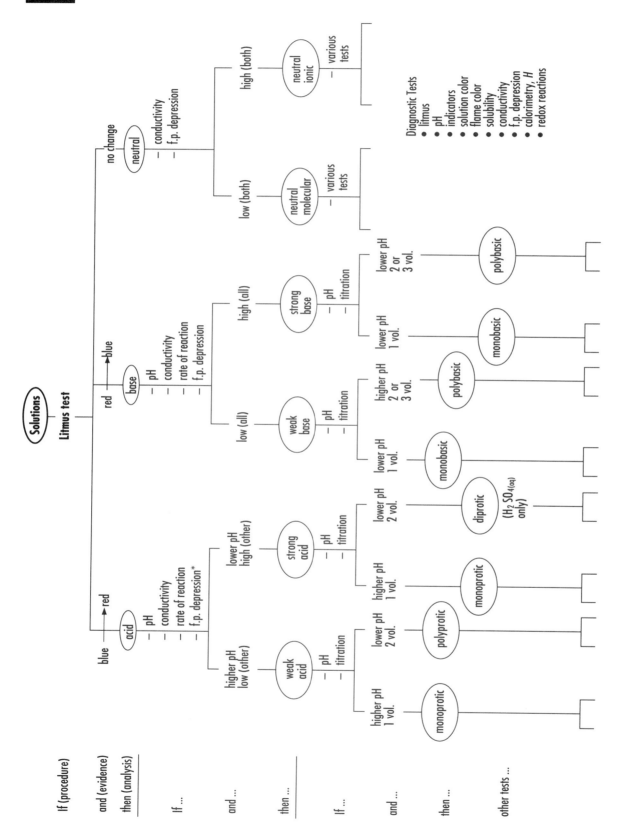

Diagnostic Tests
- litmus
- pH
- indicators
- solution color
- flame color
- solubility
- conductivity
- f.p. depression
- calorimetry, H
- redox reactions

If (procedure)

and (evidence)

then (analysis)

If ...

and ...

then ...

If ...

and ...

then ...

other tests ...

* f.p. depression — freezing point depression

67 ACID-BASE INDICATORS

(page 471)

conjugate pair

$$HIn_{(aq)} + H_2O_{(l)} \rightleftharpoons In^-_{(aq)} + H_3O^+_{(aq)}$$

acid   base

red   (litmus color)   blue

(page 471)   **Methyl Red**

$$HMr_{(aq)} + H_2O_{(l)} \rightleftharpoons H_3O^+_{(aq)} + Mr^-_{(aq)}$$

acid   base

⟵   red   orange   yellow ⟶

pH   4.8   6.0

(inside back cover)

ACID-BASE INDICATORS

Common Name of Indicator	Color of Conjugate Acid, $HIn_{(aq)}$	Approximate pH Range for Color Change	Color of Conjugate Base, $In^-_{(aq)}$
methyl violet	yellow	0.0 – 1.6	blue
thymol blue*	red	1.2 – 2.8	yellow
orange IV	red	1.4 – 2.8	yellow
benzopurpurine-48	violet	2.2 – 4.2	red
congo red	blue	3.0 – 5.0	red
methyl orange	red	3.2 – 4.4	yellow
bromocresol green	yellow	3.8 – 5.4	blue
methyl red	red	4.8 – 6.0	yellow
chlorophenol red	yellow	5.2 – 6.8	red
bromothymol blue	yellow	6.0 – 7.6	blue
litmus	red	6.0 – 8.0	blue
phenol red	yellow	6.6 – 8.0	red
metacresol purple	yellow	7.4 – 9.0	purple
thymol blue*	yellow	8.0 – 9.6	blue
phenolphthalein	colorless	8.2 – 10.0	red
thymolphthalein	colorless	9.4 – 10.6	blue
alizarin yellow R	yellow	10.1 – 12.0	red
indigo carmine	blue	11.4 – 13.0	yellow
1,3,5-trinitrobenzene	colorless	12.0 – 14.0	orange

© NELSON CANADA,
A DIVISION OF THOMSON CANADA LIMITED, 1994

68 | pH Curves

(page 480)

25.0 mL of 0.48 mol/L NaOH(aq) Titrated with 0.50 mol/L HCl(aq)

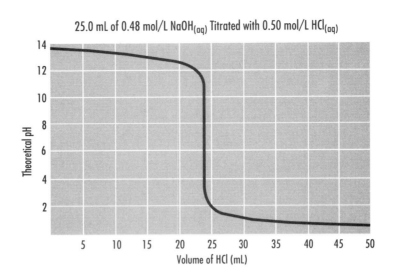

Figure 15.17
This calculated (theoretical) pH curve for the addition of 0.50 mol/L HCl$_{(aq)}$ to a 25.0 mL sample of 0.48 mol/L NaOH$_{(aq)}$ helps chemists to understand the nature of the acid-base reactions.

(page 480)

25.0 mL of 0.50 mol/L HCl(aq) Titrated with 0.48 mol/L NaOH(aq)

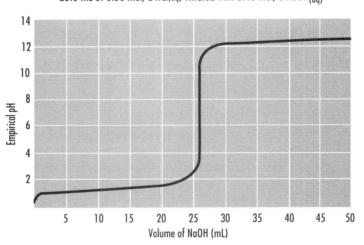

Figure 15.18
This experimentally determined (empirical) pH curve for the addition of 0.48 mol/L NaOH$_{(aq)}$ to a 25.0 mL sample of 0.50 mol/L HCl$_{(aq)}$ illustrates the typical shape of a strong acid/strong base pH curve.

Acid-Base Reactions

SA-SB
SA-WB
WA-SB
WA-WB

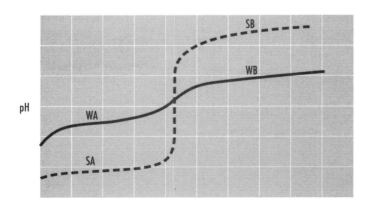

69 pH CURVES FOR POLYACIDIC AND POLYBASIC REACTIONS

(page 482)

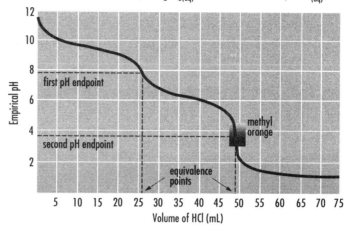

Figure 15.20
A pH curve for the addition of 0.50 mol/L $HCl_{(aq)}$ to a 25.0 mL sample of 0.50 mol/L $Na_2CO_{3(aq)}$ can be used to select an indicator for a titration.

pH Curve for $Na_2CO_{3(aq)}$

25.0 mL of 0.50 mol/L $Na_2CO_{3(aq)}$ Titrated with 0.50 mol/L $HCl_{(aq)}$

$$\underset{\underset{B}{}}{\overset{SA}{H_3O^+_{(aq)}}}, \ \underset{}{Cl^-_{(aq)}}, \ Na^+_{(aq)}, \ \underset{\underset{SB}{}}{\overset{}{CO_3^{2-}_{(aq)}}}, \ \underset{\underset{B}{}}{\overset{A}{H_2O_{(l)}}}$$

$$H_3O^+_{(aq)} + CO_3^{2-}_{(aq)} \rightarrow H_2O_{(l)} + HCO_3^-_{(aq)}$$

$$H_3O^+_{(aq)} + HCO_3^-_{(aq)} \rightarrow H_2O_{(l)} + H_2CO_{3(aq)}$$

(page 483)

25.0 mL of 0.50 mol/L $H_3PO_{4(aq)}$ Titrated with 0.48 mol/L $NaOH_{(aq)}$

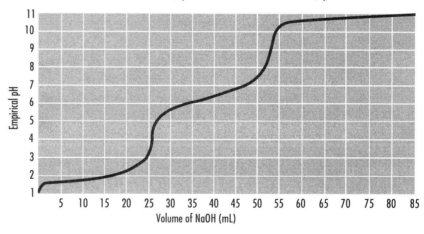

Figure 15.21
A pH curve for the addition of 0.48 mol/L $NaOH_{(aq)}$ to a 25.0 mL sample of 0.50 mol/L $H_3PO_{4(aq)}$ displays only two rapid changes in pH. This is interpreted as indicating that there are only two quantitative reactions for phosphoric acid with sodium hydroxide.

$$Na^+_{(aq)}, \ \underset{\underset{SB}{}}{\overset{}{OH^-_{(aq)}}}, \ \underset{}{\overset{SA}{\cancel{H_3PO_{4(aq)}}}} \ \underset{\underset{B}{}}{\overset{A}{H_2O_{(l)}}}, \ \underset{\underset{B}{}}{\overset{SA}{H_2PO_4^-_{(aq)}}}$$

$$OH^-_{(aq)} + H_2PO_4^-_{(aq)} \rightarrow H_2O_{(l)} + HPO_4^{2-}_{(aq)}$$

(page 483)

$$HPO_4^{2-}_{(aq)} + OH^-_{(aq)} \overset{>50\%}{\rightleftharpoons} PO_4^{3-}_{(aq)} + H_2O_{(l)}$$

As a general rule, *only quantitative reactions produce detectable endpoints in an acid-base titration.*

THE ELEMENTS

Name	Symbol	Atomic Number	Name	Symbol	Atomic Number	Name	Symbol	Atomic Number
actinium	Ac	89	helium	He	2	radium	Ra	88
aluminum	Al	13	holmium	Ho	67	radon	Rn	86
americium	Am	95	hydrogen	H	1	rhenium	Re	75
antimony	Sb	51	indium	In	49	rhodium	Rh	45
argon	Ar	18	iodine	I	53	rubidium	Rb	37
arsenic	As	33	iridium	Ir	77	ruthenium	Ru	44
astatine	At	85	iron	Fe	26	samarium	Sm	62
barium	Ba	56	krypton	Kr	36	scandium	Sc	21
berkelium	Bk	97	lanthanum	La	57	selenium	Se	34
beryllium	Be	4	lawrencium	Lr	103	silicon	Si	14
bismuth	Bi	83	lead	Pb	82	silver	Ag	47
boron	B	5	lithium	Li	3	sodium	Na	11
bromine	Br	35	lutetium	Lu	71	strontium	Sr	38
cadmium	Cd	48	magnesium	Mg	12	sulfur	S	16
calcium	Ca	20	manganese	Mn	25	tantalum	Ta	73
californium	Cf	98	mendelevium	Md	101	technetium	Tc	43
carbon	C	6	mercury	Hg	80	tellurium	Te	52
cerium	Ce	58	molybdenum	Mo	42	terbium	Tb	65
cesium	Cs	55	neodymium	Nd	60	thallium	Tl	81
chlorine	Cl	17	neon	Ne	10	thorium	Th	90
chromium	Cr	24	neptunium	Np	93	thulium	Tm	69
cobalt	Co	27	nickel	Ni	28	tin	Sn	50
copper	Cu	29	niobium	Nb	41	titanium	Ti	22
curium	Cm	96	nitrogen	N	7	unnilennium*	Une	109
dysprosium	Dy	66	nobelium	No	102	unnilhexium*	Unh	106
einsteinium	Es	99	osmium	Os	76	unnilpentium*	Unp	105
erbium	Er	68	oxygen	O	8	unnilquadium*	Unq	104
europium	Eu	63	palladium	Pd	46	unnilseptium*	Uns	107
fermium	Fm	100	phosphorus	P	15	uranium	U	92
fluorine	F	9	platinum	Pt	78	vanadium	V	23
francium	Fr	87	plutonium	Pu	94	wolfram (tungsten)	W	74
gadolinium	Gd	64	polonium	Po	84	xenon	Xe	54
gallium	Ga	31	potassium	K	19	ytterbium	Yb	70
germanium	Ge	32	praseodymium	Pr	59	yttrium	Y	39
gold	Au	79	promethium	Pm	61	zinc	Zn	30
hafnium	Hf	72	protactinium	Pa	91	zirconium	Zr	40

*These element names are derived from IUPAC prefixes for new elements. The prefix for each digit of the atomic number is combined with an -*ium* ending to form the name. The element symbol includes the first letter of each prefix in the name.

    0 - *nil*   1 - *un*   2 - *bi*   3 - *tri*   4 - *quad*   5 - *pent*   6 - *hex*   7 - *sept*   8 - *oct*   9 - *enn*

STANDARD MOLAR ENTHALPIES OF FORMATION

Chemical Name	Formula	H_f° (kJ/mol)
acetone	$(CH_3)_2CO_{(l)}$	−248.1
aluminum oxide	$Al_2O_{3(s)}$	−1675.7
ammonia	$NH_{3(g)}$	−45.9
ammonium chloride	$NH_4Cl_{(s)}$	−314.4
ammonium nitrate	$NH_4NO_{3(s)}$	−365.6
barium carbonate	$BaCO_{3(s)}$	−1216.3
barium hydroxide	$Ba(OH)_{2(s)}$	−944.7
barium oxide	$BaO_{(s)}$	−553.5
barium sulfate	$BaSO_{4(s)}$	−1473.2
benzene	$C_6H_{6(l)}$	+49.0
bromine (vapor)	$Br_{2(g)}$	+30.9
butane	$C_4H_{10(g)}$	−125.6
calcium carbonate	$CaCO_{3(s)}$	−1206.9
calcium hydroxide	$Ca(OH)_{2(s)}$	−986.1
calcium oxide	$CaO_{(s)}$	−634.9
carbon dioxide	$CO_{2(g)}$	−393.5
carbon disulfide	$CS_{2(l)}$	+89.0
carbon monoxide	$CO_{(g)}$	−110.5
chloroethene	$C_2H_3Cl_{(g)}$	+37.3
chromium(III) oxide	$Cr_2O_{3(s)}$	−1139.7
copper(I) oxide	$Cu_2O_{(s)}$	−168.6
copper(II) oxide	$CuO_{(s)}$	−157.3
copper(I) sulfide	$Cu_2S_{(s)}$	−79.5
copper(II) sulfide	$CuS_{(s)}$	−53.1
1,2-dichloroethane	$C_2H_4Cl_{2(l)}$	−126.9
ethane	$C_2H_{6(g)}$	−83.8
1,2-ethanediol	$C_2H_4(OH)_{2(l)}$	−454.8
ethanoic (acetic) acid	$CH_3COOH_{(l)}$	−432.8
ethanol	$C_2H_5OH_{(l)}$	−235.2
ethene (ethylene)	$C_2H_{4(g)}$	+52.5
ethyne (acetylene)	$C_2H_{2(g)}$	+228.2
glucose	$C_6H_{12}O_{6(s)}$	−1273.1
hexane	$C_6H_{14(l)}$	−198.7
hydrogen bromide	$HBr_{(g)}$	−36.3

Chemical Name	Formula	H_f° (kJ/mol)
hydrogen chloride	$HCl_{(g)}$	−92.3
hydrogen fluoride	$HF_{(g)}$	−273.3
hydrogen iodide	$HI_{(g)}$	+26.5
hydrogen peroxide	$H_2O_{2(l)}$	−187.8
hydrogen sulfide	$H_2S_{(g)}$	−20.6
iodine (vapor)	$I_{2(g)}$	+62.4
iron(III) oxide	$Fe_2O_{3(s)}$	−824.2
iron(II, III) oxide	$Fe_3O_{4(s)}$	−1118.4
lead(II) oxide	$PbO_{(s)}$	−219.0
lead(IV) oxide	$PbO_{2(s)}$	−277.4
magnesium carbonate	$MgCO_{3(s)}$	−1095.8
magnesium chloride	$MgCl_{2(s)}$	−641.3
magnesium hydroxide	$Mg(OH)_{2(s)}$	−924.5
magnesium oxide	$MgO_{(s)}$	−601.6
manganese(II) oxide	$MnO_{(s)}$	−385.2
manganese(IV) oxide	$MnO_{2(s)}$	−520.0
mercury(II) oxide	$HgO_{(s)}$	−90.8
mercury(II) sulfide	$HgS_{(s)}$	−58.2
methanal (formaldehyde)	$CH_2O_{(g)}$	−108.6
methane	$CH_{4(g)}$	−74.4
methanoic (formic) acid	$HCOOH_{(l)}$	−425.1
methanol	$CH_3OH_{(l)}$	−239.1
methylpropane	$C_4H_{10(g)}$	−134.2
nickel(II) oxide	$NiO_{(s)}$	−239.7
nitric acid	$HNO_{3(l)}$	−174.1
nitrogen dioxide	$NO_{2(g)}$	+33.2
nitrogen monoxide	$NO_{(g)}$	+90.2
nitromethane	$CH_3NO_{2(l)}$	−113.1
octane	$C_8H_{18(l)}$	−250.1
ozone	$O_{3(g)}$	+142.7
pentane	$C_5H_{12(l)}$	−173.5
phenylethene (styrene)	$C_6H_5CHCH_{2(l)}$	+103.8
phosphorus pentachloride	$PCl_{5(g)}$	−443.5
phosphorus trichloride (liquid)	$PCl_{3(l)}$	−319.7

Chemical Name	Formula	H_f° (kJ/mol)
phosphorus trichloride (vapor)	$PCl_{3(g)}$	−287.0
potassium chlorate	$KClO_{3(s)}$	−397.7
potassium chloride	$KCl_{(s)}$	−436.7
potassium hydroxide	$KOH_{(s)}$	−424.8
propane	$C_3H_{8(g)}$	−104.7
silicon dioxide	$SiO_{2(s)}$	−910.7
silver bromide	$AgBr_{(s)}$	−100.4
silver chloride	$AgCl_{(s)}$	−127.0
silver iodide	$AgI_{(s)}$	−61.8
sodium bromide	$NaBr_{(s)}$	−361.1
sodium chloride	$NaCl_{(s)}$	−411.2
sodium hydroxide	$NaOH_{(s)}$	−425.6
sodium iodide	$NaI_{(s)}$	−287.8
sucrose	$C_{12}H_{22}O_{11(s)}$	−2225.5
sulfur dioxide	$SO_{2(g)}$	−296.8
sulfur trioxide (liquid)	$SO_{3(l)}$	−441.0
sulfur trioxide (vapor)	$SO_{3(g)}$	−395.7
sulfuric acid	$H_2SO_{4(l)}$	−814.0
tin(II) oxide	$SnO_{(s)}$	−280.7
tin(IV) oxide	$SnO_{2(s)}$	−577.6
2,2,4-trimethylpentane	$C_8H_{18(l)}$	−259.2
urea	$CO(NH_2)_{2(s)}$	−333.5
water (liquid)	$H_2O_{(l)}$	−285.8
water (vapor)	$H_2O_{(g)}$	−241.8
zinc oxide	$ZnO_{(s)}$	−350.5
zinc sulfide	$ZnS_{(s)}$	−206.0

- Standard molar enthalpies (heats) of formation are measured at SATP (25°C and 100 kPa). The values were obtained from *The CRC Handbook of Chemistry and Physics, 71st Edition*.
- The standard molar enthalpies of elements in their standard states are defined as zero.

70 (C) DATA TABLES

REDOX HALF-REACTIONS

	Oxidizing Agents		Reducing Agents	$E°$ (V)

SOA
Strongest Oxidizing Agents

DECREASING STRENGTH OF OXIDIZING AGENTS →

Oxidizing Agents / Reducing Agents	$E°$ (V)
$F_{2(g)} + 2e^- \rightleftharpoons 2F^-_{(aq)}$	+2.87
$PbO_{2(s)} + SO_4^{2-}_{(aq)} + 4H^+_{(aq)} + 2e^- \rightleftharpoons PbSO_{4(s)} + 2H_2O_{(l)}$	+1.69
$MnO_4^-_{(aq)} + 8H^+_{(aq)} + 5e^- \rightleftharpoons Mn^{2+}_{(aq)} + 4H_2O_{(l)}$	+1.51
$Au^{3+}_{(aq)} + 3e^- \rightleftharpoons Au_{(s)}$	+1.50
$ClO_4^-_{(aq)} + 8H^+_{(aq)} + 8e^- \rightleftharpoons Cl^-_{(aq)} + 4H_2O_{(l)}$	+1.39
$Cl_{2(g)} + 2e^- \rightleftharpoons 2Cl^-_{(aq)}$	+1.36
$2HNO_{2(aq)} + 4H^+_{(aq)} + 4e^- \rightleftharpoons N_2O_{(g)} + 3H_2O_{(l)}$	+1.30
$Cr_2O_7^{2-}_{(aq)} + 14H^+_{(aq)} + 6e^- \rightleftharpoons 2Cr^{3+}_{(aq)} + 7H_2O_{(l)}$	+1.23
$O_{2(g)} + 4H^+_{(aq)} + 4e^- \rightleftharpoons 2H_2O_{(l)}$	+1.23
$MnO_{2(s)} + 4H^+_{(aq)} + 2e^- \rightleftharpoons Mn^{2+}_{(aq)} + 2H_2O_{(l)}$	+1.22
$2IO_3^-_{(aq)} + 12H^+_{(aq)} + 10e^- \rightleftharpoons I_{2(s)} + 6H_2O_{(l)}$	+1.20
$Br_{2(l)} + 2e^- \rightleftharpoons 2Br^-_{(aq)}$	+1.07
$Hg^{2+}_{(aq)} + 2e^- \rightleftharpoons Hg_{(l)}$	+0.85
$ClO^-_{(aq)} + H_2O_{(l)} + 2e^- \rightleftharpoons Cl^-_{(aq)} + 2OH^-_{(aq)}$	+0.84
$Ag^+_{(aq)} + e^- \rightleftharpoons Ag_{(s)}$	+0.80
$NO_3^-_{(aq)} + 2H^+_{(aq)} + e^- \rightleftharpoons NO_{2(g)} + H_2O_{(l)}$	+0.80
$Fe^{3+}_{(aq)} + e^- \rightleftharpoons Fe^{2+}_{(aq)}$	+0.77
$O_{2(g)} + 2H^+_{(aq)} + 2e^- \rightleftharpoons H_2O_{2(l)}$	+0.70
$I_{2(s)} + 2e^- \rightleftharpoons 2I^-_{(aq)}$	+0.54
$O_{2(g)} + 2H_2O_{(l)} + 4e^- \rightleftharpoons 4OH^-_{(aq)}$	+0.40
$Cu^{2+}_{(aq)} + 2e^- \rightleftharpoons Cu_{(s)}$	+0.34
$SO_4^{2-}_{(aq)} + 4H^+_{(aq)} + 2e^- \rightleftharpoons H_2SO_{3(aq)} + H_2O_{(l)}$	+0.17
$Sn^{4+}_{(aq)} + 2e^- \rightleftharpoons Sn^{2+}_{(aq)}$	+0.15
$1/8 S_{8(s)} + 2H^+_{(aq)} + 2e^- \rightleftharpoons H_2S_{(aq)}$	+0.14
$AgBr_{(s)} + e^- \rightleftharpoons Ag_{(s)} + Br^-_{(aq)}$	+0.07
$2H^+_{(aq)} + 2e^- \rightleftharpoons H_{2(g)}$	0.00
$Pb^{2+}_{(aq)} + 2e^- \rightleftharpoons Pb_{(s)}$	−0.13
$Sn^{2+}_{(aq)} + 2e^- \rightleftharpoons Sn_{(s)}$	−0.14
$AgI_{(s)} + e^- \rightleftharpoons Ag_{(s)} + I^-_{(aq)}$	−0.15
$Ni^{2+}_{(aq)} + 2e^- \rightleftharpoons Ni_{(s)}$	−0.26
$Co^{2+}_{(aq)} + 2e^- \rightleftharpoons Co_{(s)}$	−0.28
$PbSO_{4(s)} + 2e^- \rightleftharpoons Pb_{(s)} + SO_4^{2-}_{(aq)}$	−0.36
$Se_{(s)} + 2H^+_{(aq)} + 2e^- \rightleftharpoons H_2Se_{(aq)}$	−0.40
$Cd^{2+}_{(aq)} + 2e^- \rightleftharpoons Cd_{(s)}$	−0.40
$Cr^{3+}_{(aq)} + e^- \rightleftharpoons Cr^{2+}_{(aq)}$	−0.41
$Fe^{2+}_{(aq)} + 2e^- \rightleftharpoons Fe_{(s)}$	−0.45
$Ag_2S_{(s)} + 2e^- \rightleftharpoons 2Ag_{(s)} + S^{2-}_{(aq)}$	−0.69
$Zn^{2+}_{(aq)} + 2e^- \rightleftharpoons Zn_{(s)}$	−0.76
$Te_{(s)} + 2H^+_{(aq)} + 2e^- \rightleftharpoons H_2Te_{(aq)}$	−0.79
$2H_2O_{(l)} + 2e^- \rightleftharpoons H_{2(g)} + 2OH^-_{(aq)}$	−0.83
$Cr^{2+}_{(aq)} + 2e^- \rightleftharpoons Cr_{(s)}$	−0.91
$SO_4^{2-}_{(aq)} + H_2O_{(l)} + 2e^- \rightleftharpoons SO_3^{2-}_{(aq)} + 2OH^-_{(aq)}$	−0.93
$Al^{3+}_{(aq)} + 3e^- \rightleftharpoons Al_{(s)}$	−1.66
$Mg^{2+}_{(aq)} + 2e^- \rightleftharpoons Mg_{(s)}$	−2.37
$Na^+_{(aq)} + e^- \rightleftharpoons Na_{(s)}$	−2.71
$Ca^{2+}_{(aq)} + 2e^- \rightleftharpoons Ca_{(s)}$	−2.87
$Ba^{2+}_{(aq)} + 2e^- \rightleftharpoons Ba_{(s)}$	−2.91
$K^+_{(aq)} + e^- \rightleftharpoons K_{(s)}$	−2.93
$Li^+_{(aq)} + e^- \rightleftharpoons Li_{(s)}$	−3.04

← DECREASING STRENGTH OF REDUCING AGENTS

SRA
Strongest Reducing Agents

- All $E°$ values are reduction potentials measured relative to the standard hydrogen electrode. $E°$ values are measured at SATP using 1.0 mol/L solutions.
- Values in this table are taken from *The CRC Handbook of Chemistry and Physics*, 71st Edition.

70 (D) DATA TABLES

			ACIDS AND BASES			
Percent Reaction (%)	**Equilibrium Constant (K_a)**	**Acid**		**Conjugate Base**		
		Name	**Formula**	**Formula**	**Name**	
100	very large	perchloric acid	$HClO_{4(aq)}$	$ClO_4^-{}_{(aq)}$	perchlorate ion	
100	3.2×10^9	hydroiodic acid	$HI_{(aq)}$	$I^-{}_{(aq)}$	iodide ion	
100	1.0×10^9	hydrobromic acid	$HBr_{(aq)}$	$Br^-{}_{(aq)}$	bromide ion	
100	1.3×10^6	hydrochloric acid	$HCl_{(aq)}$	$Cl^-{}_{(aq)}$	chloride ion	
100	1.0×10^3	sulfuric acid	$H_2SO_{4(aq)}$	$HSO_4^-{}_{(aq)}$	hydrogen sulfate ion	
100	2.4×10^1	nitric acid	$HNO_{3(aq)}$	$NO_3^-{}_{(aq)}$	nitrate ion	
—	—	hydronium ion	$H_3O^+{}_{(aq)}$	$H_2O_{(l)}$	water	
51	5.4×10^{-2}	oxalic acid	$HOOCCOOH_{(aq)}$	$HOOCCOO^-{}_{(aq)}$	hydrogen oxalate ion	
30	1.3×10^{-2}	sulfurous acid ($SO_2 + H_2O$)	$H_2SO_{3(aq)}$	$HSO_3^-{}_{(aq)}$	hydrogen sulfite ion	
27	1.0×10^{-2}	hydrogen sulfate ion	$HSO_4^-{}_{(aq)}$	$SO_4^{2-}{}_{(aq)}$	sulfate ion	
23	7.1×10^{-3}	phosphoric acid	$H_3PO_{4(aq)}$	$H_2PO_4^-{}_{(aq)}$	dihydrogen phosphate ion	
8.1	7.2×10^{-4}	nitrous acid	$HNO_{2(aq)}$	$NO_2^-{}_{(aq)}$	nitrite ion	
7.8	6.6×10^{-4}	hydrofluoric acid	$HF_{(aq)}$	$F^-{}_{(aq)}$	fluoride ion	
4.2	1.8×10^{-4}	methanoic acid	$HCOOH_{(aq)}$	$HCOO^-{}_{(aq)}$	methanoate ion	
—	$\sim 10^{-4}$	methyl orange	$HMo_{(aq)}$	$Mo^-{}_{(aq)}$	methyl orange ion	
—	6.3×10^{-5}	benzoic acid	$C_6H_5COOH_{(aq)}$	$C_6H_5COO^-{}_{(aq)}$	benzoate ion	
2.3	5.4×10^{-5}	hydrogen oxalate ion	$HOOCCOO^-{}_{(aq)}$	$OOCCOO^{2-}{}_{(aq)}$	oxalate ion	
1.3	1.8×10^{-5}	ethanoic (acetic) acid	$CH_3COOH_{(aq)}$	$CH_3COO^-{}_{(aq)}$	ethanoate (acetate) ion	
—	4.4×10^{-7}	carbonic acid ($CO_2 + H_2O$)	$H_2CO_{3(aq)}$	$HCO_3^-{}_{(aq)}$	hydrogen carbonate ion	
—	$\sim 10^{-7}$	bromothymol blue	$HBb_{(aq)}$	$Bb^-{}_{(aq)}$	bromothymol blue ion	
0.10	1.1×10^{-7}	hydrosulfuric acid	$H_2S_{(aq)}$	$HS^-{}_{(aq)}$	hydrogen sulfide ion	
0.079	6.3×10^{-8}	dihydrogen phosphate ion	$H_2PO_4^-{}_{(aq)}$	$HPO_4^{2-}{}_{(aq)}$	hydrogen phosphate ion	
0.079	6.2×10^{-8}	hydrogen sulfite ion	$HSO_3^-{}_{(aq)}$	$SO_3^{2-}{}_{(aq)}$	sulfite ion	
0.054	2.9×10^{-8}	hypochlorous acid	$HClO_{(aq)}$	$ClO^-{}_{(aq)}$	hypochlorite ion	
—	$\sim 10^{-10}$	phenolphthalein	$HPh_{(aq)}$	$Ph^-{}_{(aq)}$	phenolphthalein ion	
0.0078	6.2×10^{-10}	hydrocyanic acid	$HCN_{(aq)}$	$CN^-{}_{(aq)}$	cyanide ion	
0.0076	5.8×10^{-10}	ammonium ion	$NH_4^+{}_{(aq)}$	$NH_{3(aq)}$	ammonia	
0.0076	5.8×10^{-10}	boric acid	$H_3BO_{3(aq)}$	$H_2BO_3^-{}_{(aq)}$	dihydrogen borate ion	
0.0022	4.7×10^{-11}	hydrogen carbonate ion	$HCO_3^-{}_{(aq)}$	$CO_3^{2-}{}_{(aq)}$	carbonate ion	
0.00020	4.2×10^{-13}	hydrogen phosphate ion	$HPO_4^{2-}{}_{(aq)}$	$PO_4^{3-}{}_{(aq)}$	phosphate ion	
0.00013	1.8×10^{-13}	dihydrogen borate ion	$H_2BO_3^-{}_{(aq)}$	$HBO_3^{2-}{}_{(aq)}$	hydrogen borate ion	
0.00011	1.3×10^{-13}	hydrogen sulfide ion	$HS^-{}_{(aq)}$	$S^{2-}{}_{(aq)}$	sulfide ion	
0.000040	1.6×10^{-14}	hydrogen borate ion	$HBO_3^{2-}{}_{(aq)}$	$BO_3^{3-}{}_{(aq)}$	borate ion	
—	—	water	$H_2O_{(l)}$	$OH^-{}_{(aq)}$	hydroxide ion	

SA Strongest Acids

DECREASING STRENGTH OF ACIDS

DECREASING STRENGTH OF BASES

SB Strongest Bases

- The percent reaction of acids with water is for 0.10 mol/L solutions and is only valid for concentrations close to 0.10 mol/L. All measurements of acid strengths were made at SATP. No percent reaction is given for benzoic acid or carbonic acid because these acids have molar solubilities less than 0.10 mol/L at SATP. No percent reaction is given for indicators because indicators are generally used at concentrations lower than 0.10 mol/L.

- Values in this table are taken from *Lange's Handbook of Chemistry*, 13th Edition.

87

ADDITIONAL EXERCISES

This section of *Nelson Chemistry Teacher's Resource Masters* includes Extra Practice exercises for the essential topics in *Nelson Chemistry*, Enrichment exercises that allow more able students to go beyond these topics to expand their learning, and Video exercises for some of the video resources suggested in the *Teacher's Edition*. The blackline masters are provided to save (and create) time for your productive use.

- The **Extra Practice** exercises have been created and selected to assist students with topics that are considered essential learning. Without the basics, some students may have trouble with topics that follow, and some may need more practice than others. These blackline masters should allow you to provide assistance to these students without taking extra time from your normal workload with average students. Most of the blackline masters have questions printed on the front of the page and solutions on the back. You will have to decide whether you want to provide the solutions on the back of the question sheet when you make copies for students. The advantage of printing the questions and solutions back-to-back is that the students could become independent learners and take charge of their own learning.

    One way of handling these exercises is to print class sets of them at or before the beginning of the school year and place them in a filing cabinet (or equivalent) in labelled file folders. You can then hand them out whenever you wish. Students absent from a particular class period can find the exercises themselves when they return. Alternatively, only students who need extra practice should pick up the Extra Practice exercises from your filing system. Normally, you would not have to go over these exercises in class. The solutions provided with the questions may take care of any difficulties that students have with the topic.

- The **Enrichment** exercises are on selected topics that provide extensions of core material. These exercises would not usually be handed out to all students. You may want to indicate that such exercises are available for students to work through on their own, but that the content is not going to be covered in regular tests.

- The **Video** exercises provide sample questions that can be asked about information presented in some of the videos. You could add exercises of your own for your favorite videos.

These extra exercises should provide a foundation of practice and enrichment exercises that you can build on over the years as your own resource. You might want to add lab exercises and other enrichment exercises to this section.

MASTER LIST OF ADDITIONAL EXERCISES

1 PERIODIC TABLE ASSIGNMENT

Use the _Nelson Chemistry_ periodic table to complete the following table.

	Element Name	IUPAC Symbol	Atomic Number	Group Number	Period Number	Metal (m) or Nonmetal (nm)	SATP State	Family/Series Name
1.	chlorine							
2.	magnesium							
3.			30					
4.		N						
5.				17	5			
6.			79					
7.	‸				3			alkali metals
8.	thorium			———				
9.				12			liquid	
10.		Br						
11.	argon							
12.				11	5			
13.			19					
14.	calcium							
15.				1			gas	
16.			58	———				

1 PERIODIC TABLE ASSIGNMENT

Use the *Nelson Chemistry* periodic table to complete the following table.

	Element Name	IUPAC Symbol	Atomic Number	Group Number	Period Number	Metal (m) or Nonmetal (nm)	SATP State	Family/Series Name
1.	chlorine	**Cl**	**17**	**17**	**3**	**nm**	**gas**	**halogens**
2.	magnesium	**Mg**	**12**	**2**	**3**	**m**	**solid**	**alkaline-earths**
3.	**zinc**	**Zn**	30	**12**	**4**	**m**	**solid**	**transition elements**
4.	**nitrogen**	**N**	**7**	**15**	**2**	**nm**	**gas**	**representative elements**
5.	**iodine**	**I**	**53**	17	5	**nm**	**solid**	**halogens**
6.	**gold**	**Au**	79	**11**	6	**m**	**solid**	**transition elements**
7.	**sodium**	**Na**	**11**	**1**	3	**m**	**solid**	alkali metals
8.	thorium	**Th**	**90**	——	7	**m**	**solid**	**actinides**
9.	**mercury**	**Hg**	**80**	12	6	**m**	liquid	**transition elements**
10.	**bromine**	Br	**35**	**17**	**4**	**nm**	liquid	**halogens**
11.	argon	**Ar**	**18**	**18**	**3**	**nm**	**gas**	noble gases
12.	silver	**Ag**	**47**	11	5	**m**	**solid**	**transition elements**
13.	**potassium**	**K**	19	**1**	**4**	**m**	**solid**	alkali metals
14.	calcium	**Ca**	**20**	**2**	**4**	**m**	**solid**	alkaline-earths
15.	**hydrogen**	**H**	**1**	1	**1**	**nm**	gas	representative elements
16.	**cerium**	**Ce**	58	——	6	**m**	**solid**	lanthanides

CHAPTER 2   PAGES 52–55

2 ATOMIC THEORY: ISOTOPES

Complete the following table.

	Isotope Name	Atomic Number	Mass Number	Symbol	Number of Protons	Number of Neutrons
1.	carbon-14					
2.		8	16			
3.					84	128
4.		92				146
5.	hydrogen-2					
6.		2	4			
7.					90	142
8.			12		6	
9.	lawrencium-257					
10.			1			0

2 ATOMIC THEORY: ISOTOPES

Complete the following table.

	Isotope Name	Atomic Number	Mass Number	Symbol	Number of Protons	Number of Neutrons
1.	carbon-14	6	14	$^{14}_{6}\text{C}$	6	8
2.	oxygen-16	8	16	$^{16}_{8}\text{O}$	8	8
3.	polonium-212	84	212	$^{212}_{84}\text{Po}$	84	128
4.	uranium-238	92	238	$^{238}_{92}\text{U}$	92	146
5.	hydrogen-2	1	2	$^{2}_{1}\text{H}$	1	1
6.	helium-4	2	4	$^{4}_{2}\text{He}$	2	2
7.	thorium-232	90	232	$^{232}_{90}\text{Th}$	90	142
8.	carbon-12	6	12	$^{12}_{6}\text{C}$	6	6
9.	lawrencium-257	103	257	$^{257}_{103}\text{Lr}$	103	154
10.	hydrogen-1	1	1	$^{1}_{1}\text{H}$	1	0

3 ELECTRON ENERGY-LEVEL DIAGRAMS FOR ATOMS

1								18

		13 IIIA	14 IVA	15 VA	16 VIA	17 VIIA	18 VIIIA
		$\frac{3e^-}{}$ $\frac{8e^-}{}$ $\frac{2e^-}{}$ $13p^+$ Al					

1				

	1 IA	2 IIA
2		
3		
4		

3   ELECTRON ENERGY-LEVEL DIAGRAMS FOR ATOMS

18

2
$\frac{2\,e^-}{2\,p^+}$
He

VIIIA

10
$\frac{8\,e^-}{2\,e^-}$
$10\,p^+$
Ne

18
$\frac{8\,e^-}{\frac{8\,e^-}{2\,e^-}}$
$18\,p^+$
Ar

17 VIIA

9
$\frac{7\,e^-}{2\,e^-}$
$9\,p^+$
F

17
$\frac{7\,e^-}{\frac{8\,e^-}{2\,e^-}}$
$17\,p^+$
Cl

16 VIA

8
$\frac{6\,e^-}{2\,e^-}$
$8\,p^+$
O

16
$\frac{6\,e^-}{\frac{8\,e^-}{2\,e^-}}$
$16\,p^+$
S

15 VA

7
$\frac{5\,e^-}{2\,e^-}$
$7\,p^+$
N

15
$\frac{5\,e^-}{\frac{8\,e^-}{2\,e^-}}$
$15\,p^+$
P

14 IVA

6
$\frac{4\,e^-}{2\,e^-}$
$6\,p^+$
C

14
$\frac{4\,e^-}{\frac{8\,e^-}{2\,e^-}}$
$14\,p^+$
Si

13 IIIA

5
$\frac{3\,e^-}{2\,e^-}$
$5\,p^+$
B

13
$\frac{3\,e^-}{\frac{8\,e^-}{2\,e^-}}$
$13\,p^+$
Al

1

1
$\frac{1\,e^-}{1\,p^+}$
H

2 IIA

4
$\frac{2\,e^-}{2\,e^-}$
$4\,p^+$
Be

12
$\frac{2\,e^-}{\frac{8\,e^-}{2\,e^-}}$
$12\,p^+$
Mg

20
$\frac{2\,e^-}{\frac{8\,e^-}{\frac{8\,e^-}{2\,e^-}}}$
$20\,p^+$
Ca

IA

3
$\frac{1\,e^-}{2\,e^-}$
$3\,p^+$
Li

11
$\frac{1\,e^-}{\frac{8\,e^-}{2\,e^-}}$
$11\,p^+$
Na

19
$\frac{1\,e^-}{\frac{8\,e^-}{\frac{8\,e^-}{2\,e^-}}}$
$19\,p^+$
K

1

2

3

4

4   ELECTRON ENERGY-LEVEL DIAGRAMS FOR IONS

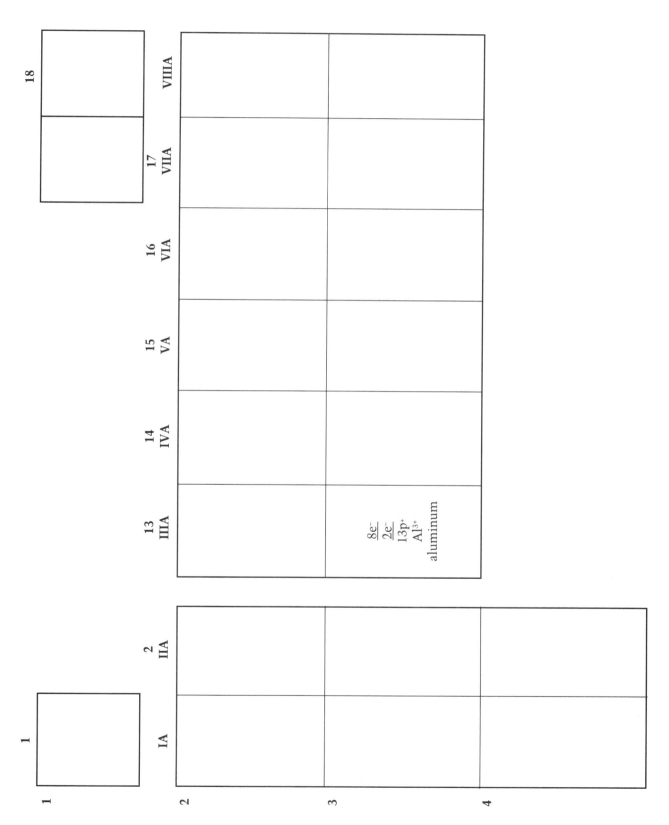

	1

	2 IIA

13 IIIA	14 IVA	15 VA	16 VIA	17 VIIA	18

8e⁻
2e⁻
13p⁺
Al³⁺
aluminum

IA	IIA		
2			
3			
4			

4 ELECTRON ENERGY-LEVEL DIAGRAMS FOR IONS

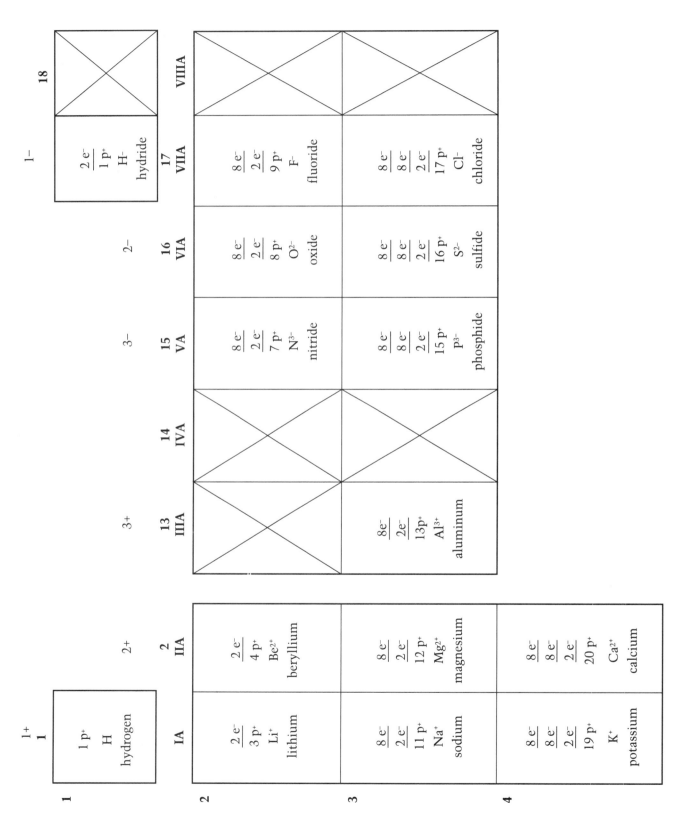

	1+ 1	2+ 2 IIA	13 IIIA 3+	14 IVA	15 VA 3−	16 VIA 2−	17 VIIA 1−	18
1	1 p⁺ H hydrogen						2 e⁻ / 1 p⁺ H⁻ hydride	
2 IA / IIA	2 e⁻ / 3 p⁺ Li⁺ lithium	2 e⁻ / 4 p⁺ Be²⁺ beryllium			8 e⁻ / 2 e⁻ / 7 p⁺ N³⁻ nitride	8 e⁻ / 2 e⁻ / 8 p⁺ O²⁻ oxide	8 e⁻ / 2 e⁻ / 9 p⁺ F⁻ fluoride	
3	8 e⁻ / 2 e⁻ / 11 p⁺ Na⁺ sodium	8 e⁻ / 2 e⁻ / 12 p⁺ Mg²⁺ magnesium	8e⁻ / 2e⁻ / 13p⁺ Al³⁺ aluminum		8 e⁻ / 8 e⁻ / 2 e⁻ / 15 p⁺ P³⁻ phosphide	8 e⁻ / 8 e⁻ / 2 e⁻ / 16 p⁺ S²⁻ sulfide	8 e⁻ / 8 e⁻ / 2 e⁻ / 17 p⁺ Cl⁻ chloride	
4	8 e⁻ / 8 e⁻ / 2 e⁻ / 19 p⁺ K⁺ potassium	8 e⁻ / 8 e⁻ / 2 e⁻ / 20 p⁺ Ca²⁺ calcium						

© NELSON CANADA,
A DIVISION OF THOMSON CANADA LIMITED, 1994

5 ATOMIC THEORY: ATOMS AND IONS

Complete the following table.

	English Name	International Symbol	Number of Protons	Number of Electrons	Number of Electrons Lost or Gained	Net Charge
1.	neon atom					
2.	lithium ion				lost 1	
3.			47			1+
4.				18		2–
5.		Si				
6.			33	36		
7.				54	lost 1	
8.			30	28		
9.				1	0	
10.		P				
11.		Ca^{2+}				
12.	selenide ion					
13.			13			3+
14.		Rb^{+}				
15.			18	18		
16.			8	10		
17.	iodine atom					
18.		Pu				
19.				54	gained 2	
20.	unnilseptium atom					

5 ATOMIC THEORY: ATOMS AND IONS

Complete the following table.

	English Name	International Symbol	Number of Protons	Number of Electrons	Number of Electrons Lost or Gained	Net Charge
1.	neon atom	Ne	10	10	0	0
2.	lithium ion	Li⁺	3	2	lost 1	1+
3.	silver ion	Ag⁺	47	46	lost 1	1+
4.	sulfide ion	S²⁻	16	18	gained 2	2−
5.	silicon atom	Si	14	14	0	0
6.	arsenide ion	As³⁻	33	36	gained 3	3−
7.	cesium ion	Cs⁺	55	54	lost 1	1+
8.	zinc ion	Zn²⁺	30	28	lost 2	2+
9.	hydrogen atom	H	1	1	0	0
10.	phosphorus atom	P	15	15	0	0
11.	calcium ion	Ca²⁺	20	18	lost 2	2+
12.	selenide ion	Se²⁻	34	36	gained 2	2−
13.	aluminum ion	Al³⁺	13	10	lost 3	3+
14.	rubidium ion	Rb⁺	37	36	lost 1	1+
15.	argon atom	Ar	18	18	0	0
16.	oxide ion	O²⁻	8	10	gained 2	2−
17.	iodine atom	I	53	53	0	0
18.	plutonium atom	Pu	94	94	0	0
19.	telluride ion	Te²⁻	52	54	gained 2	2−
20.	unnilseptium atom	Uns	107	107	0	0

6 NAMING IONS

1. Write the name of each of the ions given. Spelling is important.

Positive Ions	Negative Ions
Na^+	F^-
K^+	Cl^-
Mg^{2+}	Br^-
Ca^{2+}	I^-
Al^{3+}	O^{2-}
Ga^{3+}	S^{2-}
Fe^{2+}	N^{3-}
Fe^{3+}	P^{3-}

2. For each of the ions (except the iron ions) in the table above, label the group number using the periodic table. What pattern do you notice?

3. What problem arises when naming Fe^{2+} and Fe^{3+}? What is the solution to this problem?

4. Using your table of polyatomic ions (on the back of the periodic table sheet), name or write the formula for each of the following ions.

   (a) sulfate ion

   (b) NO_3^-

   (c) hydrogen carbonate ion

   (d) bisulfite ion

   (e) CO_3^{2-}

   (f) OH^-

6 NAMING IONS

1. Write the name of each of the ions given. Spelling is important.

Positive Ions	Negative Ions
Na^+ **sodium ion Group 1**	F^- **fluoride ion Group 17**
K^+ **potassium ion Group 1**	Cl^- **chloride ion Group 17**
Mg^{2+} **magnesium ion Group 2**	Br^- **bromide ion Group 17**
Ca^{2+} **calcium ion Group 2**	I^- **iodide ion Group 17**
Al^{3+} **aluminum ion Group 13**	O^{2-} **oxide Group 16**
Ga^{3+} **gallium ion Group 13**	S^{2-} **sulfide ion Group 16**
Fe^{2+} **iron ion**	N^{3-} **nitride ion Group 15**
Fe^{3+} **iron ion**	P^{3-} **phosphide ion Group 15**

2. For each of the ions (except the iron ions) in the table above, label the group number using the periodic table. What pattern do you notice?

   Group 1 forms 1+ ions   **Group 15 forms 3– ions**

   Group 2 forms 2+ ions   **Group 16 forms 2– ions**

   Group 13 forms 3+ ions   **Group 17 forms 1– ions**

3. What problem arises when naming Fe^{2+} and Fe^{3+}? What is the solution to this problem?

   Fe^{2+} and Fe^{3+} are different ions but are both iron ions.

   The Stock system uses a Roman numeral to distinguish these ions; Fe^{2+} is the iron(II) ion, Fe^{3+} is the iron(III) ion.

4. Using your table of polyatomic ions (on the back of the periodic table sheet), name or write the formula for each of the following ions.

   (a) sulfate ion   **SO_4^{2-}**

   (b) NO_3^- **nitrate ion**

   (c) hydrogen carbonate ion   **HCO_3^-**

   (d) bisulfite ion   **HSO_3^-**

   (e) CO_3^{2-} **carbonate ion**

   (f) OH^- **hydroxide ion**

CHAPTER 3   PAGES 80–84

7 NAMING IONIC COMPOUNDS

Write the English name of each of the compounds given. (There are no multi-valent ions in this exercise.)

	Chemical Formula	Name of Compound
1.	$AlCl_{3(s)}$	
2.	$NaI_{(s)}$	
3.	$MgO_{(s)}$	
4.	$K_2S_{(s)}$	
5.	$CaF_{2(s)}$	
6.	$GaBr_{3(s)}$	
7.	$MgCO_{3(s)}$	
8.	$Na_2SO_{4(s)}$	
9.	$CaCrO_{4(s)}$	
10.	$KOH_{(s)}$	
11.	$Zn(CH_3COO)_{2(s)}$	
12.	$Ca_3(PO_4)_{2(s)}$	
13.	$Sr(OH)_{2(s)}$	
14.	$Na_2SiO_{3(s)}$	
15.	$NH_4NO_{3(s)}$	

7 NAMING IONIC COMPOUNDS

Write the English name of each of the compounds given. (There are no multi-valent ions in this exercise.)

	Chemical Formula	Name of Compound
1.	$AlCl_{3(s)}$	**aluminum chloride**
2.	$NaI_{(s)}$	**sodium iodide**
3.	$MgO_{(s)}$	**magnesium oxide**
4.	$K_2S_{(s)}$	**potassium sulfide**
5.	$CaF_{2(s)}$	**calcium fluoride**
6.	$GaBr_{3(s)}$	**gallium bromide**
7.	$MgCO_{3(s)}$	**magnesium carbonate**
8.	$Na_2SO_{4(s)}$	**sodium sulfate**
9.	$CaCrO_{4(s)}$	**calcium chromate**
10.	$KOH_{(s)}$	**potassium hydroxide**
11.	$Zn(CH_3COO)_{2(s)}$	**zinc acetate**
12.	$Ca_3(PO_4)_{2(s)}$	**calcium phosphate**
13.	$Sr(OH)_{2(s)}$	**strontium hydroxide**
14.	$Na_2SiO_{3(s)}$	**sodium silicate**
15.	$NH_4NO_{3(s)}$	**ammonium nitrate**

8 MONO-VALENT BINARY IONIC COMPOUNDS

Complete the *Prediction* section of the following lab report.

Problem

What are the reaction products when various elements react?

Experimental Design

Various elements are reacted and the empirical formulas of the reaction products are determined.

Prediction

According to the theory of ionic compounds, the chemical *formulas* and *names* of the products formed when the following elements react are presented below.

	Reacting Elements Formula of Product	Name of Product
1.	$Mg_{(s)} + O_{2(g)} \longrightarrow$	
2.	$Al_{(s)} + Cl_{2(g)} \longrightarrow$	
3.	$Ba_{(s)} + S_{8(s)} \longrightarrow$	
4.	$Ag_{(s)} + Br_{2(l)} \longrightarrow$	
5.	$Sc_{(s)} + F_{2(g)} \longrightarrow$	
6.	$K_{(s)} + N_{2(g)} \longrightarrow$	
7.	$Ca_{(s)} + I_{2(s)} \longrightarrow$	
8.	$Ge_{(s)} + O_{2(g)} \longrightarrow$	
9.	$Cs_{(s)} + O_{2(g)} \longrightarrow$	
10.	$Zn_{(s)} + S_{8(s)} \longrightarrow$	

Translate the following equation into an English word equation.

$$Na_{(s)} + S_{8(s)} \longrightarrow Na_2S_{(s)}$$

8 MONO-VALENT BINARY IONIC COMPOUNDS

Complete the *Prediction* section of the following lab report.

Problem

What are the reaction products when various elements react?

Experimental Design

Various elements are reacted and the empirical formulas of the reaction products are determined.

Prediction

According to the theory of ionic compounds, the chemical *formulas* and *names* of the products formed when the following elements react are presented below.

	Reacting Elements Formula of Product	Name of Product
1.	$Mg_{(s)} + O_{2(g)} \rightarrow MgO_{(s)}$	magnesium oxide
2.	$Al_{(s)} + Cl_{2(g)} \rightarrow AlCl_{3(s)}$	aluminum chloride
3.	$Ba_{(s)} + S_{8(s)} \rightarrow BaS_{(s)}$	barium sulfide
4.	$Ag_{(s)} + Br_{2(l)} \rightarrow AgBr_{(s)}$	silver bromide
5.	$Sc_{(s)} + F_{2(g)} \rightarrow ScF_{3(s)}$	scandium fluoride
6.	$K_{(s)} + N_{2(g)} \rightarrow K_3N_{(s)}$	potassium nitride
7.	$Ca_{(s)} + I_{2(s)} \rightarrow CaI_{2(s)}$	calcium iodide
8.	$Ge_{(s)} + O_{2(g)} \rightarrow GeO_{2(s)}$	germanium oxide
9.	$Cs_{(s)} + O_{2(g)} \rightarrow Cs_2O_{(s)}$	cesium oxide
10.	$Zn_{(s)} + S_{8(s)} \rightarrow ZnS_{(s)}$	zinc sulfide

Translate the following equation into an English word equation.

$Na_{(s)} + S_{8(s)} \rightarrow Na_2S_{(s)}$

sodium + sulfur $\rightarrow$ sodium sulfide

9 MULTI-VALENT BINARY IONIC COMPOUNDS

Complete the *Prediction* section of the following lab report.

Problem

What are the reaction products when various elements react?

Experimental Design

Various elements are reacted and the empirical formulas of the reaction products are determined.

Prediction

Unless otherwise indicated, according to the most common charge on each metal ion, the chemical *formulas* and *names* of the products formed when the following elements react are presented below.

	Reacting Elements Formula of Product	Name of Product
1.	$Fe_{(s)} + O_{2(g)} \longrightarrow$	
2.	$Hg_{(l)} + Cl_{2(g)} \longrightarrow$	
3.	$Pb_{(s)} + Br_{2(l)} \longrightarrow$	
4.	$Co_{(s)} + I_{2(s)} \longrightarrow$	
5.	$Sn_{(s)} + F_{2(g)} \longrightarrow$	
6.	$Fe_{(s)} + S_{8(s)} \longrightarrow$	iron(II) sulfide
7.	$Cu_{(s)} + O_{2(s)} \longrightarrow$	
8.	$Pb_{(s)} + Cl_{2(g)} \longrightarrow$	lead(IV) chloride
9.	$Sn_{(s)} + Br_{2(l)} \longrightarrow$	tin(IV) bromide
10.	$Cr_{(s)} + I_{2(s)} \longrightarrow$	

Translate the following equation into an English word equation.

$Fe_{(s)} + S_{8(s)} \longrightarrow Fe_2S_{3(s)}$

9 MULTI-VALENT BINARY IONIC COMPOUNDS

Complete the *Prediction* section of the following lab report.

Problem

What are the reaction products when various elements react?

Experimental Design

Various elements are reacted and the empirical formulas of the reaction products are determined.

Prediction

Unless otherwise indicated, according to the most common charge on each metal ion, the chemical *formulas* and *names* of the products formed when the following elements react are presented below.

	Reacting Elements Formula of Product	Name of Product
1.	$Fe_{(s)} + O_{2(g)} \rightarrow$ **$Fe_2O_{3(s)}$**	**iron(III) oxide**
2.	$Hg_{(l)} + Cl_{2(g)} \rightarrow$ **$HgCl_{2(s)}$**	**mercury(II) chloride**
3.	$Pb_{(s)} + Br_{2(l)} \rightarrow$ **$PbBr_{2(s)}$**	**lead(II) bromide**
4.	$Co_{(s)} + I_{2(s)} \rightarrow$ **$CoI_{2(s)}$**	**cobalt(II) iodide**
5.	$Sn_{(s)} + F_{2(g)} \rightarrow$ **$SnF_{4(s)}$**	**tin(IV) fluoride**
6.	$Fe_{(s)} + S_{8(s)} \rightarrow$ **$FeS_{(s)}$**	iron(II) sulfide
7.	$Cu_{(s)} + O_{2(s)} \rightarrow$ **$CuO_{(s)}$**	**copper(II) oxide**
8.	$Pb_{(s)} + Cl_{2(g)} \rightarrow$ **$PbCl_{4(s)}$**	lead(IV) chloride
9.	$Sn_{(s)} + Br_{2(l)} \rightarrow$ **$SnBr_{4(s)}$**	tin(IV) bromide
10.	$Cr_{(s)} + I_{2(s)} \rightarrow$ **$CrI_{3(s)}$**	**chromium(III) iodide**

Translate the following equation into an English word equation.

$Fe_{(s)} + S_{8(s)} \rightarrow Fe_2S_{3(s)}$

iron + sulfur $\rightarrow$ iron(III) sulfide

10 MIXED IONIC COMPOUNDS

Complete the *Prediction* section of the following lab report.

Problem

What are the reaction products when various elements react?

Experimental Design

Various elements are reacted and the empirical formulas of the reaction products are determined.

Prediction

According to the theory of ionic compounds, the chemical *formulas* and *names* of the products formed when the following elements react are presented below.

	Reacting Elements Formula of Product	Name of Product
1.	$Ba_{(s)} + O_{2(g)} \longrightarrow$	
2.	$K_{(s)} + S_{8(s)} \longrightarrow$	
3.	$K_{(s)} + S_{8(s)} + O_{2(g)} \longrightarrow$	potassium sulfate
4.	$Ca_{(s)} + P_{4(s)} \longrightarrow$	
5.	$Bi_{(s)} + F_{2(g)} \longrightarrow$	
6.	$Mg_{(s)} + P_{4(s)} + O_{2(g)} \longrightarrow$	magnesium phosphate
7.	$Fe_{(s)} + Se_{(s)} \longrightarrow$	
8.	$Sr_{(s)} + N_{2(g)} \longrightarrow$	
9.	$Cr_{(s)} + Si_{(s)} + O_{2(g)} \longrightarrow Cr_2(SiO_3)_3$	
10.	$N_{2(g)} + H_{2(g)} + C_{(s)} + O_{2(g)} \longrightarrow$	ammonium carbonate

Translate the following equation into an English word equation.

$Fe_{(s)} + S_{8(s)} + O_{2(g)} \longrightarrow Fe_2(SO_4)_{3(s)}$

10   MIXED IONIC COMPOUNDS

Complete the *Prediction* section of the following lab report.

Problem

What are the reaction products when various elements react?

Experimental Design

Various elements are reacted and the empirical formulas of the reaction products are determined.

Prediction

According to the theory of ionic compounds, the chemical *formulas* and *names* of the products formed when the following elements react are presented below.

	Reacting Elements Formula of Product	Name of Product
1.	$Ba_{(s)} + O_{2(g)} \rightarrow BaO_{(s)}$	**barium oxide**
2.	$K_{(s)} + S_{8(s)} \rightarrow K_2S_{(s)}$	**potassium sulfide**
3.	$K_{(s)} + S_{8(s)} + O_{2(g)} \rightarrow K_2SO_{4(s)}$	potassium sulfate
4.	$Ca_{(s)} + P_{4(s)} \rightarrow Ca_3P_{2(s)}$	**calcium phosphide**
5.	$Bi_{(s)} + F_{2(g)} \rightarrow BiF_{3(s)}$	**bismuth(III) fluoride**
6.	$Mg_{(s)} + P_{4(s)} + O_{2(g)} \rightarrow Mg_3(PO_4)_{2(s)}$	magnesium phosphate
7.	$Fe_{(s)} + Se_{(s)} \rightarrow Fe_2Se_{3(s)}$	**iron(III) selenide**
8.	$Sr_{(s)} + N_{2(g)} \rightarrow Sr_3N_{2(s)}$	**strontium nitride**
9.	$Cr_{(s)} + Si_{(s)} + O_{2(g)} \rightarrow Cr_2(SiO_3)_3$	**chromium(III) silicate**
10.	$N_{2(g)} + H_{2(g)} + C_{(s)} + O_{2(g)} \rightarrow (NH_4)_2CO_{3(s)}$	ammonium carbonate

Translate the following equation into an English word equation.

$Fe_{(s)} + S_{8(s)} + O_{2(g)} \rightarrow Fe_2(SO_4)_{3(s)}$

iron + sulfur + oxygen $\rightarrow$ iron(III) sulfate

11 IONIC NOMENCLATURE

Write the international chemical formula or the English IUPAC name for each of the compounds given. (This exercise involves all classes of ionic compounds.)

	International Chemical Formula	IUPAC Name
1.	$SrCl_{2(s)}$	
2.	$RbBr_{(s)}$	
3.	$Na_2O_{(s)}$	
4.		aluminum sulfide
5.		zinc chloride
6.		magnesium iodide
7.	$CoCl_{2(s)}$	
8.	$TiO_{2(s)}$	
9.	$Cu_2O_{(s)}$	
10.		tin(II) sulfide
11.		chromium(III) oxide
12.		iron(II) sulfide
13.	$KC_6H_5COO_{(s)}$	
14.	$Na_2S_2O_{3(s)}$	
15.	$NH_4HCO_{3(s)}$	
16.		ammonium sulfide
17.		barium sulfite
18.		magnesium hydroxide
19.	$FeSO_4 \cdot 7H_2O_{(s)}$	
20.	$LiCl \cdot 4H_2O_{(s)}$	
21.		sodium sulfate decahydrate
22.	$Au(NO_3)_{3(s)}$	
23.		bismuth(III) sulfate
24.		lead(II) acetate-3-water
25.	$KMnO_{4(s)}$	

11 IONIC NOMENCLATURE

Write the international chemical formula or the English IUPAC name for each of the compounds given. (This exercise involves all classes of ionic compounds.)

	International Chemical Formula	IUPAC Name
1.	$SrCl_{2(s)}$	**strontium chloride**
2.	$RbBr_{(s)}$	**rubidium bromide**
3.	$Na_2O_{(s)}$	**sodium oxide**
4.	**$Al_2S_{3(s)}$**	aluminum sulfide
5.	**$ZnCl_{2(s)}$**	zinc chloride
6.	**$MgI_{2(s)}$**	magnesium iodide
7.	$CoCl_{2(s)}$	**cobalt(II) chloride**
8.	$TiO_{2(s)}$	**titanium(IV) oxide**
9.	$Cu_2O_{(s)}$	**copper(I) oxide**
10.	**$SnS_{(s)}$**	tin(II) sulfide
11.	**$Cr_2O_{3(s)}$**	chromium(III) oxide
12.	**$FeS_{(s)}$**	iron(II) sulfide
13.	$KC_6H_5COO_{(s)}$	**potassium benzoate**
14.	$Na_2S_2O_{3(s)}$	**sodium thiosulfate**
15.	$NH_4HCO_{3(s)}$	**ammonium hydrogen carbonate**
16.	**$(NH_4)_2S_{(s)}$**	ammonium sulfide
17.	**$BaSO_{3(s)}$**	barium sulfite
18.	**$Mg(OH)_{2(s)}$**	magnesium hydroxide
19.	$FeSO_4 \cdot 7H_2O_{(s)}$	**iron(II) sulfate-7-water or heptahydrate**
20.	$LiCl \cdot 4H_2O_{(s)}$	**lithium chloride-4-water or tetrahydrate**
21.	**$Na_2SO_4 \cdot 10H_2O_{(s)}$**	sodium sulfate decahydrate
22.	$Au(NO_3)_{3(s)}$	**gold(III) nitrate**
23.	**$Bi_2(SO_4)_{3(s)}$**	bismuth(III) sulfate
24.	**$Pb(CH_3COO)_2 \cdot 3H_2O_{(s)}$**	lead(II) acetate-3-water
25.	$KMnO_{4(s)}$	**potassium permanganate**

12 (ENRICHMENT) IONIC NOMENCLATURE

Write the international chemical formula or the IUPAC name for each of the following compounds. Some of the compounds listed below may require that you check the *CRC Handbook of Chemistry and Physics*, "Nomenclature of Inorganic Chemistry."

1. sodium-24 chloride

2. ammonium iodate

3. phosphonium sulfate

4. iron(II) iron(III) oxide

5. ferrous ammonium sulfate hexahydrate

6. $Hg_2Cl_{2(s)}$

7. $CuCr_2O_{4(s)}$

8. $H_3OClO_{4(s)}$

9. $CuCO_3 \cdot Cu(OH)_{2(s)}$

10. $[Co(NH_3)_6]_2(SO_4)_{3(s)}$

12 (ENRICHMENT) IONIC NOMENCLATURE

Write the international chemical formula or the IUPAC name for each of the following compounds. Some of the compounds listed below may require that you check the *CRC Handbook of Chemistry and Physics*, "Nomenclature of Inorganic Chemistry."

1. sodium-24 chloride $\qquad$ **$NaCl_{(s)}$**

2. ammonium iodate $\qquad$ **$NH_4IO_{3(s)}$**

3. phosphonium sulfate $\qquad$ **$(PH_4)_2SO_{4(s)}$**

4. iron(II) iron(III) oxide $\qquad$ **$Fe_3O_{4(s)}$ or $FeO \cdot Fe_2O_{3(s)}$**

5. ferrous ammonium sulfate hexahydrate $\qquad$ **$FeSO_4 \cdot (NH_4)_2SO_4 \cdot 6H_2O_{(s)}$**

6. $Hg_2Cl_{2(s)}$ $\qquad$ **dimercury(I) chloride**

7. $CuCr_2O_{4(s)}$ $\qquad$ **copper(II) chromium(III) oxide**

8. $H_3OClO_{4(s)}$ $\qquad$ **hydronium perchlorate**

9. $Cu(OH)_2 \cdot CuCO_{3(s)}$ $\qquad$ **copper(II) hydroxide carbonate**

10. $[Co(NH_3)_6]_2(SO_4)_{3(s)}$ $\qquad$ **hexaamminecobalt(III) sulfate**

13 MOLECULAR NOMENCLATURE

1. List the molecular prefixes from one to ten.

2. For which type of molecular substances are these prefixes used?

3. Why is memorization required for the nomenclature of many molecular substances in this unit?

	Molecular Formula (with SATP state)	English IUPAC Name
4.		oxygen
5.	$P_2O_{5(s)}$	
6.		hydrogen chloride
7.	$NH_{3(g)}$	
8.		dinitrogen tetrahydride (liquid)
9.	$ICl_{5(s)}$	
10.		methane
11.	$NI_{3(s)}$	
12.	$CH_3OH_{(l)}$	
13.		sucrose
14.	$S_4N_{2(s)}$	
15.		ethanol
16.	$CO_{(g)}$	
17.	$H_2O_{2(l)}$	
18.	$H_2S_{(g)}$	
19.		octasulfur
20.		propane

13 MOLECULAR NOMENCLATURE

1. List the molecular prefixes from one to ten.

1 mono		**6 hexa**	
2 di		**7 hepta**	
3 tri		**8 octa**	
4 tetra		**9 ennea**	
5 penta		**10 deca**	

2. For which type of molecular substances are these prefixes used?

   Molecular prefixes are used for binary molecular compounds. Some binary compounds, like water (H_2O) and ammonia (NH_3), preferentially use common names.

3. Why is memorization required for the nomenclature of many molecular substances in this unit?

   Memorization is required since neither a theory nor a complete communication system has been presented yet to predict the names and formulas for these compounds.

	Molecular Formula (with SATP state)	English IUPAC Name
4.	$O_{2(g)}$	oxygen
5.	$P_2O_{5(s)}$	**diphosphorus pentaoxide**
6.	$HCl_{(g)}$	hydrogen chloride
7.	$NH_{3(g)}$	**ammonia**
8.	$N_2H_{4(l)}$	dinitrogen tetrahydride (liquid)
9.	$ICl_{5(s)}$	**iodine pentachloride**
10.	$CH_{4(g)}$	methane
11.	$NI_{3(s)}$	**nitrogen triiodide**
12.	$CH_3OH_{(l)}$	**methanol**
13.	$C_{12}H_{22}O_{11(s)}$	sucrose
14.	$S_4N_{2(s)}$	**tetrasulfur dinitride**
15.	$C_2H_5OH_{(l)}$	ethanol
16.	$CO_{(g)}$	**carbon monoxide**
17.	$H_2O_{2(l)}$	**hydrogen peroxide**
18.	$H_2S_{(g)}$	**hydrogen sulfide**
19.	$S_{8(s)}$	octasulfur
20.	$C_3H_{8(g)}$	propane

14 BINARY MOLECULAR COMPOUNDS

Complete the *Prediction* section of the following lab report.

Problem

What are the reaction products when various elements react?

Experimental Design

Various elements are reacted and the empirical formulas of the reaction products are determined.

Prediction

According to the information provided, the chemical *formulas* and *names* of the products formed in the following reactions are presented below.

	Reacting Elements	Formula of Product	Name of Product
1.	$C_{(s)} + O_{2(g)} \rightarrow CO_{2(g)}$		
2.	$Si_{(s)} + F_{2(g)} \rightarrow$		silicon tetrafluoride (gas)
3.	$O_{2(g)} + Cl_{2(g)} \rightarrow OCl_{2(l)}$		
4.	$P_{4(s)} + O_{2(g)} \rightarrow P_4O_{10(s)}$		
5.	$B_{(s)} + H_{2(g)} \rightarrow$		diboron hexahydride (gas)
6.	$Xe_{(g)} + F_{2(g)} \rightarrow XeF_{6(s)}$		
7.	$N_{2(g)} + O_{2(g)} \rightarrow NO_{2(g)}$		
8.	$N_{2(g)} + O_{2(g)} \rightarrow N_2O_{(g)}$		
9.	$B_{(s)} + F_{2(g)} \rightarrow$		boron trifluoride (gas)
10.	$F_{2(g)} + Cl_{2(g)} \rightarrow$		chlorine monofluoride (gas)

Translate the following equation into an English word equation.

$$C_{(s)} + H_{2(g)} + O_{2(g)} \rightarrow C_2H_5OH_{(l)}$$

14 BINARY MOLECULAR COMPOUNDS

Complete the *Prediction* section of the following lab report.

Problem

What are the reaction products when various elements react?

Experimental Design

Various elements are reacted and the empirical formulas of the reaction products are determined.

Prediction

According to the information provided, the chemical *formulas* and *names* of the products formed in the following reactions are presented below.

	Reacting Elements Formula of Product	Name of Product
1.	$C_{(s)} + O_{2(g)} \rightarrow CO_{2(g)}$	**carbon dioxide**
2.	$Si_{(s)} + F_{2(g)} \rightarrow SiF_{4(g)}$	silicon tetrafluoride (gas)
3.	$O_{2(g)} + Cl_{2(g)} \rightarrow OCl_{2(l)}$	**oxygen dichloride**
4.	$P_{4(s)} + O_{2(g)} \rightarrow P_4O_{10(s)}$	**tetraphosphorus decaoxide**
5.	$B_{(s)} + H_{2(g)} \rightarrow B_2H_{6(g)}$	diboron hexahydride (gas)
6.	$Xe_{(g)} + F_{2(g)} \rightarrow XeF_{6(s)}$	**xenon hexafluoride**
7.	$N_{2(g)} + O_{2(g)} \rightarrow NO_{2(g)}$	**nitrogen dioxide**
8.	$N_{2(g)} + O_{2(g)} \rightarrow N_2O_{(g)}$	**dinitrogen oxide**
9.	$B_{(s)} + F_{2(g)} \rightarrow BF_{3(g)}$	boron trifluoride (gas)
10.	$F_{2(g)} + Cl_{2(g)} \rightarrow ClF_{(g)}$	chlorine monofluoride (gas)

Translate the following equation into an English word equation.

$C_{(s)} + H_{2(g)} + O_{2(g)} \rightarrow C_2H_5OH_{(l)}$

carbon + hydrogen + oxygen $\rightarrow$ ethanol

15 WRITING CHEMICAL EQUATIONS

Write unbalanced chemical equations for the following chemical reactions.
(Assume pure substances unless otherwise indicated. Include states of matter.)

Example: sodium metal + chlorine $\longrightarrow$ sodium chloride

Answer: $Na_{(s)} + Cl_{2(g)} \longrightarrow NaCl_{(s)}$

1. water $\longrightarrow$ hydrogen + oxygen

2. nitrogen + hydrogen $\longrightarrow$ ammonia

3. sulfuric acid + aqueous sodium hydroxide $\longrightarrow$ water + aqueous sodium sulfate

4. aluminum + aqueous copper(II) nitrate $\longrightarrow$ copper + aqueous aluminum nitrate

5. chlorine + aqueous potassium bromide $\longrightarrow$
    bromine + aqueous potassium chloride

6. lead(II) nitrate$_{(aq)}$ + sodium iodide$_{(aq)}$ $\longrightarrow$ lead(II) iodide$_{(s)}$ + sodium nitrate$_{(aq)}$

7. aqueous sodium hydroxide + aqueous aluminum sulfate $\longrightarrow$
    solid aluminum hydroxide + aqueous sodium sulfate

8. phosphorus + oxygen $\longrightarrow$ solid tetraphosphorus decaoxide

9. methanol + oxygen $\longrightarrow$ carbon dioxide + water vapor

10. nitrogen dioxide gas + water $\longrightarrow$ nitric acid + nitrogen monoxide gas

15 WRITING CHEMICAL EQUATIONS

Write unbalanced chemical equations for the following chemical reactions.
(Assume pure substances unless otherwise indicated. Include states of matter.)
Example: sodium metal + chlorine $\rightarrow$ sodium chloride
Answer: $Na_{(s)}$ + $Cl_{2(g)}$ $\rightarrow$ $NaCl_{(s)}$

1. water $\rightarrow$ hydrogen + oxygen

   $H_2O_{(l)}$ $\rightarrow$ $H_{2(g)}$ + $O_{2(g)}$

2. nitrogen + hydrogen $\rightarrow$ ammonia

   $N_{2(g)}$ + $H_{2(g)}$ $\rightarrow$ $NH_{3(g)}$

3. sulfuric acid + aqueous sodium hydroxide $\rightarrow$ water + aqueous sodium sulfate

   $H_2SO_{4(aq)}$ + $NaOH_{(aq)}$ $\rightarrow$ $HOH_{(l)}$ + $Na_2SO_{4(aq)}$

4. aluminum + aqueous copper(II) nitrate $\rightarrow$ copper + aqueous aluminum nitrate

   $Al_{(s)}$ + $Cu(NO_3)_{2(aq)}$ $\rightarrow$ $Cu_{(s)}$ + $Al(NO_3)_{3(aq)}$

5. chlorine + aqueous potassium bromide $\rightarrow$

                                             bromine + aqueous potassium chloride

   $Cl_{2(g)}$ + $KBr_{(aq)}$ $\rightarrow$ $Br_{2(l)}$ + $KCl_{(aq)}$

6. lead(II) nitrate$_{(aq)}$ + sodium iodide$_{(aq)}$ $\rightarrow$ lead(II) iodide$_{(s)}$ + sodium nitrate$_{(aq)}$

   $Pb(NO_3)_{2(aq)}$ + $NaI_{(aq)}$ $\rightarrow$ $PbI_{2(s)}$ + $NaNO_{3(aq)}$

7. aqueous sodium hydroxide + aqueous aluminum sulfate $\rightarrow$

                              solid aluminum hydroxide + aqueous sodium sulfate

   $NaOH_{(aq)}$ + $Al_2(SO_4)_{3(aq)}$ $\rightarrow$ $Al(OH)_{3(s)}$ + $Na_2SO_{4(aq)}$

8. phosphorus + oxygen $\rightarrow$ solid tetraphosphorus decaoxide

   $P_{4(s)}$ + $O_{2(g)}$ $\rightarrow$ $P_4O_{10(s)}$

9. methanol + oxygen $\rightarrow$ carbon dioxide + water vapor

   $CH_3OH_{(l)}$ + $O_{2(g)}$ $\rightarrow$ $CO_{2(g)}$ + $H_2O_{(g)}$

10. nitrogen dioxide gas + water $\rightarrow$ nitric acid + nitrogen monoxide gas

   $NO_{2(g)}$ + $H_2O_{(l)}$ $\rightarrow$ $HNO_{3(aq)}$ + $NO_{(g)}$

16 CHEMICAL NOMENCLATURE

Classify each substance given as ionic, molecular, or acid (i, m, or a). Write the international chemical formula (including state of matter at SATP) or the IUPAC name. Communicate the solubility of the substance in a water environment using (aq) for high solubility, and (s), (l), or (g) for low solubility.

	i, m, or a	Chemical Formula	Solubility	English IUPAC Name
1.		$PbI_{2(s)}$		
2.				ethanol
3.		$NaHS_{(s)}$		
4.				sulfurous acid
5.		$H_2O_{2(l)}$		
6.			(s)	titanium(IV) oxide
7.		$Co(NO_3)_2 \cdot 6H_2O_{(s)}$		
8.		$H_2S_{(g)}$		
9.				gallium sulfide
10.				sulfuric acid
11.		$CH_{4(g)}$		
12.				ammonium chromate
13.		$SO_{3(l)}$	(aq)	
14.		$HNO_{3(aq)}$		
15.			(g)	dinitrogen tetraoxide
16.		$Al_2(SO_4)_{3(s)}$		
17.		$Na_2SO_{3(s)}$		
18.				ammonia
19.				sodium thiosulfate-5-water
20.				iodine

16 CHEMICAL NOMENCLATURE

Classify each substance given as ionic, molecular, or acid (i, m, or a). Write the international chemical formula (including state of matter at SATP) or the IUPAC name. Communicate the solubility of the substance in a water environment using (aq) for high solubility, and (s), (l), or (g) for low solubility.

	i, m, or a	Chemical Formula	Solubility	English IUPAC Name
1.	i	$PbI_{2(s)}$	(s)	**lead(II) iodide**
2.	m	$C_2H_5OH_{(l)}$	(aq)	ethanol
3.	i	$NaHS_{(s)}$	(aq)	**sodium hydrogen sulfide**
4.	a	$H_2SO_{3(aq)}$	(aq)	sulfurous acid
5.	m	$H_2O_{2(l)}$	(aq)	**hydrogen peroxide**
6.	i	$TiO_{2(s)}$	(s)	titanium(IV) oxide
7.	i	$Co(NO_3)_2 \cdot 6H_2O_{(s)}$	(aq)	**cobalt(II) nitrate-6-water**
8.	m	$H_2S_{(g)}$	(aq)	**hydrogen sulfide**
9.	i	$Ga_2S_{3(s)}$	(s)	gallium sulfide
10.	a	$H_2SO_{4(aq)}$	(aq)	sulfuric acid
11.	m	$CH_{4(g)}$	(g)	**methane**
12.	i	$(NH_4)_2CrO_{4(s)}$	(aq)	ammonium chromate
13.	m	$SO_{3(l)}$	(aq)	**sulfur trioxide**
14.	a	$HNO_{3(aq)}$	(aq)	**nitric acid**
15.	m	$N_2O_{4(g)}$	(g)	dinitrogen tetraoxide
16.	i	$Al_2(SO_4)_{3(s)}$	(aq)	**aluminum sulfate**
17.	i	$Na_2SO_{3(s)}$	(aq)	**sodium sulfite**
18.	m	$NH_{3(g)}$	(aq)	ammonia
19.	i	$Na_2S_2O_3 \cdot 5H_2O_{(s)}$	(aq)	sodium thiosulfate-5-water
20.	m	$I_{2(s)}$	(s)	iodine

17 BALANCING CHEMICAL REACTION EQUATIONS

Use the Dalton theory of the conservation of atoms to balance the following chemical equations.

1. $\_Ni_{(s)} + \_HCl_{(aq)} \rightarrow \_NiCl_{2(aq)} + \_H_{2(g)}$

2. $\_Ca(OH)_{2(s)} + \_HCl_{(aq)} \rightarrow \_CaCl_{2(aq)} + \_HOH_{(l)}$

3. $\_Cl_{2(g)} + \_NaBr_{(aq)} \rightarrow \_Br_{2(l)} + \_NaCl_{(aq)}$

4. $\_Cr_2O_{3(s)} \rightarrow \_Cr_{(s)} + \_O_{2(g)}$

5. $\_Fe_{(s)} + \_HCl_{(aq)} \rightarrow \_FeCl_{3(aq)} + \_H_{2(g)}$

6. $\_C_3H_{6(g)} + \_O_{2(g)} \rightarrow \_CO_{2(g)} + \_H_2O_{(g)}$

7. $\_P_{4(s)} + \_F_{2(g)} \rightarrow \_PF_{3(l)}$

8. $\_Ca(NO_3)_{2(aq)} + \_KOH_{(aq)} \rightarrow \_Ca(OH)_{2(s)} + \_KNO_{3(aq)}$

9. $\_KHCO_{3(s)} \rightarrow \_K_2CO_{3(s)} + \_H_2O_{(l)} + \_CO_{2(g)}$

10. $\_H_3PO_{4(aq)} + \_NaOH_{(aq)} \rightarrow \_Na_3PO_{4(aq)} + \_HOH_{(l)}$

11. $\_Ca(NO_3)_{2(aq)} + \_Na_3PO_{4(aq)} \rightarrow \_Ca_3(PO_4)_{2(s)} + \_NaNO_{3(aq)}$

12. $\_Cu_{(s)} + \_HNO_{3(aq)} \rightarrow \_Cu(NO_3)_{2(aq)} + \_NO_{2(g)} + \_H_2O_{(l)}$

17 BALANCING CHEMICAL REACTION EQUATIONS

Use the Dalton theory of the conservation of atoms to balance the following chemical equations.

1. $Ni_{(s)}$ + **2** $HCl_{(aq)}$ $\rightarrow$ $NiCl_{2(aq)}$ + $H_{2(g)}$

2. $Ca(OH)_{2(s)}$ + **2** $HCl_{(aq)}$ $\rightarrow$ $CaCl_{2(aq)}$ + **2** $HOH_{(l)}$

3. $Cl_{2(g)}$ + **2** $NaBr_{(aq)}$ $\rightarrow$ $Br_{2(l)}$ + **2** $NaCl_{(aq)}$

4. **2** $Cr_2O_{3(s)}$ $\rightarrow$ **4** $Cr_{(s)}$ + **3** $O_{2(g)}$

5. **2** $Fe_{(s)}$ + **6** $HCl_{(aq)}$ $\rightarrow$ **2** $FeCl_{3(aq)}$ + **3** $H_{2(g)}$

6. **2** $C_3H_{6(g)}$ + **9** $O_{2(g)}$ $\rightarrow$ **6** $CO_{2(g)}$ + **6** $H_2O_{(g)}$

7. $P_{4(s)}$ + **6** $F_{2(g)}$ $\rightarrow$ **4** $PF_{3(l)}$

8. $Ca(NO_3)_{2(aq)}$ + **2** $KOH_{(aq)}$ $\rightarrow$ $Ca(OH)_{2(s)}$ + **2** $KNO_{3(aq)}$

9. **2** $KHCO_{3(s)}$ $\rightarrow$ $K_2CO_{3(s)}$ + $H_2O_{(l)}$ + $CO_{2(g)}$

10. $H_3PO_{4(aq)}$ + **3** $NaOH_{(aq)}$ $\rightarrow$ $Na_3PO_{4(aq)}$ + **3** $HOH_{(l)}$

11. **3** $Ca(NO_3)_{2(aq)}$ + **2** $Na_3PO_{4(aq)}$ $\rightarrow$ $Ca_3(PO_4)_{2(s)}$ + **6** $NaNO_{3(aq)}$

12. $Cu_{(s)}$ + **4** $HNO_{3(aq)}$ $\rightarrow$ $Cu(NO_3)_{2(aq)}$ + **2** $NO_{2(g)}$ + **2** $H_2O_{(l)}$

18 TRANSLATING INTO BALANCED CHEMICAL EQUATIONS

Translate each of the following chemical reactions into a complete, balanced chemical equation using international symbols and including states of matter at SATP. Assume pure states of matter unless otherwise indicated.

1. The reaction of magnesium and oxygen to form magnesium oxide is used to produce light in disposable flash bulbs.

2. Chlorine gas reacts with an aqueous solution of sodium iodide. Experimental evidence indicates that the products are solid iodine and aqueous sodium chloride.

3. Solid sodium sulfate reacts with carbon to form solid sodium sulfide (used to produce synthetic fabrics) and carbon dioxide.

4. Sulfuric acid, spilled from a battery, reacts with solid baking soda to produce aqueous sodium sulfate, carbon dioxide gas and water.

5. The roasting of zinc sulfide ore in a smelter involves the heating of the ore in the presence of oxygen to produce zinc oxide and sulfur dioxide gas.

6. Once the protective oxide coating is removed, aluminum metal reacts readily with water to form hydrogen and aluminum hydroxide.

Starting with nitrogen and hydrogen, millions of kilograms of ammonia are produced every year for use as a fertilizer. Use this information to answer to next three questions.

7. Communicate the balanced chemical equation using molecular models.

8. Communicate the balanced chemical equation using international symbols and states of matter at SATP.

9. Translate the balanced chemical equation of international symbols into a complete English sentence.

18 TRANSLATING INTO BALANCED CHEMICAL EQUATIONS

Translate each of the following chemical reactions into a complete, balanced chemical equation using international symbols and including states of matter at SATP. Assume pure states of matter unless otherwise indicated.

1. The reaction of magnesium and oxygen to form magnesium oxide is used to produce light in disposable flash bulbs.

   $2 Mg_{(s)} + O_{2(g)} \rightarrow 2 MgO_{(s)}$

2. Chlorine gas reacts with an aqueous solution of sodium iodide. Experimental evidence indicates that the products are solid iodine and aqueous sodium chloride.

   $Cl_{2(g)} + 2 NaI_{(aq)} \rightarrow I_{2(s)} + 2 NaCl_{(aq)}$

3. Solid sodium sulfate reacts with carbon to form solid sodium sulfide (used to produce synthetic fabrics) and carbon dioxide.

   $Na_2SO_{4(s)} + 2 C_{(s)} \rightarrow Na_2S_{(s)} + 2 CO_{2(g)}$

4. Sulfuric acid, spilled from a battery, reacts with solid baking soda to produce aqueous sodium sulfate, carbon dioxide gas and water.

   $H_2SO_{4(aq)} + 2 NaHCO_{3(s)} \rightarrow Na_2SO_{4(aq)} + 2 CO_{2(g)} + 2 H_2O_{(l)}$

5. The roasting of zinc sulfide ore in a smelter involves the heating of the ore in the presence of oxygen to produce zinc oxide and sulfur dioxide gas.

   $2 ZnS_{(s)} + 3 O_{2(g)} \rightarrow 2 ZnO_{(s)} + 2 SO_{2(g)}$

6. Once the protective oxide coating is removed, aluminum metal reacts readily with water to form hydrogen and aluminum hydroxide.

   $2 Al_{(s)} + 6 H_2O_{(l)} \rightarrow 3 H_{2(g)} + 2 Al(OH)_{3(s)}$

Starting with nitrogen and hydrogen, millions of kilograms of ammonia are produced every year for use as a fertilizer. Use this information to answer to next three questions.

7. Communicate the balanced chemical equation using molecular models.

   $N-N$ + $H-H$ $\rightarrow$ $N(H,H,H)$ + $N(H,H,H)$
   $H-H$
   $H-H$

8. Communicate the balanced chemical equation using international symbols and states of matter at SATP.

   $N_{2(g)} + 3 H_{2(g)} \rightarrow 2 NH_{3(g)}$

9. Translate the balanced chemical equation of international symbols into a complete English sentence.

   One molecule of nitrogen gas and three molecules of hydrogen gas react to produce two molecules of ammonia gas.

19 PREDICTING CHEMICAL REACTIONS

For each of the following questions, classify the reaction type (formation, simple decomposition, combustion, single replacement, double replacement, or other) and predict the balanced chemical equation. Provide a word equation as well.

1. $Al_{(s)} + O_{2(g)} \rightarrow$

2. $Ag_2O_{(s)} \rightarrow$

3. $Br_{2(l)} + KI_{(aq)} \rightarrow$

4. A strip of zinc metal is placed into a copper(II) nitrate solution.

5. $BaCl_{2(aq)} + Na_2SO_{4(aq)} \rightarrow$

6. Sulfuric acid is neutralized by aqueous sodium hydroxide.

7. $\rightarrow CuS_{(s)} + NaCH_3COO_{(aq)}$

8. $CuS_{(s)} + O_{2(g)} \rightarrow$

9. Propane burns in air.

10. $Na_2CO_{3(aq)} + HCl_{(aq)} \rightarrow NaCl_{(aq)} + CO_{2(g)} + H_2O_{(l)}$

© NELSON CANADA,
A DIVISION OF THOMSON CANADA LIMITED, 1994

19 PREDICTING CHEMICAL REACTIONS

For each of the following questions, classify the reaction type (formation, simple decomposition, combustion, single replacement, double replacement, or other) and predict the balanced chemical equation. Provide a word equation as well.

1. $4 Al_{(s)} + 3 O_{2(g)} \rightarrow 2 Al_2O_{3(s)}$
   formation or combustion
   aluminum + oxygen $\rightarrow$ aluminum oxide

2. $2 Ag_2O_{(s)} \rightarrow 4 Ag_{(s)} + O_{2(g)}$
   simple decomposition
   silver oxide $\rightarrow$ silver + oxygen

3. $Br_{2(l)} + 2 KI_{(aq)} \rightarrow I_{2(s)} + 2 KBr_{(aq)}$
   single replacement
   bromine + potassium iodide $\rightarrow$ iodine + potassium bromide

4. A strip of zinc metal is placed into a copper(II) nitrate solution.
   $Zn_{(s)} + Cu(NO_3)_{2(aq)} \rightarrow Cu_{(s)} + Zn(NO_3)_{2(aq)}$
   single replacement
   zinc + copper (II) nitrate $\rightarrow$ copper + zinc nitrate

5. $BaCl_{2(aq)} + Na_2SO_{4(aq)} \rightarrow BaSO_{4(s)} + 2 NaCl_{(aq)}$
   double replacement
   barium chloride + sodium sulfate $\rightarrow$ barium sulfate + sodium chloride

6. Sulfuric acid is neutralized by aqueous sodium hydroxide.
   $H_2SO_{4(aq)} + 2 NaOH_{(aq)} \rightarrow 2 HOH_{(l)} + Na_2SO_{4(aq)}$
   double replacement
   sulfuric acid + sodium hydroxide $\rightarrow$ water + sodium sulfate

7. $Na_2S_{(aq)} + Cu(CH_3COO)_{2(aq)} \rightarrow CuS_{(s)} + 2 NaCH_3COO_{(aq)}$
   double replacement
   sodium sulfate + copper (II) acetate $\rightarrow$ copper(II) sulfide + sodium acetate

8. $2 CuS_{(s)} + 3 O_{2(g)} \rightarrow 2 CuO_{(s)} + 2 SO_{2(g)}$
   combustion
   copper (II) sulfide + oxygen $\rightarrow$ copper (II) oxide + sulfur dioxide

9. Propane burns in air.
   $C_3H_{8(g)} + 5 O_{2(g)} \rightarrow 3 CO_{2(g)} + 4 H_2O_{(g)}$
   combustion
   propane + oxygen $\rightarrow$ carbon dioxide + water

10. $Na_2CO_{3(aq)} + 2 HCl_{(aq)} \rightarrow 2 NaCl_{(aq)} + CO_{2(g)} + H_2O_{(l)}$
    other
    sodium carbonate + hydrochloric acid $\rightarrow$ sodium chloride + carbon dioxide + water

20 COMMUNICATING CHEMICAL REACTION EQUATIONS

For each of the reactions described below, predict and communicate the balanced chemical equation using international symbols and including the states of matter at SATP. Write an English translation to indicate how the equation is read.

1. Sulfuric acid reacts with solid calcium phosphate.

2. Kerosene, $C_{14}H_{30(l)}$, is burned as a fuel.

3. Phosphoric acid is neutralized with a calcium hydroxide solution.

4. Aqueous ammonia and nitric acid react to form the fertilizer ammonium nitrate.

5. Carbon disulfide liquid burns in air.

6. Sodium metal reacts with water.

7. A sodium carbonate solution reacts with aqueous lead(II) nitrate to recover and dispose of an environmentally hazardous substance.

8. Aluminum and oxygen react to form a protective oxide coating.

9. Aqueous chlorine reacts with aqueous potassium iodide.

20 COMMUNICATING CHEMICAL REACTION EQUATIONS

For each of the reactions described below, predict and communicate the balanced chemical equation using international symbols and including the states of matter at SATP. Write an English translation to indicate how the equation is read.

1. Sulfuric acid reacts with solid calcium phosphate.

$$3\ H_2SO_{4(aq)} + Ca_3(PO_4)_{2(s)} \rightarrow 2\ H_3PO_{4(aq)} + 3\ CaSO_{4(s)}$$

Three moles of aqueous sulfuric acid and one mole of solid calcium phosphate react to produce two moles of aqueous phosphoric acid and three moles of solid calcium sulfate.

2. Kerosene, $C_{14}H_{30(l)}$, is burned as a fuel.

$$2\ C_{14}H_{30(l)} + 43\ O_{2(g)} \rightarrow 28\ CO_{2(g)} + 30\ H_2O_{(g)}$$

Two moles of liquid kerosene reacts with forty-three moles of oxygen gas to produce twenty-eight moles of carbon dioxide gas and thirty moles of water vapor.

3. Phosphoric acid is neutralized with a calcium hydroxide solution.

$$2\ H_3PO_{4(aq)} + 3\ Ca(OH)_{2(aq)} \rightarrow Ca_3(PO_4)_{2(s)} + 6\ HOH_{(l)}$$

Two moles of aqueous phosphoric acid and three moles of aqueous calcium hydroxide react to form one mole of solid calcium phosphate and six moles of liquid water.

4. Aqueous ammonia and nitric acid react to form the fertilizer ammonium nitrate.

$$NH_{3(aq)} + HNO_{3(aq)} \rightarrow NH_4NO_{3(aq)}$$

One mole of aqueous ammonia reacts with one mole of aqueous nitric acid to produce one mole of solid aqueous ammonium nitrate.

5. Carbon disulfide liquid burns in air.

$$CS_{2(l)} + 3\ O_{2(g)} \rightarrow CO_{2(g)} + 2\ SO_{2(g)}$$

One mole of liquid carbon disulfide and three moles of oxygen gas react to form one mole of carbon dioxide gas and two moles of sulfur dioxide gas.

6. Sodium metal reacts with water.

$$2\ Na_{(s)} + 2\ HOH_{(l)} \rightarrow 2\ NaOH_{(aq)} + H_{2(g)}$$

Two moles of solid sodium react with two moles of liquid water to produce two moles of aqueous sodium hydroxide and one mole of hydrogen gas.

7. A sodium carbonate solution reacts with aqueous lead(II) nitrate to recover and dispose of an environmentally hazardous substance.

$$Na_2CO_{3(aq)} + Pb(NO_3)_{2(aq)} \rightarrow PbCO_{3(s)} + 2\ NaNO_{3(aq)}$$

One mole of aqueous sodium carbonate and one mole of aqueous lead(II) nitrate react to produce one mole of solid lead(II) carbonate and two moles of aqueous sodium nitrate.

8. Aluminum and oxygen react to form a protective oxide coating.

$$4\ Al_{(s)} + 3\ O_{2(g)} \rightarrow 2\ Al_2O_{3(s)}$$

Four moles of solid aluminum and three moles of oxygen gas react to form two moles of solid aluminum oxide.

9. Aqueous chlorine reacts with aqueous potassium iodide.

$$Cl_{2(aq)} + 2\ KI_{(aq)} \rightarrow I_{2(s)} + 2\ KCl_{(aq)}$$

One mole of aqueous chlorine reacts with two moles of aqueous potassium iodide to produce one mole of solid iodine and two moles of aqueous potassium chloride.

21 UNDERSTANDING SOLUTIONS

List some properties that could be used to construct diagnostic tests to identify the type of solute in each of the following solutions.

1. an aqueous solution of a molecular substance

2. an aqueous solution of a neutral ionic compound

3. an aqueous solution of an acid

4. an aqueous solution of a base

Each of the following pure substances is placed in a water environment. For each mixture, list the major entities present. The solubility of ionic compounds can be obtained from the solubility table inside the back cover of the textbook and the solubility of selected molecular compounds requires memorizing the examples given in Table 4.5.

5. ammonia

6. silver sulfate

7. hydrogen chloride

8. sodium tetraborate

9. barium hydroxide

10. propane

11. aluminum

12. magnesium phosphate

13. sodium stearate

14. hydrogen nitrate

21 UNDERSTANDING SOLUTIONS

List some properties that could be used to construct diagnostic tests to identify the type of solute in each of the following solutions.

1. an aqueous solution of a molecular substance
   - **does not conduct electricity**
   - **has no effect on the color of litmus paper**

2. an aqueous solution of a neutral ionic compound
   - **conducts electricity**
   - **has no effect on the color of litmus paper**

3. an aqueous solution of an acid
   - **conducts electricity**
   - **turns blue litmus paper red**

4. an aqueous solution of a base
   - **conducts electricity**
   - **turns red litmus paper blue**

Each of the following pure substances is placed in a water environment. For each mixture, list the major entities present. The solubility of ionic compounds can be obtained from the solubility table inside the back cover of the textbook and the solubility of selected molecular compounds requires memorizing the examples given in Table 4.5.

5. ammonia

   $NH_{3(aq)}$, $H_2O_{(l)}$

6. silver sulfate

   $Ag_2SO_{4(s)}$, $H_2O_{(l)}$

7. hydrogen chloride

   $H^+_{(aq)}$, $Cl^-_{(aq)}$, $H_2O_{(l)}$

8. sodium tetraborate

   $Na^+_{(aq)}$, $B_4O_7^{2-}_{(aq)}$, $H_2O_{(l)}$

9. barium hydroxide

   $Ba^{2+}_{(aq)}$, $OH^-_{(aq)}$, $H_2O_{(l)}$

10. propane

    $C_3H_{8(g)}$, $H_2O_{(l)}$

11. aluminum

    $Al_{(s)}$, $H_2O_{(l)}$

12. magnesium phosphate

    $Mg_3(PO_4)_{2(s)}$, $H_2O_{(l)}$

13. sodium stearate

    $Na^+_{(aq)}$, $C_{17}H_{35}COO^-_{(aq)}$, $H_2O_{(l)}$

14. hydrogen nitrate

    $H^+_{(aq)}$, $NO_3^-_{(aq)}$, $H_2O_{(l)}$

22 CONCENTRATION OF A SOLUTION

Use concentration as a conversion factor to calculate the quantity requested in each question below. Communicate your problem-solving approach, including units and correct certainty.

1. Cow's milk contains 4.5 g of lactose per 100 mL of milk. What mass of lactose is present in 250 mL (one glass) of milk?

2. A 10% W/V salt solution is used for making pickles. What mass of salt is present in 750 mL of this solution?

3. A 250 mL measuring cup of cleaning solution contains 1.2 mol of dissolved ammonia. What is the molar concentration of this solution?

4. Fish require a concentration of about 4.5 mg/L of dissolved oxygen in water. What volume of water would contain 100 mg of oxygen?

5. What volume of concentrated, 14.6 mol/L, phosphoric acid would contain 2.00 mol of solute?

6. Hard water contains at least 120 ppm of dissolved minerals. If 2.0 L of hard water in a kettle is boiled to dryness, what mass of minerals would be obtained?

7. What amount of table salt is needed to prepare 12.0 L of a 5.20 mol/L solution?

8. A laboratory solution of zinc nitrate is labelled 24.0 mmol/L. What volume of this solution would contain 0.600 mol of solute?

22 CONCENTRATION OF A SOLUTION

Use concentration as a conversion factor to calculate the quantity requested in each question below. Communicate your problem-solving approach, including units and correct certainty.

1. Cow's milk contains 4.5 g of lactose per 100 mL of milk. What mass of lactose is present in 250 mL (one glass) of milk?

$$m_{lactose} = 250 \text{ mL} \times \frac{4.5 \text{ g}}{100 \text{ mL}} = 11 \text{ g}$$

2. A 10% W/V salt solution is used for making pickles. What mass of salt is present in 750 mL of this solution?

$$m_{salt} = 750 \text{ mL} \times \frac{10 \text{ g}}{100 \text{ mL}} = 75 \text{ g}$$

3. A 250 mL measuring cup of cleaning solution contains 1.2 mol of dissolved ammonia. What is the molar concentration of this solution?

$$C_{ammonia} = \frac{1.2 \text{ mol}}{0.250 \text{ L}} = 4.8 \text{ mol/L}$$

4. Fish require a concentration of about 4.5 mg/L of dissolved oxygen in water. What volume of water would contain 100 mg of oxygen?

$$v_{water} = 100 \text{ mg} \times \frac{1 \text{ L}}{4.5 \text{ mg}} = 22 \text{ L}$$

5. What volume of concentrated, 14.6 mol/L, phosphoric acid would contain 2.00 mol of solute?

$$v_{acid} = 2.00 \text{ mol} \times \frac{1 \text{ L}}{14.6 \text{ mol}} = 0.137 \text{ L}$$

6. Hard water contains at least 120 ppm of dissolved minerals. If 2.0 L of hard water in a kettle is boiled to dryness, what mass of minerals would be obtained?

$$m_{minerals} = 2.0 \text{ L} \times \frac{120 \text{ mg}}{1 \text{ L}} = 0.24 \text{ g}$$

7. What amount of table salt is needed to prepare 12.0 L of a 5.20 mol/L solution?

$$n_{salt} = 12.0 \text{ L} \times \frac{5.20 \text{ mol}}{1 \text{ L}} = 62.4 \text{ mol}$$

8. A laboratory solution of zinc nitrate is labelled 24.0 mmol/L. What volume of this solution would contain 0.600 mol of solute?

$$v_{Zn(NO_3)_2} = 0.600 \text{ mol} \times \frac{1 \text{ L}}{24.0 \text{ mmol}} \times \frac{1000 \text{ mmol}}{1 \text{ mol}} = 25.0 \text{ L}$$

23 (ENRICHMENT) CONCENTRATION OF A SOLUTION

Modern analytical chemistry uses some sophisticated technology to detect incredibly small quantities of substances. Chemical analysis is approaching one part per trillion concentrations. (The SI prefixes used in this exercise include nano (n) 10^{-9}, pico (p) 10^{-12}, femto (f) 10^{-15}, and atto (a) 10^{-18}.)

1. An ICP (inductively coupled plasma technology) vaporizes a small sample which is then analyzed by a mass spectrometer. The detection limit for lead using this method is 0.05 ng/mL.

    (a) What is the detection limit for lead in ppb?

    (b) What volume of solution would contain 1.0 g of lead?

2. A laser-based microchemical analysis has a detection limit of 9×10^{-10} mol/L using a sample volume of 0.2 pL. How many molecules are present in this sample?

3. Using a sample volume of 0.2 pL, what molar concentration of solute corresponds to the presence of one molecule?

4. The detection limit for 2,4-dinitrophenylhydrazones using capillary liquid chromatography is 15 ag in a sample volume of 0.20 pL. Convert this limit into a concentration in ppb.

23 (ENRICHMENT) CONCENTRATION OF A SOLUTION

Modern analytical chemistry uses some sophisticated technology to detect incredibly small quantities of substances. Chemical analysis is approaching one part per trillion concentrations. (The SI prefixes used in this exercise include nano (n) 10^{-9}, pico (p) 10^{-12}, femto (f) 10^{-15}, and atto (a) 10^{-18}.)

1. An ICP (inductively coupled plasma technology) vaporizes a small sample which is then analyzed by a mass spectrometer. The detection limit for lead using this method is 0.05 ng/mL.

   (a) What is the detection limit for lead in ppb?

      For dilute aqueous solutions, 1 mL is equivalent to 1 g.

      $$\frac{0.05 \text{ ng}}{1 \text{ g}} = 0.05 \times 10^{-9} = \frac{0.05}{10^9} = 0.05 \text{ ppb}$$

   (b) What volume of solution would contain 1.0 g of lead?

      $$v_{Pb} = 1.0 \text{ g} \times \frac{1 \text{ mL}}{0.05 \text{ ng}} = 20 \text{ ML}$$

2. A laser-based microchemical analysis has a detection limit of 9×10^{-10} mol/L using a sample volume of 0.2 pL. How many molecules are present in this sample?

   $$n = 0.2 \text{ pL} \times \frac{9 \times 10^{-10} \text{ mol}}{1 \text{ L}} = 2 \times 10^{-22} \text{ mol}$$

   number of molecules $= 2 \times 10^{-22} \text{ mol} \times 6.02 \times 10^{23}/\text{mol} = 1 \times 10^2$

3. Using a sample volume of 0.2 pL, what molar concentration of solute corresponds to the presence of one molecule?

   $$n = \frac{1}{6.02 \times 10^{23}/\text{mol}} = 1.66 \times 10^{-24} \text{ mol}$$

   $$C = \frac{1.66 \times 10^{-24} \text{ L}}{0.2 \times 10^{-12} \text{ L}} = 8 \times 10^{-12} \text{ mol/L}$$

4. The detection limit for 2,4-dinitrophenylhydrazones using capillary liquid chromatography is 15 ag in a sample volume of 0.20 pL. Convert this limit into a concentration in ppb.

   $$c = \frac{15 \text{ ag}}{0.20 \text{ pL}} = \frac{15 \times 10^{-9} \text{ ng}}{0.20 \times 10^{-9} \text{ mL}} = 75 \text{ ng/mL} = 75 \text{ ng/g} = 75 \text{ ppb}$$

24 DILUTION

In the following questions, "concentrated" refers to the concentration of the most common commercial reagent as listed in the table of Concentrated Reagents inside the back cover of the textbook.

1. An ammonia solution is made by diluting 150 mL of the concentrated commercial reagent until the final volume reaches 1000 mL. What is the final molar concentration?

2. What volume of a 500 ppm reagent solution is required to prepare a 2.5 L solution with a 100 ppm concentration?

3. A 500 mL bottle of concentrated acetic acid is diluted to make a 5.0% solution. Find the volume of diluted solution that is prepared.

4. In a chemical analysis, a 25.0 mL sample was diluted to 500.0 mL and analyzed. If the diluted solution had a molar concentration of 0.108 mol/L, what was the molar concentration of the original sample?

5. If a 355 mL can of soda pop is diluted to a final volume of 1.00 L, what can be said quantitatively about the concentration of the diluted solution as compared with the original solution?

24 DILUTION

In the following questions, "concentrated" refers to the concentration of the most common commercial reagent as listed in the table of Concentrated Reagents inside the back cover of the textbook.

1. An ammonia solution is made by diluting 150 mL of the concentrated commercial reagent until the final volume reaches 1000 mL. What is the final molar concentration?

$$v_i C_i = v_f C_f$$
$$150 \text{ mL} \times 14.8 \text{ mol/L} = 1000 \text{ mL} \times C_f$$
$$C_f = 2.22 \text{ mol/L}$$

2. What volume of a 500 ppm reagent solution is required to prepare a 2.5 L solution with a 100 ppm concentration?

$$v_i c_i = v_f c_f$$
$$v_i \times 500 \text{ ppm} = 2.5 \text{ L} \times 100 \text{ ppm}$$
$$v_i = 0.50 \text{ L}$$

3. A 500 mL bottle of concentrated acetic acid is diluted to make a 5.0% solution. Find the volume of diluted solution that is prepared.

$$v_i c_i = v_f c_f$$
$$500 \text{ mL} \times 99.5\% = v_f \times 5.0\%$$
$$v_f = 10 \text{ L}$$

4. In a chemical analysis, a 25.0 mL sample was diluted to 500.0 mL and analyzed. If the diluted solution had a molar concentration of 0.108 mol/L, what was the molar concentration of the original sample?

$$v_i C_i = v_f C_f$$
$$25.0 \text{ mL} \times C_i = 500.0 \text{ mL} \times 0.108 \text{ mol/L}$$
$$C_i = 2.16 \text{ mol/L}$$

5. If a 355 mL can of soda pop is diluted to a final volume of 1.00 L, what can be said quantitatively about the concentration of the diluted solution as compared with the original solution?

$$v_i c_i = v_f c_f$$
$$0.355 \text{ L} \times c_i = 1.00 \text{ L} \times c_f$$
$$c_f = 0.355 \, c_i$$

The diluted solution has a concentration 0.355 times or 35.5% of the original solution.

25 SOLUBILITY

1. A 20.0 mL sample of a saturated solution of potassium nitrate at 25°C was evaporated to produce 7.2 g of dry, solid residue. What is the solubility of potassium nitrate in grams per hundred millilitres?

2. The solubility of aluminum fluoride is 0.559 g/100 mL at 25°C.

   (a) Is it possible to dissolve 3.0 g of solid to make 500 mL of solution at 25°C? Show your calculations.

   (b) What are two possible alternatives in order to dissolve 3.0 g of aluminum fluoride?

3. Classify the following liquids as miscible or immiscible with water: gasoline, vinegar, ethanol, and cooking oil.

4. A glass of cold water left sitting on a counter at room temperature usually develops many small gas bubbles on the inside of the glass. Describe what is likely happening.

5.

Solubility of Fluoride Compounds in Water at 25°C			
Compound	**Solubility (mol/L)**	**Compound**	**Solubility (mol/L)**
ammonium fluoride	>20	magnesium fluoride	0.0012
barium fluoride	0.0068	potassium fluoride	>10
calcium fluoride	0.0020	silver fluoride	>10
cobalt(II) fluoride	1.5	sodium fluoride	1.0
copper(II) fluoride	0.46	strontium fluoride	0.0096
lithium fluoride	0.10	zinc fluoride	0.16

   (a) Classify each compound in the above table as high or low according to its solubility using the same cutoff point as the solubility table inside the back cover of the textbook.

   (b) (Enrichment) Create another category, titled F^-, for the solubility table by developing a generalization based on your answer to part (a).

25 SOLUBILITY

1. A 20.0 mL sample of a saturated solution of potassium nitrate at 25°C was evaporated to produce 7.2 g of dry, solid residue. What is the solubility of potassium nitrate in grams per hundred millilitres?

$$c_{KNO_3} = \frac{7.2 \text{ g}}{20.0 \text{ mL}} \times 100 = 36 \text{ g/100 mL}$$

2. The solubility of aluminum fluoride is 0.559 g/100 mL at 25°C.

   (a) Is it possible to dissolve 3.0 g of solid to make 500 mL of solution at 25°C? Show your calculations.

   It is not possible because only 2.80 g can be dissolved in 500 mL.

   $$m_{AlF_3} = 500 \text{ mL} \times \frac{0.559 \text{ g}}{100 \text{ mL}} = 2.80 \text{ g}$$

   (b) What are two possible alternatives in order to dissolve 3.0 g of aluminum fluoride?

   Increase the temperature or add more solvent.

3. Classify the following liquids as miscible or immiscible with water: gasoline, vinegar, ethanol, and cooking oil.

   miscible: vinegar, ethanol

   immiscible: gasoline, cooking oil

4. A glass of cold water left sitting on a counter at room temperature usually develops many small gas bubbles on the inside of the glass. Describe what is likely happening.

   Water contains dissolved air. As the temperature of the water increases, the solubility of air in water decreases. Therefore, air bubbles form in the water.

5.

Solubility of Fluoride Compounds in Water at 25°C					
Compound	Solubility (mol/L)		Compound	Solubility (mol/L)	
ammonium fluoride	>20	**high**	magnesium fluoride	0.0012	**low**
barium fluoride	0.0068	**low**	potassium fluoride	>10	**high**
calcium fluoride	0.0020	**low**	silver fluoride	>10	**high**
cobalt(II) fluoride	1.5	**high**	sodium fluoride	1.0	**high**
copper(II) fluoride	0.46	**high**	strontium fluoride	0.0096	**low**
lithium fluoride	0.10	**high**	zinc fluoride	0.16	**high**

   (a) Classify each compound in the above table as high or low according to its solubility using the same cutoff point as the solubility table inside the back cover of the textbook.

   (b) (Enrichment) Create another category, titled F⁻, for the solubility table by developing a generalization based on your answer to part (a).

	F⁻
High solubility	most
Low solubility	Group 2

26   GAS LAWS

1. What is the pressure required to compress hydrogen at 1.00 atm from 300 mL to 200 mL at a constant temperature?

2. A 400 mL sample of a gas at 10°C is warmed to 25°C at a constant pressure. Calculate the final volume.

3. A bicycle tire has a pressure of 450 kPa at 20°C. Assuming the volume does not change, what is the new pressure at 35°C?

4. Nitrogen in a 250 mL container at 65.0 kPa is transferred to a container with a volume of 600 mL.

   (a) Calculate the new pressure if the temperature is kept constant.

   (b) Calculate the new pressure if the temperature changes from 20°C to 15°C.

5. A 450 mL sample of freon gas at 1.50 atm and 15°C was compressed to 300 mL at a pressure of 2.00 atm. Calculate the final temperature in degrees Celsius.

6. A 2.75 L sample of helium gas at 99.0 kPa was heated from 21.0°C to 71.0°C and the pressure changed to 100 kPa. Calculate the final volume.

26 GAS LAWS

1. What is the pressure required to compress hydrogen at 1.00 atm from 300 mL to 200 mL at a constant temperature?

$$p_1v_1 = p_2v_2$$

$$p_2 = \frac{p_1v_1}{v_2} = \frac{1.00 \text{ atm} \times 300 \text{ mL}}{200 \text{ mL}} = 1.50 \text{ atm}$$

2. A 400 mL sample of a gas at 10°C is warmed to 25°C at a constant pressure. Calculate the final volume.

$$\frac{v_1}{T_1} = \frac{v_2}{T_2}$$

$$v_2 = \frac{v_1T_2}{T_1} = \frac{400 \text{ mL} \times 298 \text{ K}}{283 \text{ K}} = 421 \text{ mL}$$

3. A bicycle tire has a pressure of 450 kPa at 20°C. Assuming the volume does not change, what is the new pressure at 35°C?

$$\frac{p_1v_1}{T_1} = \frac{p_2v_2}{T_2} \ , \ v_2 = v_1$$

$$p_2 = \frac{p_1v_1T_2}{T_1v_2} = \frac{450 \text{ kPa} \times 308 \text{ K}}{293 \text{ K}} = 473 \text{ kPa}$$

4. Nitrogen in a 250 mL container at 65.0 kPa is transferred to a container with a volume of 600 mL.

   (a) Calculate the new pressure if the temperature is kept constant.

   $$p_1v_1 = p_2v_2$$

   $$p_2 = \frac{p_1v_1}{v_2} = \frac{65.0 \text{ kPa} \times 250 \text{ mL}}{600 \text{ mL}} = 27.1 \text{ kPa}$$

   (b) Calculate the new pressure if the temperature changes from 20°C to 15°C.

   $$\frac{p_1v_1}{T_1} = \frac{p_2v_2}{T_2}$$

   $$p_2 = \frac{p_1v_1T_2}{T_1v_2} = \frac{65.0 \text{ kPa} \times 250 \text{ mL} \times 288 \text{ K}}{293 \text{ K} \times 600 \text{ mL}} = 26.6 \text{ kPa}$$

5. A 450 mL sample of freon gas at 1.50 atm and 15°C was compressed to 300 mL at a pressure of 2.00 atm. Calculate the final temperature in degrees Celsius.

$$\frac{p_1v_1}{T_1} = \frac{p_2v_2}{T_2}$$

$$T_2 = \frac{p_2v_2T_1}{p_1v_1} = \frac{2.00 \text{ atm} \times 300 \text{ mL} \times 288 \text{ K}}{1.50 \text{ atm} \times 450 \text{ mL}} = 256 \text{ K} = -17°C$$

6. A 2.75 L sample of helium gas at 99.0 kPa was heated from 21°C to 71°C and the pressure changed to 100 kPa. Calculate the final volume.

$$\frac{p_1v_1}{T_1} = \frac{p_2v_2}{T_2}$$

$$v_2 = \frac{p_1v_1T_2}{p_2T_1} = \frac{99.0 \text{ kPa} \times 2.75 \text{ L} \times 344 \text{ K}}{100 \text{ kPa} \times 294 \text{ K}} = 3.19 \text{ L}$$

27 MOLAR MASS AND CONVERSIONS

1. Determine the molar mass of each of the following substances.

   (a) $MgI_{2(s)}$ (b) $Al(OH)_{3(s)}$

   (c) $(NH_4)_2CO_{3(s)}$ (d) $CoCl_2 \cdot 6H_2O_{(s)}$

2. Convert each of the following masses into an amount in moles of the given substance.

   (a) 8.40 g of $NaOH_{(s)}$

   (b) 4.2 kg of $H_2O_{(l)}$

3. Convert each of the following amounts into a mass in grams of the given substance.

   (a) 0.456 mol of $Al_2(SO_4)_{3(s)}$

   (b) 0.518 mmol of $CuSO_4 \cdot 5H_2O_{(s)}$

4. Complete the following table.

Substance	Molar Mass (g/mol)	Mass (g)	Amount (mol)
$CaCl_{2(s)}$		18.6	
$Al_2O_{3(s)}$			0.267
$Mg(OH)_{2(s)}$		35.00	
$Na_2CO_3 \cdot 10H_2O_{(s)}$			0.150

© NELSON CANADA,
A DIVISION OF THOMSON CANADA LIMITED, 1994

27   MOLAR MASS AND CONVERSIONS

1. Determine the molar mass of each of the following substances.

   (a) $MgI_{2(s)}$

   278.11 g/mol

   (b) $Al(OH)_{3(s)}$

   78.01 g/mol

   (c) $(NH_4)_2CO_{3(s)}$

   96.11 g/mol

   (d) $CoCl_2 \cdot 6H_2O_{(s)}$

   237.95 g/mol

2. Convert each of the following masses into an amount in moles of the given substance.

   (a) 8.40 g of $NaOH_{(s)}$

   $$n_{NaOH} = 8.40 \text{ g} \times \frac{1 \text{ mol}}{40.00 \text{ g}}$$
   $$= 0.210 \text{ mol}$$

   (b) 4.2 kg of $H_2O_{(l)}$

   $$n_{H_2O} = 4.2 \text{ kg} \times \frac{1 \text{ mol}}{18.02 \text{ g}}$$
   $$= 0.23 \text{ kmol}$$

3. Convert each of the following amounts into a mass in grams of the given substance.

   (a) 0.456 mol of $Al_2(SO_4)_{3(s)}$

   $$m_{Al_2(SO_4)_3} = 0.456 \text{ mol} \times \frac{342.14 \text{ g}}{1 \text{ mol}}$$
   $$= 156 \text{ g}$$

   (b) 0.518 mmol of $CuSO_4 \cdot 5H_2O_{(s)}$

   $$m_{CuSO_4 \cdot 5H_2O} = 0.518 \text{ mmol} \times \frac{249.71 \text{ g}}{1 \text{ mol}}$$
   $$= 0.129 \text{ g}$$

4. Complete the following table.

Substance	Molar Mass (g/mol)	Mass (g)	Amount (mol)
$CaCl_{2(s)}$	110.98	18.6	**0.168**
$Al_2O_{3(s)}$	101.96	**27.2**	0.267
$Mg(OH)_{2(s)}$	58.33	35.00	**0.6000**
$Na_2CO_3 \cdot 10H_2O_{(s)}$	286.19	42.9	0.150

28 IDEAL GAS LAW

1. What amount of air, in moles, is present in a house containing 600 m³ of air at 20°C and 98 kPa? (1 m³ = 1 kL)

2. Calculate the mass of neon gas in a neon sign with a volume of 50 L at 10°C and 3.1 kPa.

3. Calculate the volume of 8.4 g of nitrogen at 200°C and 130 kPa.

4. Hydrogen gas is generated by the decomposition of water to fill a 1.1 kL weather balloon at 20°C and 100 kPa. What is the mass of hydrogen required?

5. Calculate the volume of 16 g of oxygen at 22°C and 97.5 kPa.

28 IDEAL GAS LAW

1. What amount of air, in moles, is present in a house containing 600 m³ of air at 20°C and 98 kPa? (1 m³ = 1 kL)

$$pv = nRT$$

$$n_{air} = \frac{pv}{RT} = \frac{98 \text{ kPa} \times 600 \text{ kL}}{8.31 \text{ kPa•L/(mol•K)} \times 293 \text{ K}} = 24 \text{ kmol}$$

2. Calculate the mass of neon gas in a neon sign with a volume of 50 L at 10°C and 3.1 kPa.

$$pv = nRT$$

$$n_{Ne} = \frac{pv}{RT} = \frac{3.1 \text{ kPa} \times 50 \text{ L}}{8.31 \text{ kPa•L/(mol•K)} \times 283 \text{ K}} = 0.066 \text{ mol}$$

$$m_{Ne} = 0.066 \text{ mol} \times \frac{20.18 \text{ g}}{1 \text{ mol}} = 1.3 \text{ g}$$

3. Calculate the volume of 8.4 g of nitrogen at 200°C and 130 kPa.

$$n_{N_2} = 8.4 \text{ g} \times \frac{1 \text{ mol}}{28.02 \text{ g}} = 0.30 \text{ mol}$$

$$pv = nRT$$

$$v_{N_2} = \frac{nRT}{p} = \frac{0.30 \text{ mol} \times 8.31 \text{ kPa•L/(mol•K)} \times 473 \text{ K}}{130 \text{ kPa}} = 9.1 \text{ L}$$

4. Hydrogen gas is generated by the decomposition of water to fill a 1.1 kL weather balloon at 20°C and 100 kPa. What is the mass of hydrogen required?

$$pv = nRT$$

$$n_{H_2} = \frac{pv}{RT} = \frac{100 \text{ kPa} \times 1.1 \text{ kL}}{8.31 \text{ kPa•L/(mol•K)} \times 293 \text{ K}} = 0.045 \text{ kmol}$$

$$m_{H_2} = 0.045 \text{ kmol} \times \frac{2.02 \text{ g}}{1 \text{ mol}} = 91 \text{ g}$$

5. Calculate the volume of 16 g of oxygen at 22°C and 97.5 kPa.

$$n_{O_2} = 16 \text{ g} \times \frac{1 \text{ mol}}{32.00 \text{ g}} = 0.50 \text{ mol}$$

$$pv = nRT$$

$$v_{O_2} = \frac{nRT}{p} = \frac{0.50 \text{ mol} \times 8.31 \text{ kPa•L/(mol•K)} \times 295 \text{ K}}{97.5 \text{ kPa}} = 13 \text{ L}$$

29 (ENRICHMENT) OTHER GAS LAWS

Graham's Law of Diffusion

One of the empirical properties of gases is that they diffuse or spread out more rapidly than liquids and solids. In 1883, Scottish chemist Thomas Graham determined the relationship between the molar mass of a gas and its rate of diffusion. *At a constant temperature and pressure, the rate of diffusion of a gas varies inversely with the square root of its molar mass.*

Dalton's Law of Partial Pressures

In a mixture of non-reacting gases, each gas acts as if it were the only one present. In 1800, John Dalton was the first to recognize that *the total pressure in a gaseous, non-reacting system is equal to the sum of the pressures of the component gases, each acting separately in the system.* The pressure of one gas in a mixture of gases is known as its partial pressure and is determined by the mole fraction of that gas in the total mixture.

$$p_{total} = \Sigma p_i \quad \text{where } p_i = \frac{n_i}{n_{total}} \times p_{total}$$

1. At the same temperature and pressure, how does the rate of diffusion of ammonia gas compare with the rate for hydrogen chloride gas?

2. In order to produce the first atomic bomb and to produce enriched uranium for certain types of nuclear reactors, it is necessary to separate the two isotopes of uranium—U-235 and U-238. Uranium-235 is the reactive isotope but only about 0.7% of natural uranium is uranium-235. This separation is achieved using uranium hexafluoride gas and separating the gases by diffusion through a small opening. (In fact, this needs to be repeated thousands of times.) How does the rate of diffusion of uranium-235 hexafluoride compare with the rate for uranium-238 hexafluoride?
(Use the *CRC Handbook of Chemistry and Physics* to obtain precise values of the molar masses for the uranium isotopes and fluorine atom.)

3. A sample of dry air at SATP contains 192 L of nitrogen and 52 L of oxygen. What is the partial pressure of each gas?

4. When collecting a gas by water displacement, the pressure of the gas collected includes the partial pressure of water vapor. (The *CRC Handbook of Chemistry and Physics* has tables of water vapor pressure at different temperatures.) In an experiment, 157 mL of oxygen gas is collected over water. If the atmospheric pressure is 96.5 kPa and the temperature is 20°C, what is the partial pressure of oxygen in the gas collected?

29 (ENRICHMENT) OTHER GAS LAWS

Graham's Law of Diffusion

One of the empirical properties of gases is that they diffuse or spread out more rapidly than liquids and solids. In 1883, Scottish chemist Thomas Graham determined the relationship between the molar mass of a gas and its rate of diffusion. *At a constant temperature and pressure, the rate of diffusion of a gas varies inversely with the square root of its molar mass.*

Dalton's Law of Partial Pressures

In a mixture of non-reacting gases, each gas acts as if it were the only one present. In 1800, John Dalton was the first to recognize that *the total pressure in a gaseous, non-reacting system is equal to the sum of the pressures of the component gases, each acting separately in the system.* The pressure of one gas in a mixture of gases is known as its partial pressure and is determined by the mole fraction of that gas in the total mixture.

$$p_{total} = \Sigma p_i \quad \text{where } p_i = \frac{n_i}{n_{total}} \times p_{total}$$

1. At the same temperature and pressure, how does the rate of diffusion of ammonia gas compare with the rate for hydrogen chloride gas?

$$\frac{\text{Rate}_{NH_3}}{\text{Rate}_{HCl}} = \sqrt{\frac{M_{HCl}}{M_{NH_3}}} = \sqrt{\frac{36.46}{17.04}} = \frac{1.463}{1}$$

2. In order to produce the first atomic bomb and to produce enriched uranium for certain types of nuclear reactors, it is necessary to separate the two isotopes of uranium—U-235 and U-238. Uranium-235 is the reactive isotope but only about 0.7% of natural uranium is uranium-235. This separation is achieved using uranium hexafluoride gas and separating the gases by diffusion through a small opening. (In fact, this needs to be repeated thousands of times.) How does the rate of diffusion of uranium-235 hexafluoride compare with the rate for uranium-238 hexafluoride?
(Use the *CRC Handbook of Chemistry and Physics* to obtain precise values of the molar masses for the uranium isotopes and fluorine atom.)

   From the *CRC Handbook*, the molar mass of
   - **fluorine is 18.998 403 g/mol**
   - **uranium-235 is 235.043 924 g/mol**
   - **uranium-238 is 238.050 784 g/mol**

$$\frac{\text{Rate }_{235_{UF_6}}}{\text{Rate }_{238_{UF_6}}} = \sqrt{\frac{352.04\ 120}{349.03\ 434}} = \frac{1.004\ 298\ 2}{1}$$

3. A sample of dry air at SATP contains 192 L of nitrogen and 52 L of oxygen. What is the partial pressure of each gas?

$$n_{N_2} = 192\ \text{L} \times \frac{1\ \text{mol}}{24.8\ \text{L}} = 7.74\ \text{mol}, \quad n_{O_2} = 52\ \text{L} \times \frac{1\ \text{mol}}{24.8\ \text{L}} = 2.1\ \text{mol}$$

$$n_{total} = 7.74\ \text{mol} + 2.1\ \text{mol} = 9.8\ \text{mol}$$

$$p_{O_2} = \frac{2.1\ \text{mol}}{9.8\ \text{mol}} \times 100\ \text{kPa} = 21\ \text{kPa}, \quad p_{N_2} = \frac{7.74\ \text{mol}}{9.8\ \text{mol}} \times 100\ \text{kPa} = 79\ \text{kPa}$$

4. When collecting a gas by water displacement, the pressure of the gas collected includes the partial pressure of water vapor. (The *CRC Handbook of Chemistry and Physics* has tables of water vapor pressure at different temperatures.) In an experiment, 157 mL of oxygen gas is collected over water. If the atmospheric pressure is 96.5 kPa and the temperature is 20°C, what is the partial pressure of oxygen in the gas collected?

$$p_{H_2O} = \frac{17.535\ \text{mm Hg}}{760\ \text{mm Hg}} \times 101.325\ \text{kPa} = 2.34\ \text{kPa}$$

$$pO_2 = 96.5\ \text{kPa} - 2.34\ \text{kPa} = 94.2\ \text{kPa}$$

30 GRAVIMETRIC STOICHIOMETRY

Complete the following stoichiometric problems. Communicate your problem-solving approach, using internationally accepted symbols for elements, quantities, numbers, and units.

1. Calculate the mass of iron(III) oxide (rust) produced by the reaction of 500 g of iron with oxygen from the air.

2. What mass of precipitate should form if 2.00 g of silver nitrate in solution is reacted with excess sodium sulfide solution?

3. Determine the mass of water vapor formed when 1.00 g of butane, $C_4H_{10(g)}$, is burned in a lighter.

4. Silver metal can be recovered from waste silver nitrate solutions by reaction with copper metal. What mass of silver can be obtained using 50 g of copper?

149

30 GRAVIMETRIC STOICHIOMETRY

Complete the following stoichiometric problems. Communicate your problem-solving approach, using internationally accepted symbols for elements, quantities, numbers, and units.

1. Calculate the mass of iron(III) oxide (rust) produced by the reaction of 500 g of iron with oxygen from the air.

$$4\ Fe_{(s)}\ +\ 3\ O_{2(g)}\ \rightarrow\ 2\ Fe_2O_{3(s)}$$

500 g $\qquad\qquad\qquad$ m

55.85 g/mol $\qquad\qquad$ 159.70 g/mol

$$n_{Fe}\ =\ 500\ g\ \times\ \frac{1\ mol}{55.85\ g}\ =\ 8.95\ mol$$

$$n_{Fe_2O_3}\ =\ 8.95\ mol\ \times\ \frac{2}{4}\ =\ 4.48\ mol$$

$$m_{Fe_2O_3}\ =\ 4.48\ mol\ \times\ \frac{159.70\ g}{1\ mol}\ =\ 715\ g$$

2. What mass of precipitate should form if 2.00 g of silver nitrate in solution is reacted with excess sodium sulfide solution?

$$2\ AgNO_{3(aq)}\ +\ Na_2S_{(aq)}\ \rightarrow\ Ag_2S_{(s)}\ +\ 2\ NaNO_{3(aq)}$$

2.00 g $\qquad\qquad\qquad\qquad$ m

169.88 g/mol $\qquad\qquad\qquad$ 247.80 g/mol

$$n_{AgNO_3}\ =\ 2.00\ g\ \times\ \frac{1\ mol}{169.88\ g}\ =\ 0.0118\ mol$$

$$n_{Ag_2S}\ =\ 0.0118\ mol\ \times\ \frac{1}{2}\ =\ 0.005\ 89\ mol$$

$$m_{Ag_2S}\ =\ 0.005\ 89\ mol\ \times\ \frac{247.80\ g}{1\ mol}\ =\ 1.46\ g$$

3. Determine the mass of water vapor formed when 1.00 g of butane, $C_4H_{10(g)}$, is burned in a lighter.

$$2\ C_4H_{10(g)}\ +\ 13\ O_{2(g)}\ \rightarrow\ 8\ CO_{2(g)}\ +\ 10\ H_2O_{(g)}$$

1.00 g $\qquad\qquad\qquad\qquad\qquad$ m

58.14 g/mol $\qquad\qquad\qquad\qquad$ 18.02 g/mol

$$n_{C_4H_{10}}\ =\ 1.00\ g\ \times\ \frac{1\ mol}{58.14\ g}\ =\ 0.0172\ mol$$

$$n_{H_2O}\ =\ 0.0172\ mol\ \times\ \frac{10}{2}\ =\ 0.0860\ mol$$

$$m_{H_2O}\ =\ 0.0860\ mol\ \times\ \frac{18.02\ g}{1\ mol}\ =\ 1.55\ g$$

4. Silver metal can be recovered from waste silver nitrate solutions by reaction with copper metal. What mass of silver can be obtained using 50 g of copper?

$$2\ AgNO_{3(aq)}\ +\ Cu_{(s)}\ \rightarrow\ 2\ Ag_{(s)}\ +\ Cu(NO_3)_{2(aq)}$$

$\qquad\qquad\qquad$ 50 g $\qquad$ m

$\qquad\qquad\qquad$ 63.55 g/mol $\quad$ 107.87 g/mol

$$n_{Cu}\ =\ 50\ g\ \times\ \frac{1\ mol}{63.55\ g}\ =\ 0.79\ mol$$

$$n_{Ag}\ =\ 0.79\ mol\ \times\ \frac{2}{1}\ =\ 1.6\ mol$$

$$m_{Ag}\ =\ 1.6\ mol\ \times\ \frac{107.87\ g}{1\ mol}\ =\ 0.17\ kg$$

31 GAS STOICHIOMETRY

Complete the following stoichiometric problems. Communicate your problem-solving approach, using internationally accepted symbols for elements, quantities, numbers, and units.

1. The first step in the industrial manufacture of sulfuric acid is the complete combustion of octasulfur. What mass of octasulfur is required to produce 112 L of sulfur dioxide at STP?

2. Coal can undergo an incomplete combustion in the absence of a plentiful supply of air to produce deadly carbon monoxide gas. What volume of carbon monoxide is produced at SATP by the incomplete combustion of 150 kg of coal?

3. The first recorded observation of hydrogen gas was made by the famous alchemist Paracelsus when he added iron to sulfuric acid. Calculate the volume of hydrogen gas at 20°C and 98 kPa produced by adding 10 g of iron to an excess of sulfuric acid.

4. Ammonia reacts with sulfuric acid to form the important fertilizer, ammonium sulfate. What mass of ammonium sulfate can be produced from 75 kL of ammonia at 10°C and 110 kPa?

31 GAS STOICHIOMETRY

Complete the following stoichiometric problems. Communicate your problem-solving approach, using internationally accepted symbols for elements, quantities, numbers, and units.

1. The first step in the industrial manufacture of sulfuric acid is the complete combustion of octasulfur. What mass of octasulfur is required to produce 112 L of sulfur dioxide at STP?

$$S_{8(s)} + 8\ O_{2(g)} \rightarrow 8\ SO_{2(g)}$$

m 112 L

256.48 g/mol   22.4 L/mol

$$n_{SO_2} = 112\ L \times \frac{1\ mol}{22.4\ L} = 5.00\ mol$$

$$n_{S_8} = 5.00\ mol \times \frac{1}{8} = 0.625\ mol$$

$$m_{S_8} = 0.625\ mol \times \frac{256.48\ g}{1\ mol} = 160\ g$$

2. Coal can undergo an incomplete combustion in the absence of a plentiful supply of air to produce deadly carbon monoxide gas. What volume of carbon monoxide is produced at SATP by the incomplete combustion of 150 kg of coal?

$$2\ C_{(s)} + O_{2(g)} \rightarrow 2\ CO_{(g)}$$

150 kg v

12.01 g/mol   24.8 L/mol

$$n_C = 150\ kg \times \frac{1\ mol}{12.01\ g} = 12.5\ kmol$$

$$n_{CO} = 12.5\ kmol \times \frac{2}{2} = 12.5\ kmol$$

$$v_{CO} = 12.5\ kmol \times \frac{24.8\ L}{1\ mol} = 310\ kL$$

3. The first recorded observation of hydrogen gas was made by the famous alchemist Paracelsus when he added iron to sulfuric acid. Calculate the volume of hydrogen gas at 20°C and 98 kPa produced by adding 10 g of iron to an excess of sulfuric acid.

$$2\ Fe_{(s)} + 3\ H_2SO_{4(aq)} \rightarrow 3\ H_{2(g)} + Fe_2(SO_4)_{3(aq)}$$

55.85 g/mol v

10 g   20°C, 98 kPa

$$n_{Fe} = 10\ g \times \frac{1\ mol}{55.85\ g} = 0.18\ mol, \qquad n_{H_2} = 0.18\ mol \times \frac{3}{2} = 0.27\ mol$$

$$v_{H_2} = \frac{nRT}{p} = \frac{0.27\ mol \times 8.31\ kPa{\cdot}L/(mol{\cdot}K) \times 293\ K}{98\ kPa} = 6.7\ L$$

4. Ammonia reacts with sulfuric acid to form the important fertilizer, ammonium sulfate. What mass of ammonium sulfate can be produced from 75 kL of ammonia at 10°C and 110 kPa?

$$2\ NH_{3(g)} + H_2SO_{4(aq)} \rightarrow (NH_4)_2SO_{4(aq)}$$

75 kL m

10°C, 110 kPa   132.16 g/mol

$$n_{NH_3} = \frac{pv}{RT} = \frac{110\ kPa \times 75\ kL}{8.31\ kPa{\cdot}L/(mol{\cdot}K) \times 283\ K} = 3.5\ kmol$$

$$n_{(NH_4)_2SO_4} = 3.5\ kmol \times \frac{1}{2} = 1.8\ kmol$$

$$m_{(NH_4)_2SO_4} = 1.8\ kmol \times \frac{132.16\ g}{1\ mol} = 0.23\ Mg$$

32 SOLUTION STOICHIOMETRY

Complete the following stoichiometric problems. Communicate your problem-solving approach, using internationally accepted symbols for elements, quantities, numbers, and units.

1. What is the concentration of a $KOH_{(aq)}$ solution if 12.8 mL of this solution is required to react with 25.0 mL of 0.110 mol/L $H_2SO_{4(aq)}$?

2. What volume of 0.125 mol/L $NaOH_{(aq)}$ is required to react completely with 15.0 mL of 0.100 mol/L $Al_2(SO_4)_{3(aq)}$?

3. In a chemical analysis, a 10.0 mL sample of $H_3PO_{4(aq)}$ was reacted with 18.2 mL of 0.259 mol/L $NaOH_{(aq)}$. Calculate the concentration of the phosphoric acid.

4. The concentration of magnesium ions (assume magnesium chloride) in sea water was analyzed and found to be 50.0 mmol/L. What volume of 0.200 mol/L sodium hydroxide solution would be needed in an industrial process to precipitate all of the magnesium ions from 1.00 ML of sea water?

32 SOLUTION STOICHIOMETRY

Complete the following stoichiometric problems. Communicate your problem-solving approach, using internationally accepted symbols for elements, quantities, numbers, and units.

1. What is the concentration of a $KOH_{(aq)}$ solution if 12.8 mL of this solution is required to react with 25.0 mL of 0.110 mol/L $H_2SO_{4(aq)}$?

$$2\ KOH_{(aq)}\ +\ H_2SO_{4(aq)}\ \rightarrow\ K_2SO_{4(aq)}\ +\ 2\ HOH_{(l)}$$

12.8 mL   25.0 mL
C 0.110 mol/L

$$n_{H_2SO_4}\ =\ 25.0\ mL\ \times\ \frac{0.110\ mol}{1\ L}\ =\ 2.75\ mmol$$

$$n_{KOH}\ =\ 2.75\ mmol\ \times\ \frac{2}{1}\ =\ 5.50\ mmol$$

$$C_{KOH}\ =\ \frac{5.50\ mmol}{12.8\ mL}\ =\ 0.430\ mol/L$$

2. What volume of 0.125 mol/L $NaOH_{(aq)}$ is required to react completely with 15.0 mL of 0.100 mol/L $Al_2(SO_4)_{3(aq)}$?

$$6\ NaOH_{(aq)}\ +\ Al_2(SO_4)_{3(aq)}\ \rightarrow\ 2\ Al(OH)_{3(s)}\ +\ 3\ Na_2SO_{4(aq)}$$

v 15.0 mL
0.125 mol/L   0.100 mol/L

$$n_{Al_2(SO_4)_2}\ =\ 15.0\ mL\ \times\ \frac{0.100\ mol}{1\ L}\ =\ 1.50\ mmol$$

$$n_{NaOH}\ =\ 1.50\ mmol\ \times\ \frac{6}{1}\ =\ 9.00\ mmol$$

$$v_{NaOH}\ =\ 9.00\ mmol\ \times\ \frac{1\ L}{0.125\ mol}\ =\ 72.0\ mL$$

3. In a chemical analysis, a 10.0 mL sample of $H_3PO_{4(aq)}$ was reacted with 18.2 mL of 0.259 mol/L $NaOH_{(aq)}$. Calculate the concentration of the phosphoric acid.

$$H_3PO_{4(aq)}\ +\ 3\ NaOH_{(aq)}\ \rightarrow\ Na_3PO_{4(aq)}\ +\ 3\ HOH_{(l)}$$

10.0 mL   18.2 mL
C 0.259 mol/L

$$n_{NaOH}\ =\ 18.2\ mL\ \times\ \frac{0.259\ mol}{1\ L}\ =\ 4.71\ mmol$$

$$n_{H_3PO_4}\ =\ 4.71\ mmol\ \times\ \frac{1}{3}\ =\ 1.57\ mmol$$

$$C_{H_3PO_4}\ =\ \frac{1.57\ mmol}{10.0\ mL}\ =\ 0.157\ mol/L$$

4. The concentration of magnesium ions (assume magnesium chloride) in sea water was analyzed and found to be 50.0 mmol/L. What volume of 0.200 mol/L sodium hydroxide solution would be needed in an industrial process to precipitate all of the magnesium ions from 1.00 ML of sea water?

$$MgCl_{2(aq)}\ +\ 2\ NaOH_{(aq)}\ \rightarrow\ Mg(OH)_{2(s)}\ +\ 2\ NaCl_{(aq)}$$

1.00 ML v
50.0 mmol/L   0.200 mol/L

$$n_{MgCl_2}\ =\ 1.00\ ML\ \times\ \frac{50.0\ mmol}{1\ L}\ =\ 50.0\ kmol$$

$$n_{NaOH}\ =\ 50.0\ kmol\ \times\ \frac{2}{1}\ =\ 100\ kmol$$

$$v_{NaOH}\ =\ 100\ kmol\ \times\ \frac{1\ L}{0.200\ mol}\ =\ 500\ kL$$

33 (ENRICHMENT) STOICHIOMETRY

Complete the following stoichiometric problems. Communicate your problem-solving approach, using internationally accepted symbols for elements, quantities, numbers, and units.

1. A 6.72 g sample of zinc was placed in 100.0 mL of 1.50 mol/L hydrochloric acid. After all reaction stops, how much zinc should remain?

2. An unlabelled white solid acid $H_2X_{(s)}$ is known to react in a 1 : 2 mole ratio with sodium hydroxide. In an attempt to identify the acid, a titration provides evidence that 12.5 mL of 0.300 mol/L $NaOH_{(aq)}$ reacts with 0.169 g of the acid. What is the molar mass and possible identity of the acid?

3. What volume of hydrogen gas at STP will be produced when 100 g of aluminum is added to 4.00 L of 1.40 mol/L sulfuric acid?

33 (ENRICHMENT) STOICHIOMETRY

Complete the following stoichiometric problems. Communicate your problem-solving approach, using internationally accepted symbols for elements, quantities, numbers, and units.

1. A 6.72 g sample of zinc was placed in 100.0 mL of 1.50 mol/L hydrochloric acid. After all reaction stops, how much zinc should remain?

$$Zn_{(s)} \quad + \quad 2\,HCl_{(aq)} \quad \rightarrow \quad H_{2(g)} \quad + \quad ZnCl_{2(aq)}$$
m = 6.72 g   100.0 mL
65.38 g/mol   1.50 mol/L

$$n_{HCl} = 100.0\ mL \times \frac{1.50\ mol}{1\ L} = 150\ mmol$$

$$n_{Zn} = 150\ mmol \times \frac{1}{2} = 75.0\ mmol$$

$$m_{Zn} = 75.0\ mmol \times \frac{65.38\ g}{1\ mol} = 4.90\ g\ (reacted)$$

mass of zinc remaining = 6.72 g – 4.90 g = 1.82 g

2. An unlabelled white solid acid $H_2X_{(s)}$ is known to react in a 1 : 2 mole ratio with sodium hydroxide. In an attempt to identify the acid, a titration provides evidence that 12.5 mL of 0.300 mol/L $NaOH_{(aq)}$ reacts with 0.169 g of the acid. What is the molar mass and possible identity of the acid?

$$H_2X_{(aq)} \quad + \quad 2\,NaOH_{(aq)} \quad \rightarrow \quad 2\,HOH_{(l)} \quad + \quad Na_2X_{(aq)}$$
0.169 g   12.5 mL
M 0.300 mol/L

$$n_{NaOH} = 12.5\ mL \times \frac{0.300\ mol}{1\ L} = 3.75\ mmol$$

$$n_{H_2X} = 3.75\ mmol \times \frac{1}{2} = 1.88\ mmol$$

$$M_{H_2X} = \frac{0.169\ g}{1.88\ mmol} = 90.1\ g/mol$$

The molar mass of 90.1 g/mol is close to that of oxalic acid.

3. What volume of hydrogen gas at STP will be produced when 100 g of aluminum is added to 4.00 L of 1.40 mol/L sulfuric acid?

$$3\,H_2SO_{4(aq)} \quad + \quad 2\,Al_{(s)} \quad \rightarrow \quad 3\,H_{2(g)} \quad + \quad Al_2(SO_4)_{3(aq)}$$
4.00 L   100 g v
1.40 mol/L   26.98 g/mol   22.4 L/mol

$$n_{Al} = 100\ g \times \frac{1\ mol}{26.98\ g} = 3.71\ mol$$

If all of this aluminum reacts, then the amount of sulfuric acid required is

$$n_{H_2SO_4} = 3.71\ mol \times \frac{3}{2} = 5.56\ mol$$

The amount of sulfuric acid present is

$$n_{H_2SO_4} = 4.00\ L \times \frac{1.40\ mol}{1\ L} = 5.60\ mol$$

Therefore, excess sulfuric acid is used and aluminum is the limiting reagent.

$$n_{H_2} = 3.71\ mol \times \frac{3}{2} = 5.56\ mol$$

$$v_{H_2} = 5.56\ mol \times \frac{22.4\ L}{1\ mol} = 125\ L$$

34 SOLUTION PREPARATION

Communicate your problem-solving approach when answering the questions below.

1. Calculate the molar concentration of a solution made by dissolving 20.0 g of sodium hydroxide to make 300 mL of solution.

2. Pure sodium thiosulfate-5-water, $Na_2S_2O_3 \cdot 5H_2O_{(s)}$, is used to make 250 mL of 20.0 mmol/L solution. Find the mass of solute required.

3. What mass of copper(II) nitrate will be required to prepare 10.0 L of 0.100 mol/L solution?

4. What volume of 75 mmol/L solution can be prepared from 10 g of sodium carbonate?

5. Determine the volume of concentrated hydrochloric acid required to prepare 10.0 L of a 0.200 mol/L solution.

6. What volume of concentrated ammonia is required to prepare 2.0 L of a 1.0 mol/L solution?

34 SOLUTION PREPARATION

Communicate your problem-solving approach when answering the questions below.

1. Calculate the molar concentration of a solution made by dissolving 20.0 g of sodium hydroxide to make 300 mL of solution.

$$n_{NaOH} = 20.0 \text{ g} \times \frac{1 \text{ mol}}{40.00 \text{ g}} = 0.500 \text{ mol}$$

$$C_{NaOH} = \frac{0.500 \text{ mol}}{0.300 \text{ L}} = 1.67 \text{ mol/L}$$

2. Pure sodium thiosulfate-5-water, $Na_2S_2O_3 \bullet 5H_2O_{(s)}$, is used to make 250 mL of 20.0 mmol/L solution. Find the mass of solute required.

$$n_{Na_2S_2O_3 \bullet 5H_2O} = 0.250 \text{ L} \times \frac{20.0 \text{ mmol}}{1 \text{ L}} = 5.00 \text{ mmol}$$

$$m_{Na_2S_2O_3 \bullet 5H_2O} = 5.00 \text{ mmol} \times \frac{248.20 \text{ g}}{1 \text{ mol}} = 1.24 \text{ g}$$

3. What mass of copper(II) nitrate will be required to prepare 10.0 L of 0.100 mol/L solution?

$$n_{Cu(NO_3)_2} = 10.0 \text{ L} \times \frac{0.100 \text{ mol}}{1 \text{ L}} = 1.00 \text{ mol}$$

$$m_{Cu(NO_3)_2} = 1.00 \text{ mol} \times \frac{187.57 \text{ g}}{1 \text{ mol}} = 188 \text{ g}$$

4. What volume of 75 mmol/L solution can be prepared from 10 g of sodium carbonate?

$$n_{Na_2CO_3} = 10 \text{ g} \times \frac{1 \text{ mol}}{105.99 \text{ g}} = 94 \text{ mmol}$$

$$v_{Na_2CO_3} = 94 \text{ mmol} \times \frac{1 \text{ L}}{75 \text{ mmol}} = 1.3 \text{ L}$$

5. Determine the volume of concentrated hydrochloric acid required to prepare 10.0 L of a 0.200 mol/L solution.

$$v_i C_i = v_f C_f$$

$$v_i \times \frac{11.6 \text{ mol}}{1 \text{ L}} = 10.0 \text{ L} \times \frac{0.200 \text{ mol}}{1 \text{ L}}$$

$$v_i = 0.172 \text{ L}$$

6. What volume of concentrated ammonia is required to prepare 2.0 L of a 1.0 mol/L solution?

$$v_i C_i = v_f C_f$$

$$v_i \times \frac{14.8 \text{ mol}}{1 \text{ L}} = 2.0 \text{ L} \times \frac{1.0 \text{ mol}}{1 \text{ L}}$$

$$v_i = 0.14 \text{ L}$$

35 VOLUMETRIC STOICHIOMETRY — TITRATION

Complete the *Anyalysis* section of each of the following lab reports.

1. **Problem**

   What is the molar concentration of an unknown sodium carbonate solution?

   Evidence

Titration of 25.0 mL of $Na_2CO_{3(aq)}$ with 0.352 mol/L $HCl_{(aq)}$				
Trial	1	2	3	4
Final buret reading (mL)	16.5	31.8	47.0	16.4
Initial buret reading (mL)	0.6	16.5	31.8	1.2
Volume of $HCl_{(aq)}$ added (mL)	15.9	15.3	15.2	15.2

   Analysis

2. **Problem**

   What is the molar concentration of a potassium hydroxide solution?

   Evidence

Titration of 10.0 mL of $KOH_{(aq)}$ with 0.150 mol/L $H_2SO_{4(aq)}$				
Trial	1	2	3	4
Final buret reading (mL)	12.8	25.3	37.9	—
Initial buret reading (mL)	0.2	12.8	25.3	—
Volume of $H_2SO_{4(aq)}$ added (mL)	12.6	12.6	12.6	

   Analysis

3. **Problem**

   What is the molar concentration of a potassium permanganate solution?

   Evidence

Titration of 10.0 mL of Acidified 0.100 mol/L $Fe_2SO_{4(aq)}$ with $KMnO_{4(aq)}$				
Trial	1	2	3	4
Final buret reading (mL)	11.3	21.9	32.5	43.1
Initial buret reading (mL)	0.1	11.3	21.9	32.5
Volume of $KMnO_{4(aq)}$ added (mL)	11.2	10.5	10.6	10.6

   Analysis

   $$10\ FeSO_{4(aq)} + 2\ KMnO_{4(aq)} + 8\ H_2SO_{4(aq)} \rightarrow 5\ Fe_2(SO_4)_{3(aq)} + K_2SO_{4(aq)} + 2\ MnSO_{4(aq)} + 8\ H_2O_{(l)}$$

35 VOLUMETRIC STOICHIOMETRY — TITRATION

Complete the *Anyalysis* section of each of the following lab reports.

1. *Analysis*

$$Na_2CO_{3(aq)} + 2\,HCl_{(aq)} \rightarrow H_2CO_{3(aq)} + 2\,NaCl_{(aq)}$$

25.0 mL   15.2 mL   *(Note: 15.2 is 15.2 in the calculator)*

C 0.352 mol/L

$$n_{HCl} = 15.2\ mL \times \frac{0.352\ mol}{1\ L} = 5.36\ mmol$$

$$n_{Na_2CO_3} = 5.35\ mmol \times \frac{1}{2} = 2.68\ mmol$$

$$C_{Na_2CO_3} = \frac{2.68\ mmol}{25.0\ mL} = 0.107\ mol/L$$

According to the evidence gathered and the stoichiometric method, the molar concentration of the sodium carbonate solution is 0.107 mol/L.

2. *Analysis*

$$2\,KOH_{(aq)} + H_2SO_{4(aq)} \rightarrow 2HOH_{(l)} + K_2SO_{4(aq)}$$

10.0 mL   12.6 m   *(Note: 12.6 is 12.56 in the calculator)*

C 0.150 mol/L

$$n_{H_2SO_4} = 12.6\ mL \times \frac{0.150\ mol}{1L} = 1.89\ mmol$$

$$n_{KOH} = 1.89\ mmol \times \frac{2}{1} = 3.77\ mmol$$

$$C_{KOH} = \frac{3.77\ mmol}{10.0\ mL} = 0.377\ mol/L$$

Based on the evidence gathered and the stoichiometric method, the molar concentration of the potassium hydroxide solution is 0.377 mol/L.

3. *Analysis*

$$10\,FeSO_{4(aq)} + 2\,KMnO_{4(aq)} + 8\,H_2SO_{4(aq)} \rightarrow$$
$$5\,Fe_2(SO_4)_{3(aq)} + K_2SO_{4(aq)} + 2\,MnSO_{4(aq)} + 8\,H_2O_{(l)}$$

10.0 mL   10.6 mL

0.100 mol/L C

$$n_{FeSO_4} = 10.0\ mL \times \frac{0.100\ mol}{1\ L} = 1.00\ mmol$$

$$n_{KMnO_4} = 1.00\ mmol \times \frac{2}{10} = 0.200\ mmol$$

$$C_{KMnO_4} = \frac{0.200\ mmol}{10.6\ mL} = 0.0189\ mol/L$$

According to the evidence gathered and the method of stoichiometry, the molar concentration of the potassium permanganate solution is 0.0189 mol/L.

36 IONIC COMPOUNDS

1. Draw an electron dot diagram for each of the following atoms.

   (a) Li

   (b) Sr

   (c) Br

   (d) O

2. Classify each of the atoms in question 1 as metals or nonmetals.

3. Which combinations of atoms from question 1 would likely produce ionic compounds?

4. What is the key theoretical process used to explain the formation of ionic compounds? What is believed to be the reason for this process?

5. Use electron dot diagrams to represent the formation of lithium oxide from its constituent atoms.

6. Use electron dot diagrams to represent the formation of strontium bromide from its constituent atoms.

7. What is the most direct evidence for the presence of ions in ionic compounds?

36 IONIC COMPOUNDS

1. Draw an electron dot diagram for each of the following atoms.

   (a) Li   **Li·**   (b) Sr   **·Sr·**

   (c) Br   **:Br:**   (d) O   **·O:**

2. Classify each of the atoms in question 1 as metals or nonmetals.

   metals: Li and Sr
   nonmetals: Br and O

3. Which combinations of atoms from question 1 would likely produce ionic compounds?

   metal-nonmetal combinations: Li and Br or O; Sr and Br or O

4. What is the key theoretical process used to explain the formation of ionic compounds? What is believed to be the reason for this process?

   The formation of ionic compounds is explained by a transfer of electrons from the metal atom to the nonmetal atom. This process is thought to occur because of the relatively large difference in electronegativity of metal versus nonmetal atoms.

5. Use electron dot diagrams to represent the formation of lithium oxide from its constituent atoms.

   $$Li· \qquad ·\ddot{O}: \qquad \rightarrow \qquad Li^+_2 \; [:\ddot{O}:]^{2-}$$
   $$Li·$$

6. Use electron dot diagrams to represent the formation of strontium bromide from its constituent atoms.

   $$:\ddot{Br}:$$
   $$Sr· \qquad \rightarrow \qquad Sr^{2+} \; [:\ddot{Br}:]^-_2$$
   $$:\ddot{Br}:$$

7. What is the most direct evidence for the presence of ions in ionic compounds?

   The electrical conductivity of aqueous and molten ionic compounds is the most direct evidence for the presence of charged entities, i.e. ions.

37 MOLECULAR COMPOUNDS

1. What are the two technologies used in the analysis of the composition of molecular compounds?

2. A chemical analysis of phosgene, a poisonous gas used during World War I, gave the following evidence.

   12.1% C, 16.2% O, 71.7% Cl, and 98.8 g/mol

   Determine the molecular formula of phosgene and explain the formula using an electron dot diagram.

3. Hydrazine is important as a rocket propellant. A chemical analysis indicates that the compound has a molar mass of 32.1 g/mol and a composition of 87.4% N and 12.6% H. Calculate the molecular formula for hydrazine and explain the formula using an electron dot diagram.

For each of the following molecular compounds, explain the empirically determined formula by drawing a diagram of the structural model.

4. CH_3Cl 5. C_2H_3Cl 6. CH_3CN

For each of the following combinations of elements, theoretically predict the molecular formula of the simplest product of the formation reaction. Draw a structural diagram of the product. Do not balance the equations.

7. $Cl_{2(g)} + Br_{2(g)} \longrightarrow$
8. $N_{2(g)} + Cl_{2(g)} \longrightarrow$

9. $O_{2(g)} + F_{2(g)} \longrightarrow$

10. What evidence is there for the presence of multiple bonds in some molecular substances?

37 MOLECULAR COMPOUNDS

1. What are the two technologies used in the analysis of the composition of molecular compounds?

   A combustion analyzer and a mass spectrometer are used in the analysis of the composition of molecular compounds.

2. A chemical analysis of phosgene, a poisonous gas used during World War I, gave the following evidence.

   12.1% C, 16.2% O, 71.7% Cl, and 98.8 g/mol

   Determine the molecular formula of phosgene and explain the formula using an electron dot diagram.

   $$m_C = \frac{12.1}{100} \times 98.8\ g = 12.0\ g,\ n_C = 12.0\ g \times \frac{1\ mol}{12.01\ g} = 0.995\ mol$$

   $$m_O = \frac{16.2}{100} \times 98.8\ g = 16.0\ g,\ n_O = 16.0\ g \times \frac{1\ mol}{16.00\ g} = 1.00\ mol$$

   $$m_{Cl} = \frac{71.7}{100} \times 98.8\ g = 70.8\ g,\ n_{Cl} = 70.8\ g \times \frac{1\ mol}{35.45\ g} = 2.00\ mol$$

   According to the evidence, the molecular formula is $COCl_2$:

3. Hydrazine is important as a rocket propellant. A chemical analysis indicates that the compound has a molar mass of 32.1 g/mol and a composition of 87.4% N and 12.6% H. Calculate the molecular formula for hydrazine and explain the formula using an electron dot diagram.

   $$m_N = \frac{87.4}{100} \times 32.1\ g = 28.1\ g,\ n_N = 28.1\ g \times \frac{1\ mol}{14.01\ g} = 2.00\ mol$$

   $$m_H = \frac{12.6}{100} \times 32.1\ g = 4.04\ g,\ n_H = 4.04\ g \times \frac{1\ mol}{1.01\ g} = 4.00\ mol$$

   According to the evidence, the molecular formula is N_2H_4:

For each of the following molecular compounds, explain the empirically determined formula by drawing a diagram of the structural model.

4. CH_3Cl 5. C_2H_3Cl 6. CH_3CN

For each of the following combinations of elements, theoretically predict the molecular formula of the simplest product of the formation reaction. Draw a structural diagram of the product. Do not balance the equations.

7. $Cl_{2(g)} + Br_{2(g)} \rightarrow$ **ClBr**   Cl—Br
8. $N_{2(g)} + Cl_{2(g)} \rightarrow$ **NCl₃**   Cl—N—Cl
                                                        |
                                                        Cl
9. $O_{2(g)} + F_{2(g)} \rightarrow$ **OF₂**   O—F
                                                  |
                                                  F

10. What evidence is there for the presence of multiple bonds in some molecular substances?

    Evidence for the presence of multiple covalent bonds is the rapid change in the color of bromine as it is added to the substance.

38 (ENRICHMENT) SHAPES OF MOLECULES

Evidence indicates that molecules have definite three-dimensional shapes. These shapes appear to be important in determining the physical and chemical properties of molecular substances. Although explanations and predictions of molecular shapes can be done using the modern quantum mechanics theory, a much simpler idea is remarkably effective. This theory, called the *Valence Shell Electron Pair Repulsion* theory (or VSEPR theory) was developed in part by Ronald Gillespie of McMaster University in Hamilton, Ontario. VSEPR theory is based on the simple idea that pairs of electrons in covalent bonds or lone pairs exist as far apart as possible in order to minimize the electrostatic repulsion between groups of negatively charged particles. To predict the shape of a molecule, the number of electron pairs on the central atom is determined (from an electron dot diagram), and the electron pairs are placed in a three-dimensional space at minimum repulsion. For example, if an atom has two shared pairs of electrons and no lone pairs, a linear arrangement is expected.

$$Y:X:Y \qquad \text{or} \qquad Y—X—Y$$

This arrangement keeps the electron pairs on the opposite sides of the central atom (X) and as far away as possible from each other.

Of the atoms in the group, the central atom is the atom that has the largest number of bonding electrons. Only arrangements around the central atom can be predicted. If more than one central atom is present, the shape around each is determined separately. All electron pairs, both shared and lone pairs, need to be considered. However, the description of the shape is related to the atoms present. In the following table, "A" denotes a central atom and "X" denotes an atom bonded to "A". In the restricted VSEPR theory, lone pairs of electrons around "X" atoms do not affect the shape and are not considered.

Class	Shape	Model*	Example
AX	linear	A—X	HCl
AX_2	linear	X—A—X	HCN
AX_3	trigonal planar	 X X \ / A—A / \ X X	C_2H_4
AX_4	tetrahedral	X | A / | \ X X X	CH_4
$:AX_3$	pyramidal	A | X X X	NH_3
$:\ddot{A}X_2$	V-shaped	A / \ X X	H_2O

* Each line shown in the structural models represents a group of electrons which may include one, two, or three shared pairs.

38 (ENRICHMENT) SHAPES OF MOLECULES

For each of the following molecules, draw an electron dot diagram and use the table on the previous page to describe the shape around each central atom. Then draw a structural diagram for each. Your predictions can be tested using a molecular model kit.

1. hydrogen sulfide, H_2S

2. carbon tetrachloride, CCl_4

3. phosphine, PH_3

4. tetrachloroethylene, C_2Cl_4

5. hydrogen peroxide, H_2O_2

6. carbon dioxide, CO_2

7. hydrazine, N_2H_4

8. propane, C_3H_8

9. methylcyanide, CH_3CN

10. nitrosyl chloride, NOCl

Answers
1. V-shaped   2. tetrahedral   3. pyramidal   4. trigonal planar   5. V-shaped   6. linear   7. pyramidal   8. tetrahedral   9. tetrahedral and linear   10. pyramidal

39 ALKANES

For each of the following IUPAC names, draw a structural diagram.

1. 2-methylpentane

2. 3-ethylhexane

3. 2,3-dimethylbutane

4. octane

5. cyclobutane

6. 2,3-dimethylhexane

7. trimethylbutane

8. 3-ethyl-3-methylhexane

9. methylcyclopentane

10. 2,2,3-trimethylpentane

For each of the following questions, draw a structural diagram equation and classify the reaction type.

11. heptane burns in a fuel mixture

12. cyclooctane + hydrogen $\rightarrow$ propane + pentane

13. butane + propane $\rightarrow$ 2,3-dimethylpentane + hydrogen

39 ALKANES

For each of the following IUPAC names, draw a structural diagram.
(*You may want all hydrogen atoms shown as bonded atoms.*)

1. 2-methylpentane

$$CH_3$$
$$|$$
$$CH_3 - CH - CH_2 - CH_3$$

2. 3-ethylhexane

$$C_2H_5$$
$$|$$
$$CH_3 - CH_2 - CH - CH_2 - CH_2 - CH_3$$

3. 2,3-dimethylbutane

$$CH_3$$
$$|$$
$$CH_3 - CH - CH - CH_3$$
$$|$$
$$CH_3$$

4. octane

$$CH_3 - (CH_2)_6 - CH_3$$

5. cyclobutane

$$CH_2 - CH_2$$
$$|\qquad\;|$$
$$CH_2 - CH_2$$

6. 2,3-dimethylhexane

$$CH_3$$
$$|$$
$$CH_3 - CH - CH - CH_2 - CH_2 - CH_3$$
$$|$$
$$CH_3$$

7. trimethylbutane

$$CH_3 \quad CH_3$$
$$|\qquad\;|$$
$$CH_3 - C - CH - CH_3$$
$$|$$
$$CH_3$$

8. 3-ethyl-3-methylhexane

$$CH_3$$
$$|$$
$$CH_3 - CH_2 - C - CH_2 - CH_2 - CH_3$$
$$|$$
$$C_2H_5$$

9. methylcyclopentane

$$CH_3$$

10. 2,2,3-trimethylpentane

$$CH_3 \quad CH_3$$
$$|\qquad\;|$$
$$CH_3 - C - CH_2 - CH_2 - CH_3$$
$$|$$
$$CH_3$$

For each of the following questions, draw a structural diagram equation and classify the reaction type.

11. heptane burns in a fuel mixture
    complete combustion
    $$CH_3 - (CH_2)_5 - CH_3 + O{=}O \rightarrow O{=}C{=}O + H - O - H$$

12. cyclooctane + hydrogen $\rightarrow$ propane + pentane
    cracking

    ⬡ $+ \; H - H \rightarrow CH_3 - CH_2 - CH_3 + CH_3 - CH_2 - CH_2 - CH_2 - CH_3$

13. butane + propane $\rightarrow$ 2,3-dimethylpentane + hydrogen
    reforming

    $$CH_3 - CH_2 - CH_2 - CH_3 + CH_3 - CH_2 - CH_3 \rightarrow CH_3 - \overset{\displaystyle CH_3}{\underset{\displaystyle CH_3}{\overset{|}{\underset{|}{CH}}} - CH} - CH_2 - CH_3 + H - H$$

40 ALKENES, ALKYNES, AND AROMATICS

For each of the following IUPAC names, draw a structural diagram.

1. ethene (ethylene)

2. propyne

3. methylpropene

4. methyl-1-butyne

5. ethylbenzene

6. 3-methyl-2-pentene

For each of the following structural diagrams, write the IUPAC name.

7. $CH_3—C\equiv C—CH_3$

8. $CH_3—CH_2—CH_2—CH=CH_2$

9.
$$CH_3—\overset{\overset{\displaystyle CH_3}{|}}{C}=CH—CH_3$$

10.
$$CH_3—CH—CH_3$$

11. $CH_2=CH—CH_3$

12.
$$CH_3—C\equiv C—\overset{\overset{\displaystyle}{|}}{\underset{\underset{\displaystyle CH_3}{|}}{CH}}—CH_3$$

For each of the following questions, draw a structural diagram equation and classify the reaction type.

13. 2-methyl-1-pentene + hydrogen $\rightarrow$ 2-methylpentane

14. butane $\rightarrow$ 1-butene + 2-butene + hydrogen

15. acetylene + oxygen $\rightarrow$

40 ALKENES, ALKYNES, AND AROMATICS

For each of the following IUPAC names, draw a structural diagram.

1. ethene (ethylene)

    $CH_2{=}CH_2$

2. propyne

    $CH{\equiv}C-CH_3$

3. methylpropene

    $$\begin{matrix} & CH_3 \\ & | \\ CH_2{=}C & {-}CH_3 \end{matrix}$$

4. methyl-1-butyne

    $$\begin{matrix} & CH_3 \\ & | \\ CH{\equiv}C-CH & {-}CH_3 \end{matrix}$$

5. ethylbenzene

    C_2H_5

6. 3-methyl-2-pentene

    $$\begin{matrix} & & CH_3 \\ & & | \\ CH_3-CH{=}C & {-}CH_2{-}CH_3 \end{matrix}$$

For each of the following structural diagrams, write the IUPAC name.

7. $CH_3-C{\equiv}C-CH_3$

    2-butyne

8. $CH_3-CH_2-CH_2-CH{=}CH_2$

    1-pentene

9. $$\begin{matrix} & CH_3 \\ & | \\ CH_3-C{=}CH & {-}CH_3 \end{matrix}$$

    methyl-2-butene

10. $$\begin{matrix} CH_3-CH-CH_3 \end{matrix}$$

    2-phenylpropane

11. $CH_2{=}CH-CH_3$

    propene

12. $$\begin{matrix} CH_3-C{\equiv}C-CH-CH_3 \\ | \\ CH_3 \end{matrix}$$

    methyl-2-pentyne

For each of the following questions, draw a structural diagram equation and classify the reaction type.

13. 2-methyl-1-pentene + hydrogen $\rightarrow$ 2-methylpentane

    addition

    $$\begin{matrix} CH_3 & & & & CH_3 \\ | & & & & | \\ CH_2{=}C-CH_2-CH_2-CH_3 + H{-}H \rightarrow CH_3-CH-CH_2-CH_2-CH_3 \end{matrix}$$

14. butane $\rightarrow$ 1-butene + 2-butene + hydrogen

    cracking

    $CH_3-CH_2-CH_2-CH_3 \rightarrow CH_2{=}CH-CH_2-CH_3 + CH_3-CH{=}CH-CH_3 + H{-}H$

15. acetylene + oxygen $\rightarrow$

    complete combustion

    $CH{\equiv}CH + O{=}O \rightarrow O{=}C{=}O + H{-}O{-}H$

41 HYDROCARBON DERIVATIVES

In the following questions, the IUPAC names of a variety of hydrocarbon derivatives are provided. Draw a structural diagram for each name and identify the organic family to which the compound belongs.

1. 2-bromopentane

2. 1,4-dichlorobenzene

3. butanoic acid

4. ethyl methanoate

5. 1-butanol

6. propanal

7. 1,1-dichloro-2,2-difluoroethane

8. trimethylamine

9. 2-methyl-2-propanol

10. propanamide

Communicate acceptable English IUPAC names for the following structural models.

11. $CH_3—CH_2—CH—CH_3$
$\qquad\qquad\qquad |$
$\qquad\qquad\qquad OH$

12. $Cl—CH_2—CH_2—CH_2—Cl$

13. $CH_3—C—OH$
$\qquad\quad ||$
$\qquad\quad O$

14. $CH_3—C—O—CH_3$
$\qquad\quad ||$
$\qquad\quad O$

41 HYDROCARBON DERIVATIVES

In the following questions, the IUPAC names of a variety of hydrocarbon derivatives are provided. Draw a structural diagram for each name and identify the organic family to which the compound belongs.

1. 2-bromopentane

$$CH_3 \!-\! \overset{\overset{\displaystyle Br}{|}}{CH} \!-\! CH_2 \!-\! CH_2 \!-\! CH_3$$
organic halide

2. 1,4-dichlorobenzene

organic halide

3. butanoic acid

$$CH_3 \!-\! CH_2 \!-\! CH_2 \!-\! \overset{\overset{\displaystyle O}{||}}{C} \!-\! OH$$
carboxylic acid

4. ethyl methanoate

$$H \!-\! \overset{\overset{\displaystyle O}{||}}{C} \!-\! O \!-\! CH_2 \!-\! CH_3$$
ester

5. 1-butanol

$$CH_3 \!-\! CH_2 \!-\! CH_2 \!-\! CH_2 \!-\! OH$$
alcohol

6. propanal

$$CH_3 \!-\! CH_2 \!-\! \overset{\overset{\displaystyle O}{||}}{C} \!-\! H$$
aldehyde

7. 1,1-dichloro-2,2-difluoroethane

$$Cl \!-\! \overset{\overset{\displaystyle Cl}{|}}{CH} \!-\! \overset{\overset{\displaystyle F}{|}}{CH} \!-\! F$$
organic halide

8. trimethylamine

$$CH_3 \!-\! \overset{\overset{\displaystyle CH_3}{|}}{N} \!-\! CH_3$$
amine

9. 2-methyl-2-propanol

$$CH_3 \!-\! \overset{\overset{\displaystyle CH_3}{|}}{\underset{\underset{\displaystyle OH}{|}}{C}} \!-\! CH_3$$
alcohol

10. propanamide

$$CH_3 \!-\! CH_2 \!-\! \overset{\overset{\displaystyle O}{||}}{C} \!-\! NH_2$$
amide

Communicate acceptable English IUPAC names for the following structural models.

11. $CH_3 \!-\! CH_2 \!-\! \underset{\underset{\displaystyle OH}{|}}{CH} \!-\! CH_3$

2-butanol

12. $Cl \!-\! CH_2 \!-\! CH_2 \!-\! CH_2 \!-\! Cl$

1,3-dichloropropane

13. $CH_3 \!-\! \underset{\underset{\displaystyle O}{||}}{C} \!-\! OH$

ethanoic acid

14. $CH_3 \!-\! \underset{\underset{\displaystyle O}{||}}{C} \!-\! O \!-\! CH_3$

methyl ethanoate

42 ORGANIC REACTIONS OF HYDROCARBON DERIVATIVES

For each of the following questions, state the organic reaction type, draw structural diagrams for all reactants and products, and name all organic products. Do not balance the equations.

1. Propane reacts with fluorine.

2. Chloroethane reacts with hydroxide ions.

3. Ethanol, present in gasohol, burns in an automobile engine.

4. 1-butanol reacts in the presence of concentrated sulfuric acid.

5. Dichloromethane is produced by reacting methane and chlorine.

6. Bromine and ethene react to form an alkyl halide.

7. Hydrogen chloride and ethene react to produce an alkyl halide.

8. Acetic acid and ethanol react to produce a solvent used in nail polish remover.

© NELSON CANADA,
A DIVISION OF THOMSON CANADA LIMITED, 1994

42 ORGANIC REACTIONS OF HYDROCARBON DERIVATIVES

For each of the following questions, state the organic reaction type, draw structural diagrams for all reactants and products, and name all organic products. Do not balance the equations.

1. Propane reacts with fluorine.

   substitution

   $CH_3-CH_2-CH_3 + F-F \rightarrow F-CH_2-CH_2-CH_3 + CH_3-CH-CH_3 + H-F$

   1-fluoropropane

   $\overset{|}{F}$

   2-fluoropropane

2. Chloroethane reacts with hydroxide ions.

   elimination

   $CH_3-CH_2-Cl + OH^- \rightarrow CH_2=CH_2 + Cl^- + H-O-H$

   ethene

3. Ethanol, present in gasohol, burns in an automobile engine.

   combustion

   $CH_3-CH_2-OH + O=O \rightarrow O=C=O + H-O-H$

4. 1-butanol reacts in the presence of concentrated sulfuric acid.

   elimination

   $CH_3-CH_2-CH_2-CH_2-OH \rightarrow CH_3-CH_2-CH=CH_2 + H-O-H$

   1-butene

5. Dichloromethane is produced by reacting methane and chlorine.

   substitution

   (structural diagram: $H-\overset{H}{\underset{H}{C}}-H + Cl-Cl \rightarrow H-\overset{H}{\underset{Cl}{C}}-Cl + H-Cl$)

   dichloromethane

6. Bromine and ethene react to form an alkyl halide.

   addition

   $CH_2=CH_2 + Br-Br \rightarrow Br-CH_2-CH_2-Br$

   1,2-dibromoethane

7. Hydrogen chloride and ethene react to produce an alkyl halide.

   addition

   $CH_2=CH_2 + H-Cl \rightarrow CH_3-CH_2-Cl$

   chloroethane

8. Acetic acid and ethanol react to produce a solvent used in nail polish remover.

   condensation (esterification)

   $CH_3-\overset{O}{\overset{||}{C}}-OH + CH_3-CH_2-OH \rightarrow CH_3-\overset{O}{\overset{||}{C}}-O-CH_2-CH_3 + H-O-H$

   ethyl ethanoate

UNIT I REVIEW

CHAPTER 1

1. Distinguish between the two important types of scientific knowledge.

2. List seven ways by which empirical knowledge is communicated.

3. What are three characteristics of acceptable scientific laws and generalizations.

4. State and define three types of variables often specified in experimental designs.

5. Match the section headings used in a report of scientific work to the descriptions listed below.
   (a) a list, in numbered steps, of specific instructions to be followed
   (b) a series of judgments, with reasons, about the experimental design, procedure, and technological skills, the prediction, and the authority on which the prediction is based
   (c) a summary of the experimental method to be used to answer the problem
   (d) a proposed answer to the problem using the following format: "According to ... (some authority), ..."
   (e) an organized record of all observations—qualitative and quantitative
   (f) a question to be answered in the experiment
   (g) a complete list of all substances and equipment including sizes and quantities
   (h) includes classification, interpretation, and calculation based on the evidence obtained; provides the answer to the problem

6. The purpose of many scientific experiments is to test predictions based on some authority.
   (a) State four examples of a scientific authority.
   (b) State two examples of authorities other than scientific concepts.

7. Pure substances can be classified into two categories—elements and compounds.
   (a) How can these two types of pure substances be distinguished empirically?
   (b) According to modern theory, how are elements and compounds different?

8. Distinguish between atoms and molecules.

9. For each of the following perspectives, write a brief statement describing the focus or concern of an STS issue from that point of view: scientific, technological, economic, ecological, political.

10. Why is it important to consider different perspectives on an STS issue?

11. Give three examples of a current STS issue.

12. Classify each of the following statements using one of the issue perspectives listed in question 9. All of the statements concern sulfur dioxide emissions.
    (a) An elected representative reported to constituents that emissions of sulfur dioxide were within the limits set by environmental legislation.
    (b) Laboratory research has provided evidence that sulfur dioxide from the combustion of fossil fuels is converted in the presence of oxygen to sulfur trioxide.
    (c) The cost of ending sulfur dioxide pollution of the atmosphere will be high. The longer we delay facing the problem, the greater the cost.
    (d) Sulfur oxides and their related acids are particularly damaging to soil microbes, water lifeforms, plants, building materials, and people.

(e) One of the most promising scrubbers to remove sulfur dioxide gas from a smoke stack is the limestone-dolomite process.

CHAPTER 2

13. The purpose of this proposed investigation is to test the empirical definitions of metals and nonmetals. Write a prediction to answer the following question.

    Problem Which pure substances are classified as metals or nonmetals?

14. State the periodic law.

15. Describe the location of metals and nonmetals in the modern periodic table.

16. Complete the following table.

	Name of Element	Atomic Number	IUPAC Symbol	Group Number	Period Number	SATP State	Family/Series Name
(a)				2	3		
(b)		79					
(c)	sodium						
(d)					6		noble gases
(e)			Cl				

17. State six ways by which theoretical knowledge is communicated.

18. What are the characteristics of acceptable scientific theories?

19. If a theory is found to be unacceptable because reliable evidence contradicts the theory, what are three possible strategies employed by scientists?

20. In a few words, describe the atom associated with the Dalton model, the Thomson model, the Rutherford model, and the Bohr model.

21. Provide an analogy for each model described in the previous question.

22. The two major subatomic particles invented for the earlier atomic theories of Thomson and Rutherford are the electron and the proton. State the relative charge and location of these two subatomic particles as determined empirically by scientists.

23. Provide a theoretical definition for each of the following terms: atomic number, group number, and period number.

24. According to current atomic theory, the chemically important electrons in an atom are called the _____ electrons and are believed to exist in the _____ energy level.

25. Chemical reactions are explained as a rearrangement of _____ .

26. Draw diagrams of accepted electron energy-level models to describe potassium, carbon, oxygen, and argon atoms.

27. Do the diagrams from the previous question represent a picture of the atom? Briefly expand your answer.

28. State the theoretical rule that predicts the number of electrons accepted or donated by an atom of a representative element.

29. Using the rule from the previous question, what theoretical prediction may be made about the types of ions formed by metal and nonmetal atoms?

30. All theories, laws, and generalizations in science are restricted in their application. Which class of elements does not, in general, follow the above rule for the theoretical prediction of ions formed?

31. Predict from theory the charges on the ions formed by atoms in Groups 1 and 2, and nonmetal atoms in Groups 15, 16, and 17.

32. Using the restricted theory of quantum mechanics, draw diagrams of electron energy-level models for magnesium, sulfide, sodium, and fluoride ions.

UNIT II REVIEW

CHAPTER 3

1. By convention, ionic and molecular compounds are classified in terms of their constituent elements. Based on this convention, define ionic and molecular compounds.

2. Empirically, ionic and molecular compounds are classified in terms of their characteristic properties. Write empirical definitions for ionic and molecular compounds.

3. According to theory, how do ionic and molecular compounds differ?

4. Use the most common ion charges and electrical neutrality to *explain* each of the following empirical ionic formulas.

   (a) $CaCl_{2(s)}$
   (b) $Al_2O_{3(s)}$
   (c) $ZrO_{2(s)}$
   (d) $Ca_3(PO_4)_{2(s)}$

5. Is it possible to explain the empirically determined formula $Fe_3O_{4(s)}$ using the theory of ionic compounds? What does your answer illustrate about the theory?

6. The formulas and charges of polyatomic ions cannot be predicted or explained by the restricted theory. What way of knowing can be used to obtain the formulas and charges of polyatomic ions?

7. What is meant by a formula unit of an ionic compound?

8. Predict the ionic formula for each of the following substances. The technological use is provided in parentheses.

   (a) sodium hydroxide (lye)
   (b) potassium permanganate (fungicide)
   (c) ammonium phosphate (fertilizer)
   (d) tin(II) fluoride (toothpaste additive)
   (e) sodium hydrogen carbonate (baking soda)
   (f) aluminum chloride (deodorants)
   (g) calcium phosphate (dental powders)
   (h) aluminum sulfate (water clarifier)

9. Predict the ionic formula of the compound produced by each of the following formation reactions.

   (a) aluminum and sulfur
   (b) lithium and nitrogen
   (c) calcium and chlorine

10. Translate the following international chemical formulas into IUPAC names.

    (a) $MgI_{2(s)}$
    (b) $Na_3PO_{4(s)}$
    (c) $ZnSO_{3(s)}$
    (d) $Na_2HPO_{4(s)}$
    (e) $Na_2SO_4 \cdot 10H_2O_{(s)}$
    (f) $Fe_2O_{3(s)}$
    (g) $Cu(NO_3)_2 \cdot 6H_2O_{(s)}$
    (h) $FeS_{(s)}$
    (i) $Ti(CH_3COO)_{4(s)}$
    (j) $(NH_4)_2CO_{3(s)}$

11. The halogens and other elements with a "–gen" suffix in their names are diatomic. List the names and molecular formulas (including SATP state of matter subscripts) for the seven common diatomic elements.

12. Ozone (trioxygen), phosphorus (tetraphosphorus), and sulfur (octasulfur) are other polyatomic molecular elements. List their molecular formulas and SATP states of matter.

13. Distinguish between chloride and chlorine.

14. Translate the following English names of binary molecular compounds into international molecular formulas, including SATP states of matter.
    (a) carbon dioxide (product of combustion and respiration)
    (b) dinitrogen oxide gas (anesthetic)
    (c) sulfur trioxide gas (air pollutant)
    (d) solid diphosphorus pentaoxide (standard for rating fertilizers)

15. Translate the following international molecular formulas into English IUPAC names.
    (a) $IF_{5(l)}$
    (b) $S_2O_{7(l)}$
    (c) $N_4S_{4\ (s)}$
    (d) $OCl_{2(g)}$
    (e) $S_2Cl_{2(l)}$
    (f) $SF_{6(g)}$

16. The names and formulas of some common molecular compounds need to be memorized. Translate the English names given below into international molecular formulas, including SATP states of matter. (Memorize the names, formulas, and states of matter of these substances.)
    (a) water
    (b) hydrogen sulfide
    (c) methane
    (d) methanol
    (e) ethanol
    (f) hydrogen peroxide
    (g) ammonia
    (h) propane
    (i) sucrose
    (j) octane

17. List four characteristics of a scientific communication system acceptable to the scientific community.

Classify the substances in questions 18 to 39 as ionic (i), molecular (m), or acid (a). Complete the chemical formula by adding the SATP state of matter for each pure substance, and write the corresponding English IUPAC name for each formula.

Formula	Description
18. $NaHCO_3$	baking soda
19. $CS_{2(l)}$	toxic organic solvent
20. CH_4	natural gas
21. $Fe_2O_3 \cdot 3H_2O$	rust
22. NH_4NO_3	fertilizer
23. NH_3	important fertilizer
24. $CuSO_4 \cdot 5H_2O$	bluestone
25. $Na_2SO_4 \cdot 10H_2O$	Glauber's salt
26. $H_2SO_{4(aq)}$	battery acid
27. $H_2SO_{3(aq)}$	acid rain component
28. $C_{12}H_{22}O_{11}$	table sugar
29. $PbCrO_4$	chrome yellow pigment
30. $(NH_4)_2HPO_4$	fertilizer
31. $NaNO_2$	meat preservative

32. $Al_2O_3 \cdot 2H_2O$ bauxite ore

33. NO_2 acid-forming gas

34. TiO_2 paint pigment

35. Na_2SiO_3 water glass

36. H_2S sour gas component

37. C_2H_5OH alcoholic beverages

38. $CaCl_2$ road salt

39. $CH_3COOH_{(aq)}$ vinegar

Provide a chemical formula for each of the substances in questions 40 to 48. Include the SATP state of matter as part of the formula.

Name	Description
40. hydrochloric acid	stomach acid
41. calcium hydrogen carbonate	temporary water hardness compound
42. iron(II) sulfide	an iron ore
43. sodium carbonate decahydrate	washing soda
44. dinitrogen tetrahydride	liquid rocket fuel
45. sodium hydrogen sulfate	swimming pool chemical
46. sodium glutamate	MSG food additive
47. sulfur trioxide	acid-forming gas
48. sodium phosphate	TSP cleaner

CHAPTER 4

49. How do chemical changes differ from physical changes?

50. State the names of four diagnostic tests used to determine if a chemical reaction has occurred.

51. According to the collision-reaction theory, what are the three conditions required for a reaction to take place?

52. One large category of pollution problems relates to the combustion or burning of different chemicals. In internal combustion engines, a fuel is burned in the presence of air. The ideal hydrocarbon combustion reaction has been found to be accompanied by many side reactions, such as

    nitrogen + oxygen $\longrightarrow$ nitrogen dioxide

(a) *Problem* How does the total mass of the reactants relate to the mass of the product in the reaction of nitrogen and oxygen?
Write a prediction to answer this experimental problem.

(b) Translate the word equation into a balanced chemical equation using international chemical formulas with states of matter at SATP.

(c) Write a theoretical definition for the term coefficient, and label a coefficient in the above equation.

(d) Write a theoretical definition for the term formula subscript, and label a formula subscript in the above equation.

(e) How is the state of matter communicated in a chemical equation? Label a state of matter in the above equation.

53. A molecule or formula unit is too small a quantity of substance for laboratory work. What is the observable amount of a chemical that is convenient to measure and handle? Define this quantity.

Balance the following chemical equations in questions 54 to 61 and translate each into an English sentence using amounts in moles.

54. Cadmium pollution occurs primarily in the form of cadmium oxide. This substance is produced by the incineration of certain wastes and as a by-product of zinc refining.

$$CdS_{(s)} + O_{2(g)} \rightarrow CdO_{(s)} + SO_{2(g)}$$

55. Acceptable methods of waste disposal are important in chemistry laboratories. Waste disposal is a particular concern in laboratories that use lead compounds in reactions, e.g. the reaction below.

$$Pb(NO_3)_{2(aq)} + Cr_{(s)} \rightarrow Pb_{(s)} + Cr(NO_3)_{3(aq)}$$

56. Mercury has been known since medieval times and is widely used in industry today. There is no known cure for mercury poisoning. Mercury can be obtained by heating mercury(II) oxide.

$$HgO_{(s)} \rightarrow Hg_{(l)} + O_{2(g)}$$

57. An estimated 4000 t of arsenic and its compounds were emitted into the atmosphere in Canada during 1972. The major use of arsenic is in pesticides. One example of a pesticide, calcium arsenate, is prepared by the following industrial reaction of aqueous hydrogen arsenate.

$$H_3AsO_{4(aq)} + Ca(OH)_{2(aq)} \rightarrow HOH_{(l)} + Ca_3(AsO_4)_{2(s)}$$

58. Inadequate technology or servicing (e.g. a faulty carburettor) can cause a lack of sufficient oxygen for combustion in an automobile engine and result in the production of poisonous carbon monoxide. An ideal, complete combustion would be

$$C_8H_{18(l)} + O_{2(g)} \rightarrow CO_{2(g)} + H_2O_{(g)}$$

59. Sodium nitrite is widely used as a meat preservative in spite of the controversy associated with its use. The manufacture of sodium nitrite involves the series of reactions shown below.

(a) $N_{2(g)} + H_{2(g)} \rightarrow NH_{3(g)}$

(b) $NH_{3(g)} + O_{2(g)} \rightarrow NO_{(g)} + H_2O_{(g)}$

(c) $Na_2CO_{3(s)} + NO_{(g)} + O_{2(g)} \rightarrow NaNO_{2(s)} + CO_{2(g)}$

60. White phosphorus ignites in air and is usually stored under water.

$$P_{4(s)} + O_{2(g)} \rightarrow P_4O_{10(s)}$$

61. A convenient laboratory method for generating hydrogen is the reaction of active metals such as zinc with a strong acid.

$$Zn_{(s)} + HCl_{(aq)} \rightarrow H_{2(g)} + ZnCl_{2(aq)}$$

62. Classify the reaction types for the chemical equations in questions 54 to 61.

For each of the following: classify the reaction type, predict the balanced chemical equation, and state at least one diagnostic test to check your prediction.

63. Chlorine gas is bubbled into a sodium iodide solution.

64. Equal amounts of hydrochloric acid and aqueous sodium hydroxide are mixed.

65. Propane from a torch burns in air.

66. A piece of strontium metal is placed in water.

67. Aluminum metal is added to a copper(II) sulfate solution.

68. Nitrogen triiodide is broken down into its elements.

69. Excess sulfur in a sealed container of oxygen is heated.

70. An excess of aqueous sodium hydroxide is added to an aqueous solution of nickel(II) chloride.

UNIT III REVIEW

CHAPTER 5

1. Distinguish between aqueous electrolytes and non-electrolytes.

2. What types of substances are aqueous electrolytes and non-electrolytes?

3. Write dissociation or ionization equations for the following pure substances dissolving in water.
   (a) lithium sulfate   (b) hydrogen chloride   (c) aluminum sulfate

4. For each of the following, write the formula for the pure, undissolved substance and the formulas for the entities present when the substance is placed in water.
   (a) sodium benzoate
   (b) strontium hydroxide
   (c) hydrogen nitrate
   (d) propane
   (e) hydrogen acetate
   (f) copper(II) chloride
   (g) silver chloride
   (h) methanol

5. Write an empirical definition of an acid.

6. What theoretical idea is used to explain the acidity and aqueous conductivity of acids?

7. Predict the formula for each of the following substances.
   (a) aqueous hydrogen carbonate   (c) hydrofluoric acid
   (b) aqueous hydrogen sulfite   (d) oxalic acid

8. Write both the systematic IUPAC name and the classical name for each of the following acids.
   (a) $HNO_{3(aq)}$
   (b) $HNO_{2(aq)}$
   (c) $HI_{(aq)}$
   (d) $H_3BO_{3(aq)}$

9. For technological applications, what are three advantages of solutions?

10. Suppose you are given four, unlabelled beakers, each containing a colorless aqueous solution of one solute. The possible solutions are $NaCl_{(aq)}$, $HCl_{(aq)}$, $BaCl_{2(aq)}$, and $CH_3Cl_{(aq)}$. Write a series of diagnostic tests to distinguish each solution from the others.

11. A solution is known to contain sulfide ions and/or carbonate ions. Design an experiment to test for the presence of each ion.

12. A household cleaner contains 2.5% W/V of sodium hypochlorite. What mass of solute is present in 500 mL of this solution?

13. A calcium chloride solution has a concentration of 0.251 mol/L. What amount of calcium chloride is present in a 125 mL bottle?

14. What volume of concentrated phosphoric acid is required to prepare 250 mL of a 0.375 mol/L solution? *(See reference sheet for the initial concentration.)*

15. What is the concentration of each ion in a 2.1 mol/L solution of iron(III) chloride?

16. Classify the following solutions as acidic, basic, or neutral.

    (a) pH = 8

    (b) $[H^+_{(aq)}] = 10^{-7}$ mol/L

    (c) pH = 3

    (d) $[H^+_{(aq)}] = 10^{-10}$ mol/L

17. Complete the analysis of the following investigation report.

    Problem   What is the solubility of potassium chloride at 20°C?

    Experimental Design   Mass measurements were recorded in an experiment where 10.0 mL of a saturated potassium chloride solution at 20°C was evaporated to dryness.

    Evidence   mass of evaporating dish = 52.86 g

    mass of evaporating dish + residue = 55.24 g

18. How does the solubility of solids and gases change as the temperature increases?

19. Excess copper(II) sulfate is added to water in a closed system until no more solute dissolves at a constant temperature.

    (a) Describe some empirical properties of this mixture.

    (b) Provide a brief theoretical explanation of these properties.

CHAPTER 6

20. State three empirical properties unique to gases compared to the other states of matter.

21. Complete the following statements about gases.

    (a) At a constant temperature, as the pressure increases, the volume of gas _____ .

    (b) At a constant pressure, as the temperature decreases, the volume of gas _____ .

    (c) At a constant volume and temperature, if the amount of gas inside a container is increased, the pressure _____ .

22. A 1.5 L volume of gas is compressed at a constant temperature from 1.0 atm to 5.0 atm. What is the final volume?

23. A balloon can hold 800 mL of air before breaking. A balloon at 4.0°C containing 750 mL of air is brought into a house at 25°C. Assuming a constant pressure inside and outside the balloon, will the balloon break?

24. A sample of argon gas at 101 kPa and 22.0°C occupies a volume of 150 mL. What is the new volume if the pressure is changed to 91.5 kPa and the temperature is increased to 150°C?

25. What volume would 25.0 g of oxygen gas at 22.0°C and 98.1 kPa occupy?

26. Compare the volume that 0.278 mol of hydrogen would occupy at STP and SATP.

27. What mass of chlorine gas would occupy 86.0 L at SATP?

28. Determine the amount of propane used during a summer by a family who burns all of the 8.0 kg of propane from the cylinder in a portable barbecue.

29. A 5.0 kg bag of sugar is used in making preserves. Calculate the amount of sugar in 5.0 kg.

30. As part of a chemical analysis, 70 g of magnesium sulfate-7-water are measured. Calculate the amount used.

31. Determine the mass of sodium hydroxide that must be measured by a student to obtain 0.150 mol for an experiment.

32. An average bungalow requires about 400 kmol of methane per year for space heating.

(a) Calculate the mass of methane consumed in one year.

(b) Calculate the volume of methane at SATP used in one year.

33. To increase the cleaning power of synthetic detergents, sodium tripolyphosphate has largely replaced sodium phosphate in these detergents. Convert 1.50 mol of sodium tripolyphosphate into the corresponding mass.

CHAPTER 7

34. What are some similarities between the fields of chemistry and chemical technology?

35. What are some differences between the fields of chemistry and chemical technology?

36. List the three types of stoichiometry and how each type is recognized.

37. What mass of lead(II) iodide precipitate forms when 2.93 g of potassium iodide in solution reacts with excess lead(II) nitrate?

38. In a hard water analysis, a calcium chloride solution is reacted with excess aqueous sodium oxalate to produce 0.452 g of calcium oxalate precipitate. Determine the mass of calcium chloride present in the original solution.

39. Analysis for sulfate ions is usually done by precipitating barium sulfate from an aqueous sample of the sulfate ions. The barium sulfate precipitate is then ignited and reacted with carbon from the filter paper. Using the balanced chemical equation given below, predict the mass of carbon required to react with 1.50 g of barium sulfate precipitate.

$$BaSO_{4(s)} + 2\,C_{(s)} \longrightarrow BaS_{(s)} + 2\,CO_{2(g)}$$

40. A precipitate of aluminum hydroxide can be used to clarify water. What mass of sodium hydroxide is required to precipitate 1.50 g of aluminum sulfate in solution?

41. Powdered aluminum metal is one of the fuels used in the solid rocket boosters for the NASA Space Shuttle. What volume of oxygen at SATP is required to react completely with 100 kg of aluminum?

42. A portable hydrogen generator uses the reaction of calcium hydride and water to form calcium hydroxide and hydrogen. What volume of hydrogen at 96.5 kPa and 22°C can be produced from a 50 g cartridge of $CaH_{2(s)}$?

43. A volumetric analysis shows that it takes 32.0 mL of 2.12 mol/L $NaOH_{(aq)}$ to neutralize 10.0 mL of sulfuric acid from a car battery. Calculate the molar concentration of sulfuric acid in the battery solution.

44. Silver metal can be recycled as silver nitrate by the following reaction.

$$3\,Ag_{(s)} + 4\,HNO_{3(aq)} \longrightarrow 3\,AgNO_{3(aq)} + NO_{(g)} + 2\,H_2O_{(l)}$$

What volume of concentrated nitric acid is required to react with 1.68 kg of silver metal? (*See reference sheet for concentration of acids.*)

45. *Problem*   What is the molar solubility of calcium hydroxide in water at 23°C?

   Experimental Design   Excess solid calcium hydroxide was placed in water and stirred occasionally for a day at 23°C. A 10.0 mL sample of the saturated solution was titrated with 0.0500 mol/L hydrochloric acid solution.

   Evidence   An average of 4.2 mL of acid were required for reaction.

   (a) Complete the analysis to answer the question stated in the problem.

   (b) Suggest an alternate experimental design to answer the same problem.

46. A windshield washer solution was prepared by dissolving 100 g of methanol in water to form 2.00 L of solution. What is the molar concentration of the solution?

47. Magnesium hydroxide is a low-solubility ionic compound. What mass of solute is needed to prepare 100 mL of 0.154 mmol/L of magnesium hydroxide solution?

48. Write instructions (give the procedural steps and show required calculations) for an untrained laboratory technician to prepare 100.0 mL of 0.100 mol/L glucose solution from solid glucose.

49. A mechanic wishes to prepare 500 mL of a 3.75 mol/L solution of sulfuric acid to add to a car battery. What volume of concentrated solution is required?

50. Write instructions for a laboratory technician to prepare 100.0 mL of a 0.100 mol/L solution of glucose starting with a 2.00 mol/L solution.

51. Complete the analysis of the following investigation report.

   Problem   What is the molar concentration of an unknown sodium carboate solution?

   Experimental Design   Samples of sodium carbonate solution were titrated with a standardized hydrochloric acid solution using methyl orange as the indicator.

   Evidence

Titration of 25.0 mL Samples of $Na_2CO_{3(aq)}$ with 0.352 mol/L $HCl_{(aq)}$				
Trial	1	2	3	4
Final buret reading (mL)	16.5	31.8	47.0	16.4
Initial buret reading (mL)	0.6	16.5	31.8	1.2

52. State two alternative experimental designs to answer the problem in question 51.

UNIT IV REVIEW

CHAPTER 8

1. What is an activity series?

2. Based on general reactivity trends in the periodic table, list the elements from least active to most active for

   (a) Group 1   (b) Groups 1–6, Period 4   (c) Group 17

3. What is the octet rule and how does this relate to chemical reactivity?

4. For each of the following elements, draw an electron dot diagram, list the group numbers, and list the numbers of valence electrons, lone pairs and bonding electrons.

   (a) calcium   (d) oxygen

   (b) aluminum   (e) bromine

   (c) arsenic   (f) neon

5. (a) What types of elements would be expected to react to form compounds containing covalent bonds?

   (b) What types of elements would be expected to react to form compounds containing ionic bonds?

(c) Provide a brief explanation for (a) and (b) using the concept of electronegativity.

6. Theories are invented to explain observations. For each of the following properties of ionic compounds, write a brief explanation.

   (a) Ionic compounds are hard solids with high melting and boiling points.

   (b) Ionic compounds are electrical conductors in molten and aqueous states.

7. Using electron dot diagrams of atoms and ions, write the formation equation for each of the following compounds.

   (a) potassium bromide   (b) sodium oxide   (c) calcium fluoride

8. Use the following evidence from combustion analyzers and mass spectrometers to obtain the empirical molecular formula. Explain the formula using a structural diagram.

   (a) 15.77% C, 84.23% S, 76.14 g/mol

   (b) 9.73% H, 90.27% Si, 62.23 g/mol

9. Predict the simplest molecular formula and draw a structural diagram for each of the following formation reactions.

   (a) sulfur + chlorine $\rightarrow$

   (b) phosphorus + hydrogen $\rightarrow$

   (c) nitrogen + sulfur $\rightarrow$

10. What is the experimental evidence for double covalent bonds?

11. Compare the strength of intramolecular (covalent) bonds with that of intermolecular bonds (forces)?

12. Explain each of the following observations in terms of the characteristics of molecules and intermolecular forces.

    (a) The boiling point of fluorine is significantly less than that of chlorine.

    (b) Drops of ethanol are attracted to a charged strip.

    (c) Ice has a regular hexagonal structure.

CHAPTER 9

13. Define organic chemistry.

14. Draw complete structural diagrams for the isomers of C_3H_8O.

15. For each of the following molecular formulas, draw a structural diagram, identify the organic family, and write the IUPAC name.

    (a) CH_3OH (f) CH_3COOH

    (b) C_2H_6 (g) CH_3NH_2

    (c) $C_6H_5CH_3$ (h) CH_2Cl_2

    (d) C_3H_4 (i) CH_3COOCH_3

    (e) HCHO   (j) CH_3CONH_2

16. Identify the organic reaction type and complete the following equations using structural diagrams and IUPAC names. Do not balance the equations.

    (a)
    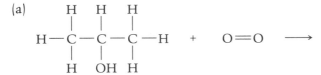

(b)

$$CH_3 - \underset{\underset{CH_3}{|}}{\overset{\overset{CH_3}{|}}{C}} - CH_2 - CH_3 \quad + \quad H - H \quad \longrightarrow \quad CH_3 - \underset{\underset{}{\overset{CH_3}{|}}}{CH} - CH_3 \quad + \quad CH_3 - CH_3$$

(c) 2-methyl-2-pentene + hydrogen $\longrightarrow$

(d) acetylene $\longrightarrow$ benzene

(e) $CH_3 - CH_2 - CH_2 - \underset{\underset{Cl}{|}}{CH} - CH_3 \quad + \quad OH^- \quad \longrightarrow$

(f) $CH_2 = CH - CH_2 - CH_3 \quad + \quad H - Cl \quad \longrightarrow$

(g) 2-phenylpropane + bromine $\longrightarrow$ 2-bromo-2-phenylpropane + hydrogen bromide

(h) 1-propanol $\longrightarrow$ propene + water

(i) butanoic acid + methanol $\longrightarrow$

17. Define a polymer.

18. Describe some similarities and differences between synthetic and natural polymers.

NELSON CHEMISTRY
TEACHER'S RESOURCE MASTERS

ANSWERS
UNIT I REVIEW

CHAPTER 1

1. Empirical knowledge is observable; theoretical knowledge is not.

2. simple descriptions, tables of evidence, graphs, empirical hypotheses, empirical definitions, generalizations, scientific laws

3. Acceptable scientific laws and generalizations describe, predict, and are simple.

4. A manipulated variable is the property that is systematically changed during an experiment. A responding variable is the property that is measured as changes are made to the manipulated variable. A controlled variable is a property that is kept constant throughout an experiment.

5. (a) procedure   (e) evidence
   (b) evaluation   (f) problem
   (c) experimental design (g) materials
   (d) prediction   (h) analysis

6. (a) theory, law, generalization, definition
   (b) reference source, manufacturer's label

7. (a) Elements are substances that cannot be broken down chemically into simpler units. Compounds can be decomposed chemically by heat and/or electricity.
   (b) According to modern theory, elements are composed of only one kind of atom and compounds are composed of two or more kinds of atoms.

8. Atoms are the smallest particles of an element, whereas molecules are larger particles composed of two or more atoms.

9. scientific—deals with research and describing and explaining natural phenomena
   technological—deals with developing and using machines, instruments, and processes
   economic—deals with money matters
   ecological—deals with relationships of living things with the environment
   political—deals with government actions and getting elected

10. Many good arguments, both for and against an issue, can be obtained by considering many perspectives. Considering only one perspective is narrrow-minded and will not solve complex problems.

11. Several current topics are acid rain, toxic chemical waste spills, nuclear energy, ozone depletion, and energy crisis.

12. (a) political   (d) ecological
    (b) scientific   (e) technological
    (c) economic

CHAPTER 2

13. *Prediction* According to the empirical definitions, metals are shiny flexible solids at SATP and are good conductors of heat and electricity. Nonmetals are non-lustrous and brittle as solids and are poor conductors of heat and electricity.

14. When elements are arranged in order of increasing atomic mass, chemical and physical properties form patterns that repeat at regular intervals.

15. Metals are located to the left of the staircase line, and nonmetals to the right.

16. (a) magnesium, 12, Mg, 2, 3, solid, alkaline-earth metals
    (b) gold, 79, Au, 11, 6, solid, transition elements
    (c) sodium, 11, Na, 1, 3, solid, alkali metals
    (d) radon, 86, Rn, 18, 6, gas, noble gases
    (e) chlorine, 17, Cl, 17, 3, gas, halogens

17. theoretical descriptions, hypotheses, and definitions, theories, analogies, models

18. Acceptable scientific theories describe, explain, predict, and are simple.

19. The theory may be restricted, revised, or replaced.

20. Dalton model: tiny indivisible sphere
    Thomson model: positively charged sphere containing negatively charged electrons
    Rutherford model: nucleus surrounded by electrons
    Bohr model: nucleus surrounded by electrons moving in fixed circular orbits

21. Dalton model: billiard ball analogy
    Thomson model: raisin bun analogy
    Rutherford model: fruit with a tiny central pit analogy or beehive analogy
    Bohr model: solar system analogy

22. Electrons are believed to have a "one negative" charge and exist in energy levels outside the nucleus. Protons are believed to have a "one positive" charge and exist inside the nucleus.

23. Atomic number represents the number of protons in the nucleus of an atom. The last digit of the group number represents the number of valence electrons. Period number represents the number of energy levels occupied by electrons.

24. valence, highest (outermost)

25. particles (atoms, ions, electrons)

26.

$1e^-$			
$8e^-$			$8e^-$
$8e^-$	$4e^-$	$6e^-$	$8e^-$
$2e^-$	$2e^-$	$2e^-$	$2e^-$
$19p^+$	$6p^+$	$8p^+$	$18p^+$
potassium atom, K	carbon atom, C	oxygen atom, O	argon atom, Ar

27. No. The restricted quantum mechanics theory does not provide locations or pictures. The diagrams list the energy levels present according to the theory.

28. Atoms of the representative elements form monatomic ions when losing or gaining electrons to form the same stable electronic structure as atoms of the nearest noble gas.

29. Metals should form positive ions and nonmetals should form negative ions by gaining electrons.

30. The transition elements, Groups 3 to 12, do not follow the rule. Their ion changes are given in the periodic table.

31. $1+, 2+, 3-, 2-, 1-$

32.

	$8e^-$		
$8e^-$	$8e^-$	$8e^-$	$8e^-$
$2e^-$	$2e^-$	$2e^-$	$2e^-$
$12p^+$	$16p^+$	$11p^+$	$9p^+$
magnesium ion, Mg^{2+}	sulfide ion, S^{2-}	sodium ion, Na^+	fluoride ion, F^-

UNIT II REVIEW

CHAPTER 3

1. Ionic compounds are metal-nonmetal combinations and molecular compounds are nonmetal-nonmetal combinations.

2. Ionic compounds are all solids at SATP and form solutions that conduct electricity. Molecular compounds are solids, liquids, or gases at SATP and form solutions that do not conduct electricity.

3. Ionic compounds contain positively and negatively charged ions, whereas molecular compounds are composed of molecules formed from nonmetal atoms.

4. (a) $CaCl_2$ $2+ + (2 \times 1-) = 0$
   (b) Al_2O_3 $(2 \times 3+) + (3 \times 2-) = 0$
   (c) ZrO_2 $4+ + (2 \times 2-) = 0$
   (d) $Ca_3(PO_4)_2$ $(3 \times 2+) + (2 \times 3-) = 0$

5. No. The theory does not allow fractional ion charges. Our theory does not work and must be restricted, revised, or replaced to solve this problem.

6. A referenced or memorized way of knowing is required for polyatomic ions.

7. A formula unit of an ionic compound is the smallest amount of the compound that has the composition given by the chemical formula.

8. (a) $NaOH_{(s)}$ (e) $NaHCO_{3(s)}$
   (b) $KMnO_{4(s)}$ (f) $AlCl_{3(s)}$
   (c) $(NH_4)_3PO_{4(s)}$ (g) $Ca_3(PO_4)_{2(s)}$
   (d) $SnF_{2(s)}$ (h) $Al_2(SO_4)_{3(s)}$

9. (a) $Al_2S_{3(s)}$
   (b) $Li_3N_{(s)}$
   (c) $CaCl_{2(s)}$

10. (a) magnesium iodide
    (b) sodium phosphate
    (c) zinc sulfite
    (d) sodium hydrogen phosphate
    (e) sodium sulfate-10-water
    (f) iron(III) oxide
    (g) copper(II) nitrate-6-water
    (h) iron(II) sulfide
    (i) titanium(IV) acetate
    (j) ammonium carbonate

11. fluorine, $F_{2(g)}$; chlorine, $Cl_{2(g)}$; bromine, $Br_{2(l)}$; iodine, $I_{2(s)}$; hydrogen, $H_{2(g)}$; nitrogen, $N_{2(g)}$; oxygen, $O_{2(g)}$

12. $O_{3(g)}$, $P_{4(s)}$, $S_{8(s)}$

13. Chloride (Cl^-) is a monatomic ion, and chlorine (Cl_2) is a diatomic molecule.

14. (a) $CO_{2(g)}$ (c) $SO_{3(g)}$
    (b) $N_2O_{(g)}$ (d) $P_2O_{5(s)}$

15. (a) iodine pentafluoride   (d) oxygen dichloride
    (b) disulfur heptaoxide   (e) disulfur dichloride
    (c) tetranitrogen tetrasulfide (f) sulfur hexafluoride

16. (a) $H_2O_{(l)}$ (f) $H_2O_{2(l)}$
    (b) $H_2S_{(g)}$ (g) $NH_{3(g)}$
    (c) $CH_{4(g)}$ (h) $C_3H_{8(g)}$
    (d) $CH_3OH_{(l)}$ (i) $C_{12}H_{22}O_{11(s)}$
    (e) $C_2H_5OH_{(l)}$ (j) $C_8H_{18(l)}$

17. The communication system must be international, logical, precise, and simple.

18. i, (s), sodium hydrogen carbonate

19. m, (l), carbon disulfide

20. m, (g), methane

21. i, (s), iron(III) oxide-3-water

22. i, (s), ammonium nitrate

23. m, (g), ammonia

24. i, (s), copper(II) sulfate-5-water

25. i, (s), sodium sulfate-10-water

26. a, (aq), sulfuric acid

27. a, (aq), sulfurous acid

28. m, (s), sucrose

29. i, (s), lead(II) chromate

30. i, (s), ammonium hydrogen phosphate

31. i, (s), sodium nitrite

32. i, (s), aluminum oxide-2-water

33. m, (g), nitrogen dioxide

34. i, (s), titanium(IV) oxide

35. i, (s), sodium silicate

36. m, (g), hydrogen sulfide

37. m, (l), ethanol

38. i, (s), calcium chloride

39. a, (aq), acetic acid (ethanoic acid)

40. $HCl_{(aq)}$

41. $Ca(HCO_3)_{2(s)}$

42. $FeS_{(s)}$

43. $Na_2CO_3 \cdot 10H_2O_{(s)}$

44. $N_2H_{4(l)}$

45. $NaHSO_{4(s)}$

46. $NaC_5H_8NO_{4(s)}$

47. $SO_{3(g)}$

48. $Na_3PO_{4(s)}$

CHAPTER 4

49. In chemical changes, new substances with different properties are formed. Physical changes do not produce new substances.

50. color change, odor change, state change, energy change

51. For a reaction to take place, particles of the reactants must collide with a certain minimum energy and with a certain orientation.

52. (a) According to the Law of Conservation of Mass, the total mass of the reactants should equal the mass of product formed.
    (b) $N_{2(g)} + 2 O_{2(g)} \longrightarrow 2 NO_{2(g)}$
    (c) A coefficient represents the number of molecules or formula units believed to react in a chemical reaction, e.g. 2 molecules of oxygen.

(d) A formula subscript represents the number of atoms or ions present in one molecule or formula unit of a substance, e.g. 2 atoms of oxygen in an oxygen molecule.

(e) The state of matter is communicated by a state of matter subscript, e.g. $_{(g)}$.

53. A mole is the amount of substance with the number of entities corresponding to Avogadro's constant,

6.02×10^{23}/mol.

54. Two moles of solid cadmium sulfide and three moles of oxygen gas react to produce two moles of solid cadmium oxide and two moles of sulfur dioxide gas.

55. Three moles of aqueous lead(II) nitrate and two moles of solid chromium react to form three moles of solid lead and two moles of aqueous chromium(III) nitrate.

56. Two moles of solid mercury(II) oxide decompose to form two moles of liquid mercury and one mole of oxygen gas.

57. Two moles of aqueous hydrogen arsenate and three moles of aqueous calcium hydroxide react to produce six moles of liquid water and one mole of solid calcium arsenate.

58. Two moles of liquid octane and twenty-five moles of oxygen gas react to form sixteen moles of carbon dioxide gas and eighteen moles of water vapor.

59. (a) One mole of nitrogen gas and three moles of hydrogen gas react to produce two moles of ammonia gas.

(b) Four moles of ammonia gas and five moles of oxygen gas react to form four moles of nitrogen monoxide gas and six moles of water vapor.

(c) Two moles of solid sodium carbonate, four moles of nitrogen monoxide gas and one mole of oxygen gas react to form four moles of solid sodium nitrite and two moles of carbon dioxide gas.

60. One mole of solid phosphorus reacts with five moles of oxygen gas to produce one mole of solid tetraphosphorus decaoxide.

61. One mole of solid zinc reacts with two moles of hydrochloric acid to produce one mole of hydrogen gas and one mole of aqueous zinc chloride.

62. reaction 54-combustion
reaction 55-single replacement
reaction 56-simple decomposition
reaction 57-double replacement
reaction 58-complete combustion
reaction 59(a)-formation
reaction 59(b)-other or incomplete combustion
reaction 59(c)-other
reaction 60-formation or combustion
reaction 61-single replacement

63. single replacement

$Cl_{2(g)} + 2\ NaI_{(aq)} \longrightarrow I_{2(s)} + 2\ NaCl_{(aq)}$

If a few millilitres of a chlorinated hydrocarbon is added to the mixture, with shaking, and the color of the solvent appears purple or violet, then iodine is likely present.

64. double replacement (neutralization)

$HCl_{(aq)} + NaOH_{(aq)} \longrightarrow NaCl_{(aq)} + HOH_{(l)}$

If strips of red and blue litmus are dipped into the final solution, and neither litmus changes color, then a neutral solution has formed. If a thermometer is used to measure the initial and final temperatures, and the temperature increases, then a neutralization has likely occurred.

65. complete combustion

$C_3H_{8(g)} + 5\ O_{2(g)} \longrightarrow 3\ CO_{2(g)} + 4\ H_2O_{(g)}$

If cobalt(II) chloride paper is exposed to the final products, and the paper turns from blue to pink, then water is likely present. If the gases produced are bubbled into a limewater solution, and the limewater turns cloudy, then carbon dioxide is likely present.

66. single replacement

$Sr_{(s)} + 2\ HOH_{(l)} \longrightarrow H_{2(g)} + Sr(OH)_{2(aq)}$

If a flame is inserted into the gas produced, and a squeak or pop sound is heard, then hydrogen is likely present. If strips of red and blue litmus are dipped into the final solution, and the red litmus turns blue, then a base is likely present. If a sample of the solution is placed into a flame, and a bright red flame color appears, then strontium ions are likely present.

67. single replacement

$2\ Al_{(s)} + 3\ CuSO_{4(aq)} \longrightarrow 3\ Cu_{(s)} + Al_2(SO_4)_{3(aq)}$

If the final mixture is observed, and a red-brown solid appears, then copper metal is likely present. If the color of the solution is observed, and the color changes from blue to colorless, then copper(II) sulfate has likely reacted and aluminum sulfate formed.

68. simple decomposition

$2\ NI_{3(s)} \longrightarrow N_{2(g)} + 3\ I_{2(s)}$

If a few millilitres of a chlorinated hydrocarbon is added to the mixture, with shaking, and the color of the solvent appears purple or violet, then iodine is likely present.

69. formation (or combustion)

$S_{8(s)} + 8\ O_{2(g)} \longrightarrow 8\ SO_{2(g)}$

If a glowing splint is inserted into the container after the reaction, and the splint does not glow brighter or relight, then oxygen has likely reacted. If the gas produced is bubbled into water and tested with litmus paper, and the blue litmus turns red, then sulfur dioxide gas is likely present. (precipitation)

70. double replacement (precipitation)

$2\ NaOH_{(aq)} + NiCl_{2(aq)} \longrightarrow Ni(OH)_{2(s)} + 2\ NaCl_{(aq)}$

If the final mixture is observed, and a solid is formed, then nickel(II) hydroxide is likely present. If the color of the solution is observed, and the original green color disappears, then nickel(II) chloride has likely reacted.

UNIT III REVIEW

CHAPTER 5

1. Electrolytes are soluble compounds whose aqueous solutions conduct electricity. Solutions of non-electrolytes do not conduct electricity.

2. electrolytes: acids, bases, ionic compounds
non-electrolytes: molecular compounds

3. (a) $Li_2S_{(s)} \longrightarrow 2\ Li^+_{(aq)} + S^{2-}_{(aq)}$
(b) $HCl_{(g)} \longrightarrow H^+_{(aq)} + Cl^-_{(aq)}$
(c) $Al_2(SO_4)_{3(s)} \longrightarrow 2\ Al^{3+}_{(aq)} + 3\ SO_4^{2-}_{(aq)}$

4. (a) $NaC_6H_5COO_{(s)}$; $Na^+_{(aq)}$, $C_6H_5COO^-_{(aq)}$, $H_2O_{(l)}$
    (b) $Sr(OH)_{2(s)}$; $Sr^{2+}_{(aq)}$, $OH^-_{(aq)}$, $H_2O_{(l)}$
    (c) $HNO_{3(l)}$; $H^+_{(aq)}$, $NO_3^-_{(aq)}$, $H_2O_{(l)}$
    (d) $C_3H_{8(g)}$; $C_3H_{8(g)}$, $H_2O_{(l)}$
    (e) $CH_3COOH_{(l)}$; $CH_3COOH_{(aq)}$, $H_2O_{(l)}$
    (f) $CuCl_{2(s)}$; $Cu^{2+}_{(aq)}$, $Cl^-_{(aq)}$, $H_2O_{(l)}$
    (g) $AgCl_{(s)}$; $AgCl_{(s)}$, $H_2O_{(l)}$
    (h) $CH_3OH_{(l)}$; $CH_3OH_{(aq)}$, $H_2O_{(l)}$

5. An acid is a substance whose aqueous solution turns blue litmus red, conducts electricity and neutralizes bases.

6. Acidity is explained by the presence of hydrogen ions and conductivity is explained by the presence of both hydrogen ions and anions of the acid.

7. (a) $H_2CO_{3(aq)}$
    (b) $H_2SO_{3(aq)}$
    (c) $HF_{(aq)}$
    (d) $HOOCCOOH_{(aq)}$

8. (a) aqueous hydrogen nitrate; nitric acid
    (b) aqueous hydrogen nitrite; nitrous acid
    (c) aqueous hydrogen iodide; hydroiodic acid
    (d) aqueous hydrogen borate; boric acid

9. Solutions make it easy to handle chemicals, react chemicals and control reactions.

10. (1) If all four solutions are tested for conductivity, and one of them does not conduct, then that solution is $CH_3Cl_{(aq)}$.
    (2) If the remaining three solutions are tested with red and blue litmus paper, and one of the solutions turns blue litmus red, then that solution is $HCl_{(aq)}$.
    (3) If the remaining two solutions are tested with a flame test, a bright yellow flame indicates $NaCl_{(aq)}$, and a yellow-green flame indicates $BaCl_{2(aq)}$.
    or If the remaining two solutions are tested by adding a few millilitres of sodium sulfate solution, and one of the solutions forms a precipitate, then that solution is $BaCl_{2(aq)}$.

11. If calcium chloride solution is added to the unknown and a precipitate forms, then carbonate ions are present. After filtering, if necessary, a copper(II) nitrate solution is added to the filtrate and a precipitate forms, then sulfide ions are likely present. *Other reagents can be used to test for the presence of each ion.*

12. 13 g

13. 31.4 mmol

14. 6.42 mL

15. 2.1 mol/L $Fe^{3+}_{(aq)}$
    6.3 mol/L $Cl^-_{(aq)}$

16. (a) basic   (c) acidic
    (b) neutral   (d) basic

17. 23.8 g/100 mL or 3.19 mol/L

18. The solubility of solids usually increases as the temperature increases and the solubility of gases always decreases with temperature.

19. (a) The quantity of solid, the intensity of the blue color, and the electrical conductivity remain constant.
    (b) According to the theory of dynamic equilibrium, these properties are constant because the rate of crystallizing has become equal to the rate of dissolving.

CHAPTER 6

20. Gases always expand to fill their containers. Gases are highly compressible and diffuse very quickly.

21. (a) decreases   (c) increases
    (b) decreases

22. 0.30 L

23. yes (v_f = 807 mL)

24. 237 mL

25. 19.5 L

26. 6.23 L at STP; 6.89 L at SATP

27. 246 g

28. 0.18 kmol

29. 28 mol

30. 0.28 mol

31. 6.00 g

32. (a) 6.42 Mg
    (b) 9.92 ML

33. 552 g

CHAPTER 7

34. Both fields deal with chemicals and their reactions. Both use empirical methods.

35. Chemistry is more theoretical and international; chemical technology is more empirical and localized. Chemistry emphasizes concepts; chemical technology emphasizes methods and practical applications.

36. gravimetric: deals with masses
    gas: deals with pressure, volume and temperature of gases
    solution: deals with volume and concentration of a solution

37. 4.07 g

38. 0.392 g

39. 0.154 g

40. 1.05 g

41. 68.9 kL

42. 60 L

43. 3.39 mol/L

44. 1.35 L

45. (a) 11 mmol/L
    (b) Excess calcium hydroxide was placed in water and stirred occasionally for a day at 23°C. The solute of a measured volume of the saturated solution was crystallized by evaporating and the mass determined. (Alternatively, the mass of precipitate obtained by adding excess sodium carbonate to a measured volume of the saturated solution was determined after filtration.)

46. 1.56 mol/L

47. 0.898 mg

48. (1) Obtain 1.80 g of glucose in a clean, dry 150 mL beaker.
    (2) Dissolve the solid in about 50 mL of distilled water.
    (3) Transfer the solution to a 100 mL volumetric flask.
    (4) Add distilled water to the calibration line.
    (5) Stopper the flask and mix the contents thoroughly.

49. 105 mL

50. (1) Using a volumetric pipet transfer 5.00 mL of the concentrated solution into 50 mL of distilled water in a 100 mL volumetric flask.
    (2) Add distilled water to the calibration line on the flask.
    (3) Stopper the flask and mix the contents thoroughly.

51. 0.107 mol/L

52. crystallization (evaporating solvent)
    filtration (low solubility product)
    gas collection (CO_2 gas from decomposition)

UNIT IV REVIEW

CHAPTER 8

1. An activity series is an order of reactivity of elements.
2. (a) Li, Na, K, Rb, Cs, F
   (b) Cr, V, Ti, Sc, Ca, K
   (c) At_2, I_2, Br_2, Cl_2, F_2
3. The octet rule states that a maximum of eight electrons can occupy orbitals in the valence level of an atom. When four valence orbitals are completely filled (8 electrons), the valence electronic structure apears very stable, i.e. non-reactive.
4. (a) ·Ca·   2, 2, 0, 2
   (b) ·Al·   13, 3, 0, 3
   (c) :As·   15, 5, 1, 3
   (d) :O·   16, 6, 2, 2
   (e) :Br: 17, 7, 3, 1
   (f) :Ne: 18, 8, 4, 0
5. (a) Nonmetals react to form molecular compounds.
   (b) Metals and nonmetals react to form ionic compounds.
   (c) Nonmetals tend to have high electronegativities. Therefore, in the competition for electrons, neither atom wins and bonding electrons are shared. Metals tend to have low electronegativities. Therefore, in the competition for electrons, metals tend to lose electrons and non-metals tend to gain electrons forming ions.
6. (a) Ionic bonds are strong and hold ions in a rigid structure. Strong bonds would mean considerable energy is required to overcome these bonds to melt or boil the sample.
   (b) Positively and negatively charged particles (ions) are free to move in molten and aqueous states and are able to conduct an electric current.
7. (a) $K\cdot \; + \; \cdot \ddot{\underset{..}{Br}}: \; \rightarrow \; K^+ \; [:\ddot{\underset{..}{Br}}:]^-$
   (b) $2\,[Na\cdot] \; + \; \cdot \ddot{O}: \; \rightarrow \; Na^+_2 \; [:\ddot{\underset{..}{O}}:]^{2-}$
   (c) $\cdot Ca\cdot \; + \; 2\,[\cdot \ddot{\underset{..}{F}}:] \; \rightarrow \; Ca^{2+} \; [:\ddot{\underset{..}{F}}:]^-_2$
8. (a) CS_2 S=C=S
   (b) Si_2H_6 (H—Si—Si—H with H's above and below each Si)
9. (a) SCl_2 Cl—S—Cl

(b) PH_3 (H—P—H with H above P)
(c) N_2S_2 S=N—N=S
10. A rapid decolorization of bromine when added to a sample is the evidence for double covalent bonds.
11. Intermolecular bonds between molecules are generally weaker than the intramolecular bonds inside molecules.
12. (a) Fluorine molecules have fewer electrons than chlorine molecules and therefore weaker London forces.
   (b) Ethanol molecules are polar.
   (c) Hydrogen bonding among water molecules determines the structure of ice.

CHAPTER 9

13. Organic chemistry is the study of the molecular compounds of carbon.
14. (structural formulas of three organic molecules)
15. (a) $CH_3—OH$ alcohol   methanol
    (b) $CH_3—CH_3$ alkane   ethane
    (c) CH_3 (on benzene ring)   aromatic   methylbenzene or phenylmethane
    (d) $CH\equiv C—CH_3$ alkyne   propyne
    (e) (H—C(=O)—H)   aldehyde   methanal (formaldehyde)
    (f) ($CH_3—C(=O)—OH$)   carboxylic acid   ethanoic acid (acetic acid)
    (g) ($CH_3—N(H)—H$)   amine   methylamine
    (h) $Cl—CH_2—Cl$ organic halide   dichloromethane
    (i) ($CH_3—C(=O)—O—CH_3$)   ester   methyl ethanoate

(j)

$$CH_3-\overset{\overset{\displaystyle O}{\|}}{C}-O-\overset{\overset{\displaystyle H}{|}}{N}-H \qquad amide \qquad ethanamide$$

16. (a) complete combustion

$$\rightarrow \quad O=C=O \ + \ H-O-H$$

2-propanol + oxygen → carbon dioxide + water

(b) cracking

2,2-dimethylbutane + hydrogen →
methylpropane + ethane

(c) addition (hydrogenation)

$$CH_3-\overset{\overset{\displaystyle CH_3}{|}}{C}=CH-CH_2-CH_3 \ + \ H-H \ \rightarrow$$

$$CH_3-\overset{\overset{\displaystyle CH_3}{|}}{C}H-CH_2-CH_2-CH_3$$

2-methylpentane

(d) reforming

$$CH\equiv CH \quad \rightarrow \quad \hexagon$$

(e) elimination

2-chloropentane + hydroxide ion →

$$CH_3-CH_2-CH_2-CH=CH_2$$

1-pentene

$$+ \ CH_3-CH_2-CH=CH-CH_3$$

2-pentene

$$+ \ Cl^- \ + \ H-O-H$$

chloride ion   water

(f) addition

1-butene + hydrogen chloride →

$$CH_2Cl-CH_2-CH_2-CH_3$$

1-chlorobutane

$$+ \ CH_3-CHCl-CH_2-CH_3$$

2-chlorobutane

(g) substitution

$$CH_3-CH-CH_3 \ + \ Br-Br \ \rightarrow$$

$$CH_3-CBr-CH_3 \qquad + \qquad H-Br$$

(h) elimination

$$CH_3-CH_2-CH_2-OH \ \rightarrow$$

$$CH_3-CH=CH_2 \ + \ H-O-H$$

(i) condensation (esterification)

$$CH_3-CH_2-CH_2-\overset{\overset{\displaystyle O}{\|}}{C}-OH \ + \ CH_3OH \ \rightarrow$$

$$CH_3-CH_2-CH_2-\overset{\overset{\displaystyle O}{\|}}{C}-O-CH_3 \ + \ H-O-H$$

methyl butanoate   water

17. A polymer is a substance whose molecules are made up of many similar small molecules linked together in long chains.

18. Both synthetic and natural polymers may be addition or condensation polymers. Synthetic polymers generally do not occur in nature and are manufactured for specific purposes. Therefore, synthetic polymers usually have some unique properties not found in natural polymers.

43 HEAT CALCULATIONS

1. Calculate the quantity of heat required to warm 1.25 L of water from 22.0°C to 98.0°C in an electric kettle.

2. Assuming a volumetric heat capacity the same as pure water, calculate the heat that must be released from a soft drink in a 2.00 L container when the soft drink is cooled from 22.0°C to 10.0°C.

3. What mass of aluminum in a car engine will absorb 1.00 MJ of heat when the temperature rises from 22°C to 102°C after the car is started?

4. Assume the liquid coolant in a car engine has a volumetric heat capacity of 3.88 kJ/(L•°C). Determine the volume of coolant that will absorb 1.00 MJ of heat during a temperature rise from 22°C to 102°C.

5. In a laboratory experiment, 2.00 kJ of heat flowed to a 100 g sample of a liquid solvent, causing a temperature increase from 15.40°C to 21.37°C. Calculate the specific heat capacity of the liquid.

6. A human body loses about 360 kJ of heat every hour. Assuming that an average human body is equivalent to about 60 kg of water, what temperature decrease would this heat transfer cause? (Of course, this heat is replaced by body metabolism.)

43 HEAT CALCULATIONS

1. Calculate the quantity of heat required to warm 1.25 L of water from 22.0°C to 98.0°C in an electric kettle.

$$q = vc\Delta t$$

$$= 1.25 \text{ L} \times \frac{4.19 \text{ kJ}}{\text{L} \cdot °\text{C}} \times (98.0 - 22.0)°\text{C}$$

$$= 398 \text{ kJ}$$

2. Assuming a volumetric heat capacity the same as pure water, calculate the heat that must be released from a soft drink in a 2.00 L container when the soft drink is cooled from 22.0°C to 10.0°C.

$$q = vc\Delta t$$

$$= 2.00 \text{ L} \times \frac{4.19 \text{ kJ}}{\text{L} \cdot °\text{C}} \times (22.0 - 10.0)°\text{C}$$

$$= 101 \text{ kJ}$$

3. What mass of aluminum in a car engine will absorb 1.00 MJ of heat when the temperature rises from 22°C to 102°C after the car is started?

$$q = mc\Delta t$$

$$1.00 \text{ MJ} = m \times \frac{0.900 \text{ J}}{\text{g} \cdot °\text{C}} \times (102 - 22)°\text{C}$$

$$m = 14 \text{ kg}$$

4. Assume the liquid coolant in a car engine has a volumetric heat capacity of 3.88 kJ/(L•°C). Determine the volume of coolant that will absorb 1.00 MJ of heat during a temperature rise from 22°C to 102°C.

$$q = vc\Delta t$$

$$1.00 \text{ MJ} = v \times \frac{3.88 \text{ kJ}}{\text{L} \cdot °\text{C}} \times (102 - 22)°\text{C}$$

$$v = 3.2 \text{ L}$$

5. In a laboratory experiment, 2.00 kJ of heat flowed to a 100 g sample of a liquid solvent, causing a temperature increase from 15.40°C to 21.37°C. Calculate the specific heat capacity of the liquid.

$$q = mc\Delta t$$

$$2.00 \text{ kJ} = 100 \text{ g} \times c \times (21.37 - 15.40)°\text{C}$$

$$c = 3.35 \text{ J/(g} \cdot °\text{C)}$$

6. A human body loses about 360 kJ of heat every hour. Assuming that an average human body is equivalent to about 60 kg of water, what temperature decrease would this heat transfer cause? (Of course, this heat is replaced by body metabolism.)

$$q = mc\Delta t$$

$$360 \text{ kJ} = 60 \text{ kg} \times \frac{4.19 \text{ J}}{\text{g} \cdot °\text{C}} \times \Delta t$$

$$\Delta t = 1.4°\text{C}$$

44 ENTHALPY CHANGES

1. Enthalpy changes may be classified into three types — phase, chemical, and nuclear.
   (a) How are these types different empirically?

   (b) How are they similar theoretically?

2. Calculate the enthalpy change for the melting of a 30 g ice cube.

3. A reference gives a value of +39.23 kJ/mol for the molar enthalpy of vaporization for methanol. What enthalpy change occurs in the evaporation of 10.0 g of methanol?

4. Given H_{vap} is +1.37 kJ/mol for NH_3 (page 293), find the mass of ammonia that can be condensed from vapor to liquid (with no temperature change) during an enthalpy change of 10.0 kJ.

5. An experiment produces evidence that the evaporation of 4.00 g of liquid butane, $C_4H_{10(l)}$, requires a gain in enthalpy of 1.67 kJ. Find the molar enthalpy of vaporization for butane from this evidence.

44 ENTHALPY CHANGES

1. Enthalpy changes may be classified into three types — phase, chemical, and nuclear.

   (a) How are these types different empirically?

   Phase changes involve only a change of state. Chemical changes result in new chemical substances being formed. In nuclear changes, new elements or sub-atomic particles are produced.

   (b) How are they similar theoretically?

   Phase, chemical and nuclear changes all involve a change in potential energy as a result of changes in bonding.

2. Calculate the enthalpy change for the melting of a 30 g ice cube.

   $\Delta H_{fus} = nH_{fus}$

   $$= 30 \text{ g} \times \frac{1 \text{ mol}}{18.02 \text{ g}} \times \frac{6.03 \text{ kJ}}{\text{mol}}$$

   $$= 10 \text{ kJ}$$

3. A reference gives a value of +39.23 kJ/mol for the molar enthalpy of vaporization for methanol. What enthalpy change occurs in the evaporation of 10.0 g of methanol?

   $\Delta H_{vap} = nH_{vap}$

   $$= 10.0 \text{ g} \times \frac{1 \text{ mol}}{32.05 \text{ g}} \times \frac{39.23 \text{ kJ}}{\text{mol}}$$

   $$= 12.2 \text{ kJ}$$

4. Given H_{vap} is +1.37 kJ/mol for NH_3 (page 293), find the mass of ammonia that can be condensed from vapor to liquid (with no temperature change) during an enthalpy change of 10.0 kJ.

   $\Delta H_{cond} = nH_{cond}$

   $$10.0 \text{ kJ} = m \times \frac{1 \text{ mol}}{17.04 \text{ g}} \times \frac{1.37 \text{ kJ}}{\text{mol}}$$

   $$m = 124 \text{ g}$$

5. An experiment produces evidence that the evaporation of 4.00 g of liquid butane, $C_4H_{10(l)}$, requires a gain in enthalpy of 1.67 kJ. Find the molar enthalpy of vaporization for butane from this evidence.

   $\Delta H_{vap} = nH_{vap}$

   $$1.67 \text{ kJ} = 4.00 \text{ g} \times \frac{1 \text{ mol}}{58.14 \text{ g}} \times H_{vap}$$

   $$H_{vap} = 24.3 \text{ kJ/mol}$$

45 TOTAL ENERGY CHANGES

1. Find the total energy lost by the contents of a styrofoam picnic cooler when a 3.60 kg block of ice placed inside the cooler completely melts (at 0°C). The ice had an initial temperature of –10.0°C.

2. The specific heat capacity of solid lead is listed in the *Nelson Chemistry* reference tables, the molar enthalpy of fusion is 5.0 kJ/mol, and the melting point of lead is listed in the *Nelson Chemistry* periodic table. Calculate the total energy required to change 100 g of lead at 25°C to molten lead at its melting point.

3. What total energy must be absorbed (from the environment) to change 1.00 kg of ice at –30.0°C on a lake surface in February to water at 20.0°C in August.

4. Energy is released when 100 kg of steam at 150.0°C and standard pressure changes to water at 60.0°C.
   (a) Sketch a cooling curve.   (b) Calculate the total energy change.

5. A 500 kg steel boiler is used to convert 200 kg of water into steam. Calculate the total energy required to heat both the steel and its contents from 20.0°C to 100.0°C. The specific heat capacity of the steel is 0.528 J/(g•°C).

45 TOTAL ENERGY CHANGES

1. $\Delta E_{total} = \underset{\text{(ice warming)}}{q} + \underset{\text{(ice melting)}}{\Delta H_{fus}}$

   $= mc\Delta t + nH_{fus}$

   $= 3.60 \text{ kg} \times \dfrac{2.01 \text{ J}}{\text{g}\bullet°\text{C}} \times (0 - -10.0)°\text{C} + 3.60 \text{ kg} \times \dfrac{1 \text{ mol}}{18.02 \text{ g}} \times \dfrac{6.03 \text{ kJ}}{\text{mol}}$

   $= 72.4 \text{ kJ} + 1.20 \text{ MJ}$

   $= 1.28 \text{ MJ}$

2. $\Delta E_{total} = \underset{\text{(lead warming)}}{q} + \underset{\text{(lead melting)}}{\Delta H_{fus}}$

   $= mc\Delta t + nH_{fus}$

   $= 100 \text{ g} \times \dfrac{0.159 \text{ J}}{\text{g}\bullet°\text{C}} \times (328 - 25)°\text{C} + 100 \text{ g} \times \dfrac{1 \text{ mol}}{207.20 \text{ g}} \times \dfrac{5.0 \text{ kJ}}{\text{mol}}$

   $= 4.82 \text{ kJ} + 2.4 \text{ kJ}$

   $= 7.2 \text{ kJ}$

3. $\Delta E_{total} = \underset{\text{(ice warming)}}{q} + \underset{\text{(ice melting)}}{\Delta H_{fus}} + \underset{\text{(water warming)}}{q}$

   $= mc\Delta t + nH_{fus} + mc\Delta t$

   $= 1.00 \text{ kg} \times \dfrac{2.01 \text{ J}}{\text{g}\bullet°\text{C}} \times (0 - -30.0)°\text{C} + 1.00 \text{ kg} \times \dfrac{1 \text{ mol}}{18.02 \text{ g}} \times \dfrac{6.03 \text{ kJ}}{\text{mol}}$

   $+ 1.00 \text{ kg} \times \dfrac{4.19 \text{ J}}{\text{g}\bullet°\text{C}} \times (20.0 - 0.0)°\text{C}$

   $= 60.3 \text{ kJ} + 335 \text{ kJ} + 83.8 \text{ kJ}$
   $= 479 \text{ kJ}$

4. (a)

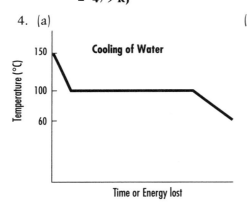

   (b) $\Delta E_{total} = \underset{\text{(steam cooling)}}{q} + \underset{\text{(steam condensing)}}{\Delta H_{cond}} + \underset{\text{(water cooling)}}{q}$

   $= mc\Delta t + nH_{cond} + mc\Delta t$

   $= 100 \text{ kg} \times \dfrac{2.01 \text{ J}}{\text{g}\bullet°\text{C}} \times (150 - 100)°\text{C}$

   $+ 100 \text{ kg} \times \dfrac{1 \text{ mol}}{18.02 \text{ g}} \times \dfrac{40.8 \text{ kJ}}{\text{mol}}$

   $+ 100 \text{ kg} \times \dfrac{4.19 \text{ J}}{\text{g}\bullet°\text{C}} \times (100.0 - 60.0)°\text{C}$

   $= 10 \text{ MJ} + 226 \text{ MJ} + 16.8 \text{ MJ}$
   $= 253 \text{ MJ}$

5. $\Delta E_{total} = \underset{\text{(water warming)}}{q} + \underset{\text{(water boiling)}}{\Delta H_{vap}} + \underset{\text{(steel warming)}}{q}$

   $= mc\Delta t + nH_{vap} + mc\Delta t$

   $= 200 \text{ kg} \times \dfrac{4.19 \text{ J}}{\text{g}\bullet°\text{C}} \times (100.0 - 20.0)°\text{C} + 200 \text{ kg} \times \dfrac{1 \text{ mol}}{18.02 \text{ g}} \times \dfrac{40.8 \text{ kJ}}{\text{mol}}$

   $+ 500 \text{ kg} \times \dfrac{0.528 \text{ J}}{\text{g}\bullet°\text{C}} \times (100.0 - 20.0)°\text{C}$

   $= 67.0 \text{ MJ} + 453 \text{ MJ} + 21.1 \text{ MJ}$
   $= 541 \text{ MJ}$

46 CALORIMETRY

1. A student mixed 100.0 mL of 1.50 mol/L sulfuric acid with 200.0 mL of 1.50 mol/L sodium hydroxide. Both solutions were at 19.67°C initially and the highest temperature reached by the reaction mixture was 34.06°C. Calculate the molar enthalpy of neutralization for sulfuric acid.

2. A calorimeter has a heat capacity of 40.00 kJ/°C. Complete combustion of 1.00 g of hydrogen in this calorimeter causes a temperature increase of 3.54°C. Calculate the molar enthalpy of combustion for hydrogen from this evidence.

3. A reference gives the molar enthalpy of combustion for methane as –803 kJ/mol. What minimum mass of methane must be burned to warm 4.00 L of water from 22.4°C to 87.6°C, assuming no heat losses?

4. Combustion of 3.50 g of ethanol, $C_2H_5OH_{(l)}$, in a calorimeter with a heat capacity of 15.2 kJ/°C causes a temperature increase from 19.88°C to 26.18°C. Find the molar enthalpy of combustion for ethanol from this evidence.

5. Find the temperature increase expected for 1.00 L of water when it absorbs all of the energy from the combustion of 1.00 g of acetylene, $C_2H_{2(g)}$. The molar enthalpy of combustion for acetylene is –1.29 MJ/mol.

46 CALORIMETRY

1. A student mixed 100.0 mL of 1.50 mol/L sulfuric acid with 200.0 mL of 1.50 mol/L sodium hydroxide. Both solutions were at 19.67°C initially and the highest temperature reached by the reaction mixture was 34.06°C. Calculate the molar enthalpy of neutralization for sulfuric acid.

$$\Delta H_n \underset{H_2SO_4}{} = q_{(mixture)}$$

$$nH_n = vc\Delta t$$

$$100.0 \text{ mL} \times \frac{1.50 \text{ mol}}{1 \text{ L}} \times H_n = 300.0 \text{ mL} \times \frac{4.19 \text{ kJ}}{L \cdot °C} \times (34.06 - 19.67)°C$$

$$H_n \underset{H_2SO_4}{} = 121 \text{ kJ/mol}$$

The molar enthalpy of neutralization for sulfuric acid is reported as –121 kJ/mol.

2. A calorimeter has a heat capacity of 40.00 kJ/°C. Complete combustion of 1.00 g of hydrogen in this calorimeter causes a temperature increase of 3.54°C. Calculate the molar enthalpy of combustion for hydrogen from this evidence.

$$\Delta H_c \underset{H_2}{} = q_{(calorimeter)}$$

$$nH_c = C\Delta t$$

$$1.00 \text{ g} \times \frac{1 \text{ mol}}{2.02 \text{ g}} \times H_c = \frac{40.00 \text{ kJ}}{°C} \times 3.54°C$$

$$H_c = 286 \text{ kJ/mol}$$

The molar enthalpy of combustion for hydrogen is reported as –286 kJ/mol.

3. A reference gives the molar enthalpy of combustion for methane as –803 kJ/mol. What minimum mass of methane must be burned to warm 4.00 L of water from 22.4°C to 87.6°C, assuming no heat losses?

$$\Delta H_c \underset{CH_4}{} = q_{(water)}$$

$$nH_c = vc\Delta t$$

$$m \times \frac{1 \text{ mol}}{16.05 \text{ g}} \times \frac{803 \text{ kJ}}{\text{mol}} = 4.00 \text{ L} \times \frac{4.19 \text{ kJ}}{L \cdot °C} \times (87.6 - 22.4)°C$$

$$m = 21.8 \text{ g}$$

The mass of methane that must be burned is 21.8 g.

4. Combustion of 3.50 g of ethanol, $C_2H_5OH_{(l)}$, in a calorimeter with a heat capacity of 15.2 kJ/°C causes a temperature increase from 19.88°C to 26.18°C. Find the molar enthalpy of combustion for ethanol from this evidence.

$$\Delta H_c \underset{C_2H_5OH}{} = q_{(calorimeter)}$$

$$nH_c = C\Delta t$$

$$3.50 \text{ g} \times \frac{1 \text{ mol}}{46.08 \text{ g}} \times H_c = \frac{15.2 \text{ kJ}}{°C} \times (26.18 - 19.88)°C$$

$$H_c = 1.26 \text{ MJ/mol}$$

The molar enthalpy of combustion for ethanol is reported as –1.26 MJ/mol.

5. Find the temperature increase expected for 1.00 L of water when it absorbs all of the energy from the combustion of 1.00 g of acetylene, $C_2H_{2(g)}$. The molar enthalpy of combustion for acetylene is –1.29 MJ/mol.

$$\Delta H_c \underset{(acetylene)}{} = q_{(water)}$$

$$nH_c = vc\Delta t$$

$$1.00 \text{ g} \times \frac{1 \text{ mol}}{26.04 \text{ g}} \times \frac{1.29 \text{ MJ}}{\text{mol}} = 1.00 \text{ L} \times \frac{4.19 \text{ kJ}}{L \cdot °C} \times \Delta t$$

$$\Delta t = 11.8°C$$

47 COMMUNICATING ENTHALPY CHANGES

1. Iron(II) sulfide ore is roasted according to the following chemical equation.

   $$4 \text{ FeS}_{(s)} + 7 \text{ O}_{2(g)} \longrightarrow 2 \text{ Fe}_2\text{O}_{3(s)} + 4 \text{ SO}_{2(g)} \qquad\qquad \Delta H_c = -2456 \text{ kJ}$$

   (a) Rewrite this chemical equation including the energy as a term in the balanced equation.

   (b) What is the molar enthalpy for iron(II) sulfide in this reaction?

   (c) What is the molar enthalpy for iron(III) oxide in this reaction?

2. Boron reacts with hydrogen to form diboron hexahydride (diborane) gas. The molar enthalpy of reaction for boron is +15.7 kJ/mol.

   (a) Write the balanced chemical equation using whole number coefficients and including the energy change as a ΔH_r.

   (b) Write the balanced chemical equation using whole number coefficients and including the energy change as a term in the balanced equation.

3. The molar enthalpy of combustion for octane, $C_8H_{18(l)}$, is reported to be –1.3 MJ/mol.

   (a) Write the balanced chemical equation using whole number coefficients and including the energy change as a ΔH_r.

   (b) Write the balanced chemical equation using whole number coefficients and including the energy change as a term in the balanced equation.

4. Draw potential energy diagrams to communicate the following chemical reactions. Assume SATP conditions.

   (a) the formation of chromium(III) oxide

   (b) the simple decomposition of silver iodide

   (c) the formation of carbon disulfide

47 COMMUNICATING ENTHALPY CHANGES

1. Iron(II) sulfide ore is roasted according to the following chemical equation.

$$4\ FeS_{(s)}\ +\ 7\ O_{2(g)}\ \rightarrow\ 2\ Fe_2O_{3(s)}\ +\ 4\ SO_{2(g)} \qquad \Delta H_c = -2456\ kJ$$

   (a) Rewrite this chemical equation including the energy as a term in the balanced equation.
   $$4\ FeS_{(s)}\ +\ 7\ O_{2(g)}\ \rightarrow\ 2\ Fe_2O_{3(s)}\ +\ 4\ SO_{2(g)}\ +\ 2456\ kJ$$

   (b) What is the molar enthalpy for iron(II) sulfide in this reaction?
   $$H_c \atop FeS = \frac{-2456\ kJ}{4\ mol} = -614\ kJ/mol$$

   (c) What is the molar enthalpy for iron(III) oxide in this reaction?
   $$H_c \atop Fe_2O_3 = \frac{-2456\ kJ}{2\ mol} = -1228\ kJ/mol$$

2. Boron reacts with hydrogen to form diboron hexahydride (diborane) gas. The molar enthalpy of reaction for boron is +15.7 kJ/mol.

   (a) Write the balanced chemical equation using whole number coefficients and including the energy change as a ΔH_r.
   $$2\ B_{(s)}\ +\ 3\ H_{2(g)}\ \rightarrow\ B_2H_{6(g)} \qquad \Delta H_r = 2\ mol \times +15.7\ kJ/mol = +31.4\ kJ$$

   (b) Write the balanced chemical equation using whole number coefficients and including the energy change as a term in the balanced equation.
   $$2\ B_{(s)}\ +\ 3\ H_{2(g)}\ +\ 31.4\ kJ\ \rightarrow\ B_2H_{6(g)}$$

3. The molar enthalpy of combustion for octane, $C_8H_{18(l)}$, is reported to be –1.3 MJ/mol.

   (a) Write the balanced chemical equation using whole number coefficients and including the energy change as a ΔH_r.
   $$2\ C_8H_{18(l)}\ +\ 25\ O_{2(g)}\ \rightarrow\ 16\ CO_{2(g)}\ +\ 18\ H_2O_{(g)}$$
   $$\Delta H_r = 2\ mol \times -1.3\ MJ/mol = -2.6\ MJ$$

   (b) Write the balanced chemical equation using whole number coefficients and including the energy change as a term in the balanced equation.
   $$2\ C_8H_{18(l)}\ +\ 25\ O_{2(g)}\ \rightarrow\ 16\ CO_{2(g)}\ +\ 18\ H_2O_{(g)}\ +\ 2.6\ MJ$$

4. (a) **The Formation of Chromium(III) Oxide** (b) **The Simple Decomposition of Silver Iodide**

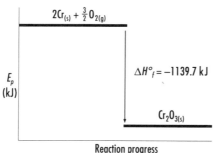

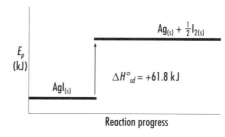

   (c) **The Formation of Carbon Disulfide**

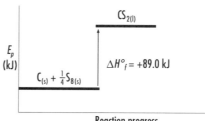

48 PREDICTING ΔH_r USING HESS'S LAW

1. The enthalpy changes for the formation of two wolfram bromides are shown below.

   $W_{(s)} + 2\,Br_{2(l)} \rightarrow WBr_{4(s)}$ $\qquad\qquad\qquad\qquad\qquad\qquad$ $\Delta H°_r = -146.7\ kJ$
   $W_{(s)} + 3\,Br_{2(l)} \rightarrow WBr_{6(s)}$ $\qquad\qquad\qquad\qquad\qquad\qquad$ $\Delta H°_r = -184.4\ kJ$

   Calculate the standard enthalpy change for the following reaction.

   $Br_{2(l)} + WBr_{4(s)} \rightarrow WBr_{6(s)}$

2. Given: $\qquad\qquad\qquad\qquad$ $N_2O_{4(g)} \rightarrow 2\,NO_{2(g)}$ $\qquad\qquad\qquad\qquad$ $\Delta H°_r = +58\ kJ$

   $\qquad\qquad\qquad$ $NO_{(g)} + \dfrac{1}{2}O_{2(g)} \rightarrow NO_{2(g)}$ $\qquad\qquad\qquad\qquad$ $\Delta H°_r = -56\,kJ$

   Calculate the standard enthalpy change for the following reaction.

   $2\,NO_{(g)} + O_{2(g)} \rightarrow N_2O_{4(g)}$

3. Use the following reactions and enthalpy changes to predict the standard enthalpy change for

   $2\,NO_{2(g)} + 2\,H_2O_{(g)} \rightarrow 3\,O_{2(g)} + N_2H_{4(g)}$

   $\dfrac{1}{2}N_{2(g)} + O_{2(g)} \rightarrow NO_{2(g)}$ $\qquad\qquad\qquad\qquad\qquad\qquad$ $\Delta H°_f = +33.2\ kJ$

   $H_{2(g)} + \dfrac{1}{2}O_{2(g)} \rightarrow H_2O_{(g)}$ $\qquad\qquad\qquad\qquad\qquad\qquad$ $\Delta H°_f = -241.8\ kJ$

   $N_{2(g)} + 2\,H_{2(g)} \rightarrow N_2H_{4(g)}$ $\qquad\qquad\qquad\qquad\qquad\qquad$ $\Delta H°_r = +47.6\ kJ$

4. Use the following formation reaction evidence to calculate the standard enthalpy change for the complete combustion of cycloheptane.

   $C_{(s)} + O_{2(g)} \rightarrow CO_{2(g)}$ $\qquad\qquad\qquad\qquad\qquad\qquad\qquad$ $\Delta H°_f = -393.5\ kJ$

   $H_{2(g)} + \dfrac{1}{2}O_{2(g)} \rightarrow H_2O_{(g)}$ $\qquad\qquad\qquad\qquad\qquad\qquad$ $\Delta H°_f = -241.8\ kJ$

   $7\,C_{(s)} + 7\,H_{2(g)} \rightarrow C_7H_{14(l)}$ $\qquad\qquad\qquad\qquad\qquad\qquad$ $\Delta H°_f = +115.0\ kJ$

48 PREDICTING ΔH_r USING HESS'S LAW

1. The enthalpy changes for the formation of two wolfram bromides are shown below.

$W_{(s)} + 2 Br_{2(l)} \rightarrow WBr_{4(s)}$ $\Delta H°_r = -146.7$ kJ
$W_{(s)} + 3 Br_{2(l)} \rightarrow WBr_{6(s)}$ $\Delta H°_r = -184.4$ kJ

Calculate the standard enthalpy change for the following reaction.

$Br_{2(l)} + WBr_{4(s)} \rightarrow WBr_{6(s)}$

$\quad\quad WBr_{4(s)} \rightarrow W_{(s)} + 2 Br_{2(l)}$ $\Delta H° = +146.7$ kJ
$\underline{W_{(s)} + 3 Br_{2(l)} \rightarrow WBr_{6(s)}}$ $\Delta H° = -184.4$ kJ
$Br_{2(l)} + WBr_{4(s)} \rightarrow WBr_{6(s)}$ $\Delta H°_r = -37.7$ kJ

2. Given:

$N_2O_{4(g)} \rightarrow 2 NO_{2(g)}$ $\Delta H°_r = +58$ kJ

$NO_{(g)} + \frac{1}{2}O_{2(g)} \rightarrow NO_{2(g)}$ $\Delta H°_r = -56$ kJ

Calculate the standard enthalpy change for the following reaction.

$2 NO_{(g)} + O_{2(g)} \rightarrow N_2O_{4(g)}$

$\quad\quad 2 NO_{2(g)} \rightarrow N_2O_{4(g)}$ $\Delta H° = -58$ kJ
$\underline{2 NO_{(g)} + O_{2(g)} \rightarrow 2 NO_{2(g)}}$ $\Delta H° = -112$ kJ
$2 NO_{(g)} + O_{2(g)} \rightarrow N_2O_{4(g)}$ $\Delta H°_r = -170$ kJ

3. Use the following reactions and enthalpy changes to predict the standard enthalpy change for

$2 NO_{2(g)} + 2 H_2O_{(g)} \rightarrow 3 O_{2(g)} + N_2H_{4(g)}$

$\frac{1}{2}N_{2(g)} + O_{2(g)} \rightarrow NO_{2(g)}$ $\Delta H°_f = +33.2$ kJ

$H_{2(g)} + \frac{1}{2}O_{2(g)} \rightarrow H_2O_{(g)}$ $\Delta H°_f = -241.8$ kJ

$N_{2(g)} + 2 H_{2(g)} \rightarrow N_2H_{4(g)}$ $\Delta H°_r = +47.6$ kJ

$\quad\quad 2 NO_{2(g)} \rightarrow N_{2(g)} + 2 O_{2(g)}$ $\Delta H° = -66.4$ kJ
$\quad\quad 2 H_2O_{(l)} \rightarrow 2 H_{2(g)} + O_{2(g)}$ $\Delta H° = +483.6$ kJ
$\underline{N_{2(g)} + 2 H_{2(g)} \rightarrow N_2H_{4(g)}}$ $\Delta H° = +47.6$ kJ
$2 NO_{2(g)} + 2 H_2O_{(g)} \rightarrow 3 O_{2(g)} + N_2H_{4(g)}$ $\Delta H°_r = +464.8$ kJ

4. Use the following formation reaction evidence to calculate the standard enthalpy change for the complete combustion of cycloheptane.

$C_{(s)} + O_{2(g)} \rightarrow CO_{2(g)}$ $\Delta H°_f = -393.5$ kJ

$H_{2(g)} + \frac{1}{2}O_{2(g)} \rightarrow H_2O_{(g)}$ $\Delta H°_f = -241.8$ kJ

$7 C_{(s)} + 7 H_{2(g)} \rightarrow C_7H_{14(l)}$ $\Delta H°_f = +115.0$ kJ

$\quad\quad 14 C_{(s)} + 14 O_{2(g)} \rightarrow 14 CO_{2(g)}$ $\Delta H° = -5509.2$ kJ
$\quad\quad 14 H_{2(g)} + 7 O_{2(g)} \rightarrow 14 H_2O_{(l)}$ $\Delta H° = -3385.2$ kJ
$\quad\quad 2 C_7H_{14(l)} \rightarrow 14 C_{(s)} + 14 H_{2(g)}$ $\Delta H° = -230.0$ kJ
$2 C_7H_{14(l)} + 21 O_{2(g)} \rightarrow 14 CO_{2(g)} + 14 H_2O_{(g)}$ $\Delta H°_c = -9124.2$ kJ

49 PREDICTING ΔH°_r USING STANDARD MOLAR ENTHALPIES OF FORMATION

1. Laboratory quantities of ethylene can be prepared by an elimination reaction of ethanol using an acid catalyst. Calculate the standard enthalpy change for the conversion of ethanol into ethylene and water.

2. Calculate the standard molar enthalpy of combustion for acetic acid.

3. An initial step in the production of iron in a blast furnace involves the conversion of iron(III) oxide and carbon monoxide into iron(II,III) oxide and carbon dioxide. Calculate the standard molar enthalpy of reaction for iron(III) oxide.

4. The fertilizer urea is produced along with liquid water by the reaction of ammonia and carbon dioxide. Calculate the standard molar enthalpy of reaction for ammonia.

49 PREDICTING ΔH°_r USING STANDARD MOLAR ENTHALPIES OF FORMATION

1. Laboratory quantities of ethylene can be prepared by an elimination reaction of ethanol using an acid catalyst. Calculate the standard enthalpy change for the conversion of ethanol into ethylene and water.

$$C_2H_5OH_{(l)} \rightarrow C_2H_{4(g)} + H_2O_{(l)}$$

$$\begin{aligned}
\Delta H^\circ_r &= \Sigma nH^\circ_{fp} - \Sigma nH^\circ_{fr} \\
&= (1\ mol \times +52.5\ kJ/mol + 1\ mol \times -285.8\ kJ/mol) \\
&\quad - (1\ mol \times -235.2\ kJ/mol) \\
&= -233.3\ kJ - (-235.2\ kJ) \\
&= +1.9\ kJ
\end{aligned}$$

2. Calculate the standard molar enthalpy of combustion for acetic acid.

$$CH_3COOH_{(l)} + 2\ O_{2(g)} \rightarrow 2\ CO_{2(g)} + 2\ H_2O_{(g)}$$

$$\begin{aligned}
\Delta H^\circ_c &= \Sigma nH^\circ_{fp} - \Sigma nH^\circ_{fr} \\
&= (2\ mol \times -393.5\ kJ/mol + 2\ mol \times -241.8\ kJ/mol) \\
&\quad - (1\ mol \times -432.8\ kJ/mol + 2\ mol \times 0\ kJ/mol) \\
&= -1270.6\ kJ - (-432.8\ kJ) \\
&= -837.8\ kJ
\end{aligned}$$

$$H^\circ_{c\ CH_3COOH} = \frac{-837.8\ kJ}{1\ mol} = -837.8\ kJ/mol$$

3. An initial step in the production of iron in a blast furnace involves the conversion of iron(III) oxide and carbon monoxide into iron(II,III) oxide and carbon dioxide. Calculate the standard molar enthalpy of reaction for iron(III) oxide.

$$3\ Fe_2O_{3(s)} + CO_{(g)} \rightarrow 2\ Fe_3O_{4(s)} + CO_{2(g)}$$

$$\begin{aligned}
\Delta H^\circ_r &= \Sigma nH^\circ_{fp} - \Sigma nH^\circ_{fr} \\
&= (2\ mol \times -1118.4\ kJ/mol + 1\ mol \times -393.5\ kJ/mol) \\
&\quad - (3\ mol \times -824.2\ kJ/mol + 1\ mol \times -110.5\ kJ/mol) \\
&= -2630.3\ kJ - (-2583.1\ kJ) \\
&= -47.2\ kJ
\end{aligned}$$

$$H^\circ_{r\ Fe_2O_3} = \frac{-47.2\ kJ}{3\ mol} = -15.7\ kJ/mol$$

4. The fertilizer urea is produced along with liquid water by the reaction of ammonia and carbon dioxide. Calculate the standard molar enthalpy of reaction for ammonia.

$$2\ NH_{3(g)} + CO_{2(g)} \rightarrow CO(NH_2)_{2(s)} + H_2O_{(l)}$$

$$\begin{aligned}
\Delta H^\circ_r &= \Sigma nH^\circ_{fp} - \Sigma nH^\circ_{fr} \\
&= (1\ mol \times -333.5\ kJ/mol + 1\ mol \times -285.8\ kJ/mol) \\
&\quad - (2\ mol \times -45.9\ kJ/mol + 1\ mol \times -393.5\ kJ/mol) \\
&= -619.3\ kJ - (-485.3\ kJ) \\
&= -134.0\ kJ
\end{aligned}$$

$$H^\circ_{r\ NH_3} = \frac{-134.0\ kJ}{2\ mol} = -67.0\ kJ/mol$$

50 MULTI-STEP ENERGY CALCULATIONS

1. For the following combustion, what mass of carbon dioxide is produced when 1500 kJ of energy is released?

$$2\,C_2H_{6(g)} \;+\; 7\,O_{2(g)} \;\longrightarrow\; 4\,CO_{2(g)} \;+\; 6\,H_2O_{(g)} \;+\; 2502\ \text{kJ}$$

2. How much energy is released when 1.00 t of sulfur trioxide is produced by the following reaction?

$$2\,SO_{2(g)} \;+\; O_{2(g)} \;\longrightarrow\; 2\,SO_{3(g)} \qquad\qquad \Delta H_r = -192.8\ \text{kJ}$$

3. In respiration, glucose is oxidized by oxygen gas to produce carbon dioxide gas, liquid water, and energy. What is the energy released when 18.0 g of glucose is consumed?

4. Methanol is burned in a bomb calorimeter. Liquid water is formed as a product. If 3.40 g of methanol reacts, what is the expected temperature change in a calorimeter with a heat capacity of 6.75 kJ/°C?

5. A waste heat exchanger is used to absorb the energy from the complete combustion of hydrogen sulfide gas. What volume of water undergoing a temperature change of 64°C is required to absorb all of the energy from the burning of 15 kg of hydrogen sulfide?

50   MULTI-STEP ENERGY CALCULATIONS

1. $H_c = \dfrac{-2502\ kJ}{4\ mol} = -626\ kJ/mol$
   $_{CO_2}$

   $\Delta H_c = nH_c$
   $1500\ kJ = m \times \dfrac{1\ mol}{44.01\ g} \times 626\ kJ/mol$

   $m = 106\ g$

2. $H_r = \dfrac{-192.8\ kJ}{2\ mol} = -96.4\ kJ/mol$
   $_{SO_3}$

   $\Delta H_c = nH_c$
   $= 1.00\ Mg \times \dfrac{1\ mol}{80.06\ g} \times \dfrac{96.4\ kJ}{mol} = 1.20\ GJ$

3. $C_6H_{12}O_{6(s)} + 6\ O_{2(g)} \longrightarrow 6\ CO_{2(g)} + 6\ H_2O_{(l)}$

   $\Delta H°_r = \sum nH°_{fp} - \sum nH°_{fr}$
   $\quad = (6\ mol \times -393.5\ kJ/mol + 6\ mol \times -285.8\ kJ/mol)$
   $\quad\quad - (1\ mol \times -1273.1\ kJ/mol + 6\ mol \times 0\ kJ/mol)$
   $\quad = -2802.7\ kJ$

   $H_r = \dfrac{-2802.7\ kJ}{1\ mol} = -2802.7\ kJ/mol$
   $_{C_6H_{12}O_6}$

   $\Delta H_r = nH_r$
   $\quad = 18.0\ g \times \dfrac{1\ mol}{180.18\ g} \times \dfrac{2802.7\ kJ}{mol} = 280\ kJ$

4. $CH_3OH_{(l)} + \dfrac{3}{2}\ O_{2(g)} \longrightarrow CO_{2(g)} + 2\ H_2O_{(l)}$

   $\Delta H_c° = \sum nH°_{fp} - \sum nH°_{fr}$
   $\quad = (1\ mol \times -393.5\ kJ/mol + 2\ mol \times -285.8\ kJ/mol)$
   $\quad\quad - (1\ mol \times -239.1\ kJ/mol + \dfrac{3}{2}\ mol \times 0\ kJ/mol)$
   $\quad = -726.0\ kJ$

   $H_c = \dfrac{726.0\ kJ}{1\ mol} = -726.0\ kJ/mol$
   $_{CH_3OH}$

   $\begin{array}{cc} nH_c & = & C\Delta t \\ \text{(methanol)} & & \text{(calorimeter)} \end{array}$

   $3.40\ g \times \dfrac{1\ mol}{32.05\ g} \times \dfrac{726.0\ kJ}{mol} = \dfrac{6.75\ kJ}{°C} \times \Delta t$

   $\Delta t = 11.4°C$

5. $H_2S_{(g)} + \dfrac{3}{2}\ O_{2(g)} \longrightarrow SO_{2(g)} + H_2O_{(g)}$

   $\Delta H_c° = \sum nH°_{fp} - \sum nH°_{fp}$
   $\quad = (1\ mol \times -296.8\ kJ/mol + 1\ mol \times -241.8\ kJ/mol)$
   $\quad\quad - (1\ mol \times -20.6\ kJ/mol + \dfrac{3}{2}\ mol \times 0\ kJ/mol)$
   $\quad = -518.0\ kJ$

   $H_c = \dfrac{-518.0\ kJ}{1\ mol} = -518.0\ kJ/mol$
   $_{H_2S}$

   $\begin{array}{cc} \Delta H_c & = & q \\ {}_{H_2S} & & \text{(water)} \end{array}$

   $nH_c = vc\Delta t$

   $15\ kg \times \dfrac{1\ mol}{34.08\ g} \times \dfrac{518.0\ kJ}{mol} = v \times \dfrac{4.19\ kJ}{L•°C} \times 64°C$

   $v = 0.85\ kL$

51 (ENRICHMENT) $H°_f$ FROM THE FORMATION METHOD

The formation method discussed previously provides the relationship, *representing the law of conservation of energy* to calculate a molar enthalpy of formation from a known enthalpy change, $\Delta H°_r$.

$$\Delta H°_r = \Sigma n H°_{fp} - \Sigma n H°_{fr}$$

As long as the enthalpy change of a reaction is known, this mathematical formula may be used to calculate an unknown molar enthalpy of formation for a compound that appears in the balanced chemical equation. Owing to their technological importance, combustion reactions have been extensively studied and considerable data are available on molar enthalpies of combustion, e.g. in the *CRC Handbook of Chemistry and Physics*. The molar enthalpies of formation of most organic compounds are obtained from combustion data and the relationship given above.

The formation method can be applied to reactions other than combustion. The following example illustrates the formation method of calculating $H°_f$ using a well-known addition reaction to provide the empirical data for the calculation.

EXAMPLE

Ethylene is reacted with chlorine in a calorimeter to form 1,2-dichloroethane. The molar enthalpy of reaction for ethylene is –217.5 kJ/mol. Using this information, calculate the standard molar enthalpy of formation for 1,2-dichloroethane.

$C_2H_{4(g)} + Cl_{2(g)} \rightarrow C_2H_4Cl_{2(l)}$ $\Delta H°_r = 1 \text{ mol} \times -179.4 \text{ kJ/mol} = -179. \text{ kJ}$

$\Delta H°_r = \Sigma n H°_{fp} - \Sigma n H°_{fr}$

$-179.4 \text{ kJ} = (1 \text{ mol} \times H°_f) - (1 \text{ mol} \times +52.5 \text{ kJ/mol} + 1 \text{ mol} \times 0 \text{ kJ/mol})$

$H°_f = -126.9 \text{ kJ/mol}$
$C_2H_4Cl_2$

1. Sodium metal and excess fluorine gas were placed in a bomb calorimeter that has a heat capacity of 4.76 kJ/°C. After the reaction was complete the temperature had risen from 20.67°C to 23.97° and 1.15 g of sodium fluoride had formed. Calculate the molar enthalpy of formation of sodium fluoride from this evidence.

2. The industrial production of stone, glass, clay, and concrete ranks number five in energy use. In several Canadian provinces limestone is mined and then heated in a kiln to produce lime for making mortar and concrete. Using only the following information, what is the standard molar enthalpy of formation of calcium carbonate?
   (1) $CaCO_{3(s)} \rightarrow CaO_{(s)} + CO_{2(g)}$ $\Delta H°_r = +178.5$ kJ
   (2) $2 \ Ca_{(s)} + O_{2(g)} \rightarrow 2 \ CaO_{(s)}$ $\Delta H°_f = -1269.8$ kJ
   (3) $C_{(s)} + O_{2(g)} \rightarrow CO_{2(g)}$ $\Delta H°_f = -393.5$ kJ

3. The *CRC Handbook of Chemistry and Physics* lists the molar heat (enthalpy) of combustion for hexane as –4163.1 kJ/mol with initial and final measurements made at SATP. Carbon dioxide gas and liquid water are the final products. Predict the standard molar enthalpy of formation for hexane, $C_6H_{14(l)}$.

4. The production of chemicals is the second major use of energy by industry. Chemicals, such as fertilizers, are an integral part of our technological lifestyle. Superphosphate fertilizer is produced by the following reaction.
   $Ca_3(PO_4)_{2(s)} + 2 \ H_2SO_{4(l)} \rightarrow Ca(H_2PO_4)_{2(s)} + 2 \ CaSO_{4(s)}$
   The standard molar enthalpy of formation for calcium dihydrogen phosphate is –3104.7 kJ/mol, the molar enthalpy of formation for calcium sulfate is –1434.1 kJ/mol, and the enthalpy change of the given equation is –224.1 kJ. Use the given information to calculate the standard molar enthalpy of formation for calcium phosphate.

51 (ENRICHMENT) $H°_f$ FROM THE FORMATION METHOD

1. $Na_{(s)} + \frac{1}{2} F_{2(g)} \longrightarrow NaF_{(s)}$

   (*Since this is a formation reaction, $\Delta H°_f = \Delta H_r$.*)

   $$\Delta H_f = q$$
   $$nH_f = C\Delta t$$

   $$1.15 \text{ g} \times \frac{1 \text{ mol}}{41.99 \text{ g}} \times H_f = \frac{4.76 \text{ kJ}}{°C} \times (23.77 - 20.67)°C$$

   $$H_f = -574 \text{ kJ/mol}$$
   $$\text{NaF}$$

   From calorimetry, the molar enthalpy of formation of sodium fluoride is −574 kJ/mol.

2. from (1) $CaO_{(s)} + CO_{2(g)} \longrightarrow CaCO_{3(s)}$ $\Delta H° = -178.5 \text{ kJ}$

   from (2) $Ca_{(s)} + \frac{1}{2} O_{2(g)} \longrightarrow CaO_{(s)}$ $\Delta H° = -634.9 \text{ kJ}$

   from (3) $C_{(s)} + O_{2(g)} \longrightarrow CO_{2(g)}$ $\Delta H° = -393.5 \text{ kJ}$

   net $Ca_{(s)} + C_{(s)} + \frac{3}{2} O_{2(g)} \longrightarrow CaCO_{3(s)}$ $\Delta H° = -1206.9 \text{ kJ}$

   $$H°_f = \frac{-1206.9 \text{ kJ}}{1 \text{ mol}} = -1206.9 \text{ kJ/mol}$$
   $$\text{CaCO}_3$$

   From Hess's law, the standard molar enthalpy of formation of calcium carbonate is −1206.9 kJ/mol.

3. $C_6H_{14(l)} + \frac{19}{2} O_{2(g)} \longrightarrow 6 CO_{2(g)} + 7 H_2O_{(l)}$ $\Delta H°_c = -4163.1 \text{ kJ}$

   (*For simplicity, the combustion equation is balanced for one mole of hexane.*)

   $$\Delta H°_c = \Sigma nH°_{fp} - \Sigma nH°_{fr}$$
   $$-4163.1 \text{ kJ} = (6 \text{ mol} \times -393.5 \text{ kJ/mol} + 7 \text{ mol} \times -285.8 \text{ kJ/mol})$$
   $$- (1 \text{ mol} \times H°_f + \frac{19}{2} \text{ mol} \times 0 \text{ kJ/mol})$$

   $$H°_f = -198.5 \text{ kJ/mol}$$
   $$\text{C}_6\text{H}_{14}$$

   From the formation method, the standard molar enthalpy of formation of hexane is −198.5 kJ/mol.

4. $Ca_3(PO_4)_{2(s)} + 2 H_2SO_{4(l)} \longrightarrow Ca(H_2PO_4)_{2(s)} + 2 CaSO_{4(s)}$ $\Delta H_r = -224.1 \text{ kJ}$

   $$\Delta H°_r = \Sigma nH°_{fp} - \Sigma nH°_{fr}$$
   $$-224.1 \text{ kJ} = (1 \text{ mol} \times -3104.7 \text{ kJ/mol} + 2 \text{ mol} \times -1434.1 \text{ kJ/mol})$$
   $$- (1 \text{ mol} \times H°_f + 2 \text{ mol} \times -814.0 \text{ kJ/mol})$$

   $$H°_f = -4120.8 \text{ kJ/mol}$$
   $$\text{Ca}_3(\text{PO}_4)_2$$

   From the formation method, the standard molar enthalpy of formation of calcium phosphate is −4120.8 kJ/mol.

52 (ENRICHMENT) BOND ENERGIES

According to current theories, chemical reactions are believed to involve the breaking and forming of chemical bonds. Based on the theoretical concept of a bond, energy is required to break bonds holding particles together and energy is released when new chemical bonds are formed between two particles. This concept is summarized by the following word equation.

$$\text{bonded particles} + \text{energy} \rightleftharpoons \text{separated particles}$$

This concept is supported by evidence of bond dissociation energies. A standard molar enthalpy or *molar bond energy*, H°_{be}, is defined as the energy required to break one mole of bonds in a substance at SATP. Average molar bond energies are available from the study of elements and simple compounds. (See table to the right.) Assuming a chemical reaction is a series of reactions involving the breaking of bonds in the reactants (an endothermic process) followed by the forming of new bonds in the products (an exothermic process), Hess's law may be expressed as follows.

$$\Delta H^\circ_r = \Sigma \Delta H^\circ_{be}$$

where $\Delta H^\circ_{be} = n H^\circ_{be}$ for each bond type

Using average bond energies from a reference table, a good approximation of the enthalpy change of a given chemical reaction may be obtained.

AVERAGE STANDARD MOLAR BOND ENERGIES	
Bond Type	**Molar Bond Energy, H°_{be},(kJ/mol)***
H — H	436
C — H	415
C — C	348
C = C	612
C ≡ C	838
H — Cl	432
Cl — Cl	243
O = O	499
O — H	460
C = O	799 (in CO_2)

* at SATP relative to atoms

Although this method relies on empirical molar bond energies, it represents a combination of empirically-determined and theoretical concepts.

An explanation of a simple chemical reaction involves changes in potential energy as bonds are broken and formed. A potential graph is often used to represent the energy changes during the reaction. The x-axis or reaction coordinate represents the progress of the reaction from the initial reactants to hypothetical, intermediate species to the final products.

For example, the formation of hydrogen chloride is believed to involve the breaking of the bonds of the diatomic reactant molecules to form hypothetical hydrogen and chlorine atoms. These atoms then release energy as they recombine to form hydrogen-chlorine bonds in the product.

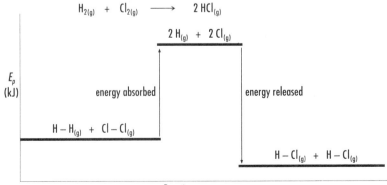

The Reaction of Hydrogen and Chlorine

52 (ENRICHMENT) BOND ENERGIES

According to current bonding theories, energy is absorbed to break the bonds in one mole of hydrogen and one mole of chlorine molecules. Note that $H°_{be}$ is positive.

$H_{2(g)} \rightarrow H_{(g)} + H_{(g)}$ $\Delta H°_{be} = 1 \text{ mol} \times +436 \text{ kJ/mol} = +436 \text{ kJ}$

$Cl_{2(g)} \rightarrow Cl_{(g)} + Cl_{(g)}$ $\Delta H°_{be} = 1 \text{ mol} \times +243 \text{ kJ/mol} = +243 \text{ kJ}$

According to current bonding theories, energy is released when hydrogen and chlorine atoms bond to form hydrogen chloride molecules. Note that $H°_{be}$ is negative.

$2 H_{(g)} + 2 Cl_{(g)} \rightarrow 2 HCl_{(g)}$ $\Delta H°_{be} = 2 \text{ mol} \times -432 \text{ kJ/mol} = -864 \text{ kJ}$

According to Hess's law, the enthalpy change of the net reaction can be calculated by adding the enthalpy changes of the individual bond-breaking and bond-forming reactions.

$H_{2(g)} + Cl_{2(g)} \rightarrow 2 HCl_{(g)}$ $\Delta H_f = \Sigma \Delta H°_{be} = +436 \text{ kJ} + 234 \text{ kJ} + -864 \text{ kJ} = -185 \text{ kJ}$

The molar enthalpy of formation for hydrogen chloride can now be calculated.

$$H°_f = \frac{-185 \text{ kJ}}{2 \text{ mol}} = -93 \text{ kJ/mol}$$

This molar enthalpy of formation from the enthalpy change using bond energies agrees well with the data table value, i.e. −92.3 kJ/mol, obtained using a direct calorimetric measurement.

EXAMPLE

The combustion of methane in natural gas is a common energy source for heating homes. Use bond energies to predict the molar enthalpy of combustion of methane.

$CH_{4(g)} + 2 O_{2(g)} \rightarrow CO_{2(g)} + 2 H_2O_{(g)}$

$$\begin{aligned}
\Delta H°_c = \Sigma \Delta H°_{be} &= \Sigma n H°_{be} \\
&= 4 \text{ mol} \times 415 \text{ kJ/mol} + 2 \text{ mol} \times 499 \text{ kJ/mol} \\
&\quad + 2 \text{ mol} \times -799 \text{ kJ/mol} + 4 \text{ mol} \times -460 \text{ kJ/mol} \\
&= -780 \text{ kJ}
\end{aligned}$$

$$H°_c = \frac{\Delta H°_c}{n} = \frac{-780 \text{ kJ}}{1 \text{ mol}} = -780 \text{ kJ/mol}$$

From bond energies, according to the bond energies method, the molar enthalpy of combustion of methane is −780 kJ/mol.

1. Propane gas is used for heating homes and for fueling automobiles and barbecues. Use bond energies to predict the molar enthalpy of combusion of propane.

   (−1963 kJ/mol)

2. Acetylene gas is used in oxy-acetylene torches to cut metals. Use bond energies to predict the molar enthalpy of combustion of acetylene.

   (−1200 kJ/mol)

3. Hydrogen gas can be used to fuel rockets and automobiles. Predict the molar enthalpy of combustion of hydrogen by using bond energies.

   (−235 kJ/mol)

53 REDOX TABLE BUILDING

1. Consider the following redox equations, which represent spontaneous reactions in an experiment similar to Investigations 12.2, i.e. metals reacting with cations. From this evidence, set up a table of relative strengths of oxidizing and reducing agents. Write half-reaction equations and label the strongest oxidizing agent and reducing agent.

$$Co_{(s)} + Pd^{2+}_{(aq)} \xrightarrow{spont.} Co^{2+}_{(aq)} + Pd_{(s)}$$

$$Pd_{(s)} + Pt^{2+}_{(aq)} \xrightarrow{spont.} Pd^{2+}_{(aq)} + Pt_{(s)}$$

$$Mg_{(s)} + Co^{2+}_{(aq)} \xrightarrow{spont.} Mg^{2+}_{(aq)} + Co_{(s)}$$

2. The following equations are interpretations of the evidence from the reactions of four metals with various cation solutions. Make a table of redox half-reactions and arrange the four metallic ions and the hydrogen ion in order of their decreasing tendency to react.

$$Cd_{(s)} + 2 H^+_{(aq)} \xrightarrow{spont.} Cd^{2+}_{(aq)} + H_{2(g)}$$

$$Hg_{(s)} + 2 H^+_{(aq)} \xrightarrow{non\text{-}spont.} Hg^{2+}_{(aq)} + H_{2(g)}$$

$$Be_{(s)} + Cd^{2+}_{(aq)} \xrightarrow{spont.} Be^{2+}_{(aq)} + Cd_{(s)}$$

$$Ca^{2+}_{(aq)} + Be_{(s)} \xrightarrow{non\text{-}spont.} Ca_{(s)} + Be^{2+}_{(aq)}$$

3. In an experiment similar to Investigation 12.2, four metals were placed into test tubes containing various ion solutions. Their resulting behavior is communicated by the equations below. Create a half-reaction table and order the metallic ions and the hydrogen ion according to their tendency to react. Label the strongest oxidizing agent and reducing agent.

$$Pt_{(s)} + 2 H^+_{(aq)} \xrightarrow{non\text{-}spont.} Pt^{2+}_{(aq)} + H_{2(g)}$$

$$2 Ce_{(s)} + 3 Ni^{2+}_{(aq)} \xrightarrow{spont.} 2 Ce^{3+}_{(aq)} + 3 Ni_{(s)}$$

$$3 Sr_{(s)} + 2 Ce^{3+}_{(aq)} \xrightarrow{spont.} 3 Sr^{2+}_{(aq)} + 2 Ce_{(s)}$$

$$Ni_{(s)} + 2 H^+_{(aq)} \xrightarrow{spont.} Ni^{2+}_{(aq)} + H_{2(g)}$$

4. A student is required to store an aqueous solution of iron(III) nitrate. She has a choice of a copper, tin, iron, or silver container. Use the table of redox half-reactions on page 552 of *Nelson Chemistry* and an appropriate generalization to predict which container would be most suitable for storing the solution?

5. An analytical chemist reacts an unknown metal X with a copper(II) sulfate solution, plating out copper metal. Metal X does not react with aqueous zinc nitrate. What is the order for these metal ions in decreasing tendency to react? What groups of metals are eliminated as a possible identity of the unknown metal? What other solutions might next be chosen to help identify the metal?

© NELSON CANADA,
A DIVISION OF THOMSON CANADA LIMITED, 1994

53 REDOX TABLE BUILDING

1. According to the evidence and the spontaneity rule, a table of oxidizing and reducing agents in order of strength is developed as follows.

SOA $Pt^{2+}_{(aq)} + 2e^- \rightleftharpoons Pt_{(s)}$
  $Pd^{2+}_{(aq)} + 2e^- \rightleftharpoons Pd_{(s)}$
  $Co^{2+}_{(aq)} + 2e^- \rightleftharpoons Co_{(s)}$
  $Mg^{2+}_{(aq)} + 2e^- \rightleftharpoons Mg_{(s)}$ SRA

2. According to the evidence and the spontaneity rule, the following table of redox half-reactions is developed.

SOA $Hg^{2+}_{(aq)} + 2e^- \rightleftharpoons Hg_{(s)}$
  $2 H^+_{(aq)} + 2e^- \rightleftharpoons H_{2(g)}$
  $Cd^{2+}_{(aq)} + 2e^- \rightleftharpoons Cd_{(s)}$
  $Be^{2+}_{(aq)} + 2e^- \rightleftharpoons Be_{(s)}$
  $Ca^{2+}_{(aq)} + 2e^- \rightleftharpoons Ca_{(s)}$ SRA

3. According to the evidence and the spontaneity rule, the following table of redox half-reactions is created.

SOA $Pt^{2+}_{(aq)} + 2e^- \rightleftharpoons Pt_{(s)}$
  $2 H^+_{(aq)} + 2e^- \rightleftharpoons H_{2(g)}$
  $Ni^{2+}_{(aq)} + 2e^- \rightleftharpoons Ni_{(s)}$
  $Ce^{3+}_{(aq)} + 3e^- \rightleftharpoons Ce_{(s)}$
  $Sr^{2+}_{(aq)} + 2e^- \rightleftharpoons Sr_{(s)}$ SRA

4. Based upon the spontaneity rule, she should choose the silver container for storing the iron(III) nitrate solution. This is, of course, from a scientific perspective. From an economic perspective, one might be concerned about the cost of using a silver container to store an inexpensive solution which could also be stored in a glass container.

5. According to the evidence and the spontaneity rule, the metal ions in order of decreasing tendency to react are listed below.

$Cu^{2+}_{(aq)} + 2e^- \rightleftharpoons Cu_{(s)}$
$X^{2+}_{(aq)} + 2e^- \rightleftharpoons X_{(s)}$
$Zn^{2+}_{(aq)} + 2e^- \rightleftharpoons Zn_{(s)}$

Metal X is between copper and zinc in a table of redox half-reactions. This eliminates Groups 1, 2, and 3 metals as a possible identity for Metal X. Groups 1, 2, and 3 metals are stronger reducing agents than zinc. The aqueous solutions of metal ions between copper and zinc ions in the table of redox half-reactions on page 552 may be used to help narrow down the possibility of what Metal X may be. Some possible solutions that can be used are lead(II), nickel(II), cobalt(II), cadmium(II), chromium(III), and iron(III).

54 PREDICTING REDOX REACTIONS

For each of the following questions, use the five-step method outlined in the *Nelson Chemistry* text to predict or describe and communicate the most likely redox reaction.

1. An aqueous solution of potassium permanganate was reacted with an acidic solution of sodium bromide and an orange-brown substance was formed.

2. A strip of silver metal is placed in a solution of aqueous nickel(II) chloride.

3. Liquid mercury is mixed with a paste of acidic manganese(IV) oxide.

4. Hydrogen peroxide and silver nitrate are mixed.

5. Potassium metal is placed into some water.

6. In a car battery lead and lead(IV) oxide electrodes are exposed to a sulfuric acid electrolyte. (Assume that the sulfuric acid ionizes to hydrogen and sulfate ions.)

54 PREDICTING REDOX REACTIONS

For each of the following questions, use the five-step method outlined in the *Nelson Chemistry* text to predict or describe and communicate the most likely redox reaction. (**Only the net redox reaction is shown below.**)

1. An aqueous solution of potassium permanganate was reacted with an acidic solution of sodium bromide and an orange-brown substance was formed.

$$2\ MnO_4^-{}_{(aq)}\ +\ 16\ H^+{}_{(aq)}\ +\ 10\ Br^-{}_{(aq)}\ \overset{spont.}{\longrightarrow}\ 5\ Br_2{}_{(l)}\ +\ 2\ Mn^{2+}{}_{(aq)}\ +\ 8\ H_2O_{(l)}$$

2. A strip of silver metal is placed in a solution of aqueous nickel(II) chloride.

$$2\ Ag_{(s)}\ +\ Ni^{2+}{}_{(aq)}\ \overset{non\text{-}spont.}{\longrightarrow}\ Ni_{(s)}\ +\ 2\ Ag^+{}_{(aq)}$$

3. Liquid mercury is mixed with a paste of acidic manganese(IV) oxide.

$$MnO_2{}_{(s)}\ +\ 4\ H^+{}_{(aq)}\ +\ Hg_{(l)}\ \overset{spont.}{\longrightarrow}\ Hg^{2+}{}_{(aq)}\ +\ Mn^{2+}{}_{(aq)}\ +\ 2\ H_2O_{(l)}$$

4. Hydrogen peroxide and silver nitrate are mixed.

$$H_2O_2{}_{(l)}\ +\ 2\ Ag^+{}_{(aq)}\ \overset{spont.}{\longrightarrow}\ 2\ Ag_{(s)}\ +\ O_2{}_{(g)}\ +\ 2\ H^+{}_{(aq)}$$

5. Potassium metal is placed into some water.

$$2\ K_{(s)}\ +\ 2\ H_2O_{(l)}\ \overset{spont.}{\longrightarrow}\ H_2{}_{(g)}\ +\ 2\ K^+{}_{(aq)}\ +\ 2\ OH^-{}_{(aq)}$$

6. In a car battery lead and lead(IV) oxide electrodes are exposed to a sulfuric acid electrolyte. (Assume that the sulfuric acid ionizes to hydrogen and sulfate ions.)

$$Pb_{(s)}\ +\ PbO_2{}_{(s)}\ +\ 4\ H^+{}_{(aq)}\ +\ 2\ SO_4^{2-}{}_{(aq)}\ \overset{spont.}{\longrightarrow}\ 2\ PbSO_4{}_{(s)}\ +\ 2\ H_2O_{(l)}$$

55 REDOX TITRATION

Complete the Analysis for each of the following redox titrations.

Problem 1
What is the concentration of the chromium(II) ions in a solution obtained from the analysis of a stainless steel fork?

Experimental Design
A 0.0114 mol/L potassium permanganate solution is titrated to oxidize 25.0 mL of chromium(II) ions to chromium(III) ions in an acidic solution.

Evidence

VOLUMES OF POTASSIUM PERMANGANATE SOLUTION REQUIRED				
Trial	1	2	3	4
Final buret reading (mL)	14.2	25.4	36.5	47.6
Initial buret reading (mL)	2.7	14.2	25.4	36.5

Analysis

Problem 2
What is the concentration of the nickel(II) ions in a solution obtained in the analysis of nichrome wire?

Experimental Design
A 0.0105 mol/L potassium permanganate solution is titrated to oxidize 10.0 mL of nickel(II) ions to nickel(III) ions in an acidic solution.

Evidence

VOLUMES OF POTASSIUM PERMANGANATE SOLUTION REQUIRED				
Trial	1	2	3	4
Final buret reading (mL)	15.8	28.1	40.4	42.9
Initial buret reading (mL)	2.7	15.8	28.1	30.7

Analysis

55 REDOX TITRATION

Problem 3
What is the concentration of the tin(II) ions in a solution obtained from the analysis of a solder wire?

Experimental Design
An acidic iron(II) ammonium sulfate-6-water $(Fe(NH_4)_2(SO_4)_2 \cdot 6H_2O)$ solution was prepared as a primary standard and 10.0 mL of the solution is titrated with a potassium dichromate solution. The standardized potassium dichromate solution is titrated to oxidize 10.0 mL of tin(II) ions in an acidic solution.

Evidence
Mass of primary standard used to prepare 100.0 mL of solution = 1.87 g

VOLUMES OF POTASSIUM DICHROMATE SOLUTION REQUIRED FOR PRIMARY STANDARD				
Trial	1	2	3	4
Final buret reading (mL)	12.5	23.5	34.5	45.5
Initial buret reading (mL)	1.2	12.5	23.5	34.5

VOLUMES OF POTASSIUM DICHROMATE SOLUTION REQUIRED FOR TIN(II) IONS				
Trial	1	2	3	4
Final buret reading (mL)	11.6	22.1	32.6	43.1
Initial buret reading (mL)	0.8	11.6	22.1	32.6

Analysis

56 OXIDATION NUMBERS

1. Assign oxidation numbers to chlorine in each of the following chemicals.

   $HCl_{(aq)}$, $Cl_{2(g)}$, $NaClO_{(s)}$, $Cl^-_{(aq)}$, $HClO_{3(aq)}$, $ClO_3^-_{(aq)}$, $KClO_{2(s)}$, $ClO_{2(g)}$, $HClO_{4(aq)}$

2. Assign oxidation numbers to manganese in each of the following chemicals.

   $MnO_{2(s)}$, $KMnO_{4(s)}$, $Mn_{(s)}$, $MnO_4^{2-}_{(aq)}$, $MnCl_{2(s)}$, $Mn_2O_{7(s)}$, $Mn^{2+}_{(aq)}$

For the following reaction equations, use oxidation numbers to identify the oxidizing agent, the reducing agent, and to balance the equation. Show your complete problem solving approach as outlined in *Nelson Chemistry*.

3. $AsO_3^{3-}_{(aq)}$ + $IO_3^-_{(aq)}$ $\rightarrow$ $AsO_4^{3-}_{(aq)}$ + $I^-_{(aq)}$

4. $CuO_{(s)}$ + $NH_{3(g)}$ $\rightarrow$ $N_{2(g)}$ + $H_2O_{(l)}$ + $Cu_{(s)}$

5. $MnO_4^-_{(aq)}$ + $H_2Se_{(g)}$ + $H^+_{(aq)}$ $\rightarrow$ $Se_{(s)}$ + $Mn^{2+}_{(aq)}$ + $H_2O_{(l)}$

6. $PbO_{2(s)}$ + $Pb_{(s)}$ + $H_2SO_{4(aq)}$ $\rightarrow$ $PbSO_{4(s)}$ + $H_2O_{(l)}$

7. $Cl_{2(g)}$ + $OH^-_{(aq)}$ $\rightarrow$ $ClO_3^-_{(aq)}$ + $Cl^-_{(aq)}$ + $H_2O_{(l)}$

56 OXIDATION NUMBERS

1. Assign oxidation numbers to chlorine in each of the following chemicals.

$$\overset{-1}{HCl_{(aq)}}, \quad \overset{0}{Cl_{2(g)}}, \quad \overset{+1}{NaClO_{(s)}}, \quad \overset{-1}{Cl^-_{(aq)}}, \quad \overset{+5}{HClO_{3(aq)}}, \quad \overset{+5}{ClO_3^-_{(aq)}}, \quad \overset{+3}{KClO_{2(s)}}, \quad \overset{+4}{ClO_{2(g)}}, \quad \overset{+7}{HClO_{4(aq)}}$$

2. Assign oxidation numbers to manganese in each of the following chemicals.

$$\overset{+4}{MnO_{2(s)}}, \quad \overset{+7}{KMnO_{4(s)}}, \quad \overset{0}{Mn_{(s)}}, \quad \overset{+6}{MnO_4^{2-}_{(aq)}}, \quad \overset{+2}{MnCl_{2(s)}}, \quad \overset{+7}{Mn_2O_{7(s)}}, \quad \overset{+2}{Mn^{2+}_{(aq)}}$$

For the following reaction equations, use oxidation numbers to identify the oxidizing agent, the reducing agent, and to balance the equation. Show your complete problem solving approach as outlined in *Nelson Chemistry*.

3. $3 \overset{+3\ -2}{AsO_3^{3-}_{(aq)}} + \overset{+5-2}{IO_3^-_{(aq)}} \rightarrow 3 \overset{+5-2}{AsO_4^{3-}_{(aq}} + \overset{-1}{I^-_{(aq)}}$

$\quad\ \ \overset{+3}{} \qquad\qquad\quad \overset{+5}{} \qquad\qquad\ \overset{+5}{} \qquad\qquad \overset{-1}{}$

$2e^-/As \qquad\qquad 6e^-/I$

$2e^-/AsO_3^{3-} \qquad 6e^-/IO_3^-$

$\mathbf{RA} \qquad\qquad\qquad \mathbf{OA}$

4. $3 \overset{+2\ -2}{CuO_{(s)}} + 2 \overset{-3+1}{NH_{3(g)}} \rightarrow \overset{0}{N_{2(g)}} + 3 \overset{+1\ -2}{H_2O_{(l)}} + 3 \overset{0}{Cu_{(s)}}$

$\quad\ \ \overset{+2}{} \qquad\qquad \overset{-3}{} \qquad\qquad \overset{0}{} \qquad\qquad\qquad\qquad \overset{0}{}$

$2e^-/Cu \qquad\qquad 3e^-/N$

$2e^-/CuO \qquad\qquad 3e^-/NH_3$

$\mathbf{OA} \qquad\qquad\qquad \mathbf{RA}$

5. $2 \overset{+7\ -2}{MnO_4^-_{(aq)}} + 5 \overset{+1\ -2}{H_2Se_{(g)}} + 6 \overset{+1}{H^+_{(aq)}} \rightarrow 5 \overset{0}{Se_{(s)}} + 2 \overset{+2}{Mn^{2+}_{(aq)}} + 8 \overset{+1\ -2}{H_2O_{(l)}}$

$\quad\ \ \overset{+7}{} \qquad\qquad\quad \overset{-2}{} \qquad\qquad\qquad\quad \overset{0}{} \qquad\quad \overset{+2}{}$

$5e^-/Mn \qquad\qquad 2e^-/Se$

$5e^-/MnO_4^- \qquad\quad 2e^-/H_2Se$

$\mathbf{OA} \qquad\qquad\qquad \mathbf{RA}$

6. $\overset{+4\ -2}{PbO_{2(s)}} + \overset{0}{Pb_{(s)}} + 2 \overset{+1\ +6\ -2}{H_2SO_{4(aq)}} \rightarrow 2 \overset{+2\ +6-2}{PbSO_{4(s)}} + 2 \overset{+1\ -2}{H_2O_{(l)}}$

$\quad\ \ \overset{+4}{} \qquad\quad \overset{0}{} \qquad\qquad\qquad\qquad\qquad \overset{+2}{}$

$2e^-/Pb \qquad\quad 2e^-/Pb$

$2e^-/PbO_2 \qquad 2e^-/Pb$

$\mathbf{OA} \qquad\qquad \mathbf{RA}$

(Note that the lead(IV) ion and the lead atom both are converted into lead(II) ions.)

7. $6 \overset{0}{Cl_{2(g)}} + 12 \overset{-2+1}{OH^-_{(aq)}} 2 \overset{+5\ -2}{ClO_3^-_{(aq)}} + 10 \overset{-1}{Cl^-_{(aq)}} + 6 \overset{+1\ -2}{H_2O_{(l)}}$

$\quad \overset{0}{} \qquad\qquad\qquad\qquad \overset{+5}{} \qquad\qquad \overset{-1}{}$

$5e^-/Cl \qquad\qquad 1e^-/Cl$

$10e^-/Cl_2 \qquad\qquad 1e^-/Cl_2$

$\mathbf{RA} \qquad\qquad\qquad \mathbf{OA}$

(You might find it easier to work on this question if you write $Cl_{2(g)}$ twice and then recombine the two later. Note that in basic solutions, it is most efficient to balance electrons lost and gained, balance oxidizing and reducing agents, and then balance charge with the hydroxide ions before balancing oxygen and hydrogen atoms.)

57 (ENRICHMENT) BALANCING "SKELETON" REDOX REACTION EQUATIONS

Skeleton equations for redox half-reactions can be balanced by adding electrons, $H^+_{(aq)}$ or $OH^-_{(aq)}$, and $H_2O_{(l)}$. After balancing electrons lost and gained, the half-reactions can be added to obtain a balanced net reaction equation.

One procedure is:

1. Write the "skeleton" equation to represent the reactants and products of the net reaction.

2. Split the net reaction into two half-reactions, and balance atoms other than H and O.

3. Use oxidation numbers to add electrons to the appropriate side of the reduction and oxidation half-reaction equations.

4. If necessary, balance charge for each half-reaction equation by adding $H^+_{(aq)}$ for acidic solutions or $OH^-_{(aq)}$ for basic solutions.

5. If necessary, balance the hydrogen and oxygen atoms by adding $H_2O_{(l)}$ to each half-reaction equation.

6. Multiply the two half-reaction equations by appropriate coefficients to balance electrons gained and lost.

7. Add the two half-reactions and cancel electrons (and other entities where appropriate).

EXAMPLE

The alcohol $(C_2H_5OH_{(aq)})$ content of blood can be determined by adding an acidic dichromate solution to a blood sample. The main products are acetic acid and chromium(III) ions. Write a balanced, net redox reaction equation.

$$Cr_2O_7^{2-}{}_{(aq)} + C_2H_5OH_{(aq)} \longrightarrow CH_3COOH_{(aq)} + Cr^{3+}_{(aq)}$$

$$2\,[14H^+_{(aq)} + 6\,e^- + \overset{+6\ -2}{Cr_2O_7^{2-}{}_{(aq)}} \longrightarrow 2\,\overset{+3}{Cr^{3+}_{(aq)}} + 7\,H_2O_{(l)}]$$
$$\underset{+6}{}\qquad\underset{+3}{}$$

$$3\,[H_2O_{(l)} + \overset{-2\ +1\ -2\ +1}{C_2H_5OH_{(aq)}} \longrightarrow \overset{0\ +1\ 0\ -2\ -2\ +1}{CH_3COOH_{(aq)}} + 4\,e^- + 4\,H^+_{(aq)}]$$
$$\underset{-2}{}\qquad\underset{0\quad 0}{}$$

$$2\,Cr_2O_7^{2-}{}_{(aq)} + 16\,H^+_{(aq)} + 3\,C_2H_5OH_{(aq)} \longrightarrow 3\,CH_3COOH_{(aq)} + 4\,Cr^{3+}_{(aq)} + 11\,H_2O_{(l)}$$

Balance the following redox equations.

1. $Sn^{2+}_{(aq)} + Fe^{3+}_{(aq)} \longrightarrow Fe^{2+}_{(aq)} + Sn^{4+}_{(aq)}$

2. $CH_3OH_{(aq)} + MnO_4^-{}_{(aq)} \longrightarrow Mn^{2+}_{(aq)} + CH_2O_{(aq)}$ (acidic)

3. $Fe^{2+}_{(aq)} + Cr_2O_7^{2-}{}_{(aq)} \longrightarrow Cr^{3+}_{(aq)} + Fe^{3+}_{(aq)}$ (acidic)

4. $CH_3OH_{(aq)} + MnO_4^-{}_{(aq)} \longrightarrow MnO_4^{2-}{}_{(aq)} + CO_3^{2-}{}_{(aq)}$ (basic)

5. $MnO_4^-{}_{(aq)} + SO_3^{2-}{}_{(aq)} \longrightarrow SO_4^{2-}{}_{(aq)} + MnO_{2(s)}$ (basic)

57 (ENRICHMENT) BALANCING "SKELETON" REDOX REACTION EQUATIONS

1. $2 [Fe^{3+}_{(aq)} + e^- \rightarrow Fe^{2+}_{(aq)}]$
          $Sn^{2+}_{(aq)} \rightarrow Sn^{4+}_{(aq)} + 2 e^-$
    _____
    $2 Fe^{3+}_{(aq)} + Sn^{2+}_{(aq)} \rightarrow Sn^{4+}_{(aq)} + 2 Fe^{2+}_{(aq)}$

2. $2 [8 H^+_{(aq)} + 5 e^- + MnO_4^-_{(aq)} \rightarrow Mn^{2+}_{(aq)} + 4 H_2O_{(l)}]$
          $5 [CH_3OH_{(aq)} \rightarrow CH_2O_{(aq)} + 2 e^- + 2 H^+_{(aq)}]$
    _____
    $6 H^+_{(aq)} + 2 MnO_4^-_{(aq)} + 5 CH_3OH_{(aq)} \rightarrow 5 CH_2O_{(aq)} + 2 Mn^{2+}_{(aq)} + 8 H_2O_{(l)}$

3. $14 H^+_{(aq)} + 6 e^- + Cr_2O_7^{2-}_{(aq)} \rightarrow 2 Cr^{3+}_{(aq)} + 7 H_2O_{(l)}$
          $6 [Fe^{2+}_{(aq)} \rightarrow Fe^{3+}_{(aq)} + e^-]$
    _____
    $14 H^+_{(aq)} + Cr_2O_7^{2-}_{(aq)} + 6 Fe^{2+}_{(aq)} \rightarrow 6 Fe^{3+}_{(aq)} + 2 Cr^{3+}_{(aq)} + 7 H_2O_{(l)}$

4. $6 [MnO_4^-_{(aq)} + e^- \rightarrow MnO_4^{2-}_{(aq)}]$
        $8 OH^-_{(aq)} + CH_3OH_{(aq)} \rightarrow CO_3^{2-}_{(aq)} + 6 e^- + 6 H_2O_{(l)}$
    _____
    $6 MnO_4^-_{(aq)} + CH_3OH_{(aq)} + 8 OH^-_{(aq)} \rightarrow CO_3^{2-}_{(aq)} + 6 H_2O_{(l)} + 6 MnO_4^{2-}_{(aq)}$

5. $2 [2 H_2O_{(l)} + 3 e^- + MnO_4^-_{(aq)} \rightarrow MnO_{2(s)} + 4 OH^-_{(aq)}]$
          $3 [2 OH^-_{(aq)} + SO_3^{2-}_{(aq)} \rightarrow SO_4^{2-}_{(aq)} + 2 e^- + H_2O_{(l)}]$
    _____
    $H_2O_{(l)} + 2 MnO_4^-_{(aq)} + 3 SO_3^{2-}_{(aq)} \rightarrow 3 SO_4^{2-}_{(aq)} + 2 MnO_{2(s)} + 2 OH^-_{(aq)}$

58 VOLTAIC CELLS

For each of the following cells, use the given cell notation to identify the strongest oxidizing and reducing agents. Write chemical equations to represent the cathode, anode and net cell reactions. Label electrodes, electrolytes, electron flow, and ion movement. Predict the cell potential.

1. $Cd_{(s)}$ | $Cd(NO_3)_{2(aq)}$ || $AgNO_{3(aq)}$ | $Ag_{(s)}$

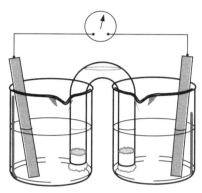

2. $Pt_{(s)}$ | $IO_3^-{}_{(aq)}$, $H^+{}_{(aq)}$ || $Zn^{2+}{}_{(aq)}$ | $Zn_{(s)}$

58 VOLTAIC CELLS

For each of the following cells, use the given cell notation to identify the strongest oxidizing and reducing agents. Write chemical equations to represent the cathode, anode and net cell reactions. Label electrodes, electrolytes, electron flow, and ion movement. Predict the cell potential.

1. $Cd_{(s)} \mid Cd(NO_3)_{2(aq)} \parallel AgNO_{3(aq)} \mid Ag_{(s)}$

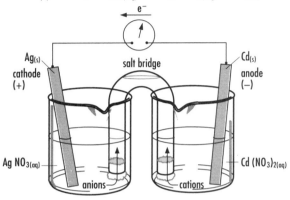

	OA	SOA	
$Cd_{(s)}$	$\mid Cd(NO_3)_{2(aq)} \parallel$	$AgNO_{3(aq)} \mid Ag_{(s)}$	
SRA		RA	
cathode		$2\,[\,Ag^+_{(aq)} + e^- \rightarrow Ag_{(s)}\,]$	$E°_r = +0.80$ V
anode		$Cd_{(s)} \rightarrow Cd^{2+}_{(aq)} + 2\,e^-$	$E°_r = -0.40$ V
net		$2\,Ag^+_{(aq)} + Cd_{(s)} \rightarrow 2\,Ag_{(s)} + Cd^{2+}_{(aq)}$	$\Delta E°_r = +1.20$ V

2. $Pt_{(s)} \mid IO_3^-_{(aq)}, H^+_{(aq)} \parallel Zn^{2+}_{(aq)} \mid Zn_{(s)}$

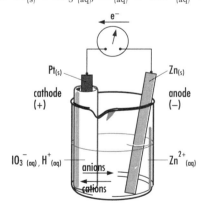

	SOA	OA	
$Pt_{(s)}$	$\mid IO_3^-_{(aq)}, H^+_{(aq)} \parallel$	$Zn^{2+}_{(aq)} \mid Zn_{(s)}$	
RA		SRA	
cathode		$2\,IO_3^-_{(aq)} + 12\,H^+_{(aq)} + 10\,e^- \rightarrow I_{2(s)} + 6\,H_2O_{(l)}$	$E°_r = +1.20$ V
anode		$5\,[\,Zn_{(s)} \rightarrow Zn^{2+}_{(aq)} + 2\,e^-\,]$	$E°_r = -0.76$ V
net		$2\,IO_3^-_{(aq)} + 12\,H^+_{(aq)} + 5\,Zn_{(s)} \rightarrow$	$\Delta E° = +1.96$ V
		$I_{2(s)} + 6\,H_2O_{(l)} + 5\,Zn^{2+}_{(aq)}$	

59 STANDARD CELLS AND CELL POTENTIALS

1. Assume that the reference half-cell is changed to a standard mercury-mercury(II) half cell.
   (a) What would be the reduction potential of a standard chlorine half-cell?
   (b) What would be the reduction potential of a standard nickel half-cell?
   (c) What would be the net cell potential of a standard chlorine-nickel cell?
   (d) Why is the answer to part C the same as the answer obtained using the standard hydrogen half-cell as the reference?

2. For each of the following standard cells, write the cell notation, label electrodes and determine the cell potential.
   (a) cobalt-hydrogen standard cell

   (b) zinc-aluminum standard cell

   (c) tin(IV)-zinc standard cell

3. For each of the following cell notations, write the cathode, anode, and net cell reaction equations and calculate the cell potential.
   (a) $Pb_{(s)}$ | $Pb^{2+}_{(aq)}$ || $Ni^{2+}_{(aq)}$ | $Ni_{(s)}$

   (b) $Pt_{(s)}$ | $SO_4^{2-}_{(aq)}$, $H^+_{(aq)}$, $H_2SO_{3(aq)}$ || $Ag^+_{(aq)}$ | $Ag_{(s)}$

   (c) $Cd_{(s)}$ | $Cd^{2+}_{(aq)}$ || $ClO_4^-_{(aq)}$, $H^+_{(aq)}$, $Cl^-_{(aq)}$ | $C_{(s)}$

4. Use the standard cell described below to determine the standard reduction potential of the gallium half-cell.

   $Cu_{(s)}$ | $Cu^{2+}_{(aq)}$ || $Ga^{3+}_{(aq)}$ | $Ga_{(s)}$ $\Delta E° = +0.90$ V
   cathode   anode

© NELSON CANADA,
A DIVISION OF THOMSON CANADA LIMITED, 1994

225

59 STANDARD CELLS AND CELL POTENTIALS

1. (a) +0.51 V
   (b) −1.11 V
   (c) +1.62 V
   (d) **A cell potential is the difference in reduction potentials. This difference (how far apart the half-cells are in the table) does not depend on the individual reduction potentials.**

2. (a) $Co_{(s)}$ | $Co^{2+}_{(aq)}$ ‖ $H^+_{(aq)}$, $H_{2(g)}$ | $Pt_{(s)}$ $\Delta E° = +0.28$ V
       anode   cathode

   (b) $Zn_{(s)}$ | $Zn^{2+}_{(aq)}$ ‖ $Al^{3+}_{(aq)}$ | $Al_{(s)}$ $\Delta E° = +0.90$ V
       cathode   anode

   (c) $C_{(s)}$ | $Sn^{4+}_{(aq)}$, $Sn^{2+}_{(aq)}$ ‖ $Zn^{2+}_{(aq)}$ | $Zn_{(s)}$ $\Delta E° = +0.91$ V
       cathode   anode

3. (a) cathode $Pb^{2+}_{(aq)} + 2\,e^- \rightarrow Pb_{(s)}$ $E_r° = -0.13$ V
          anode $Ni_{(s)} \rightarrow Ni^{2+}_{(aq)} + 2\,e^-$ $E_r° = -0.26$ V

          net $Pb^{2+}_{(aq)} + Ni_{(s)} \rightarrow Pb_{(s)} + Ni^{2+}_{(aq)}$ $\Delta E° = +0.13$ V

   (b) cathode $2\,[\,Ag^+_{(aq)} + e^- \rightarrow Ag_{(s)}\,]$ $E°_r = +0.80$ V
          anode $H_2SO_{3(aq)} + H_2O_{(l)} \rightarrow$
                                 $SO_4^{2-}_{(aq)} + 4\,H^+_{(aq)} + 2\,e^-$ $E°_r = +0.17$ V

          net $H_2SO_{3(aq)} + H_2O_{(l)} + 2\,Ag^+_{(aq)} \rightarrow$ $\Delta E° = +0.63$ V
                              $SO_4^{2-}_{(aq)} + 4\,H^+_{(aq)} + 2\,Ag_{(s)}$

   (c) cathode $ClO_4^-_{(aq)} + 8\,H^+_{(aq)} + 8\,e^- \rightarrow Cl^-_{(aq)} + H_2O_{(l)}$ $E°_r = +1.39$ V
          anode $4\,[\,Cd_{(s)} \rightarrow Cd^{2+}_{(aq)} + 2\,e^-\,]$ $E°_r = -0.40$ V

          net $ClO_4^-_{(aq)} + 8\,H^+_{(aq)} + 4\,Cd_{(s)} \rightarrow$ $\Delta E° = +1.79$ V
                         $Cl^-_{(aq)} + H_2O_{(l)} + 4\,Cd^{2+}_{(aq)}$

4. $3\,[\,Cu^{2+}_{(aq)} + 2\,e^- \rightarrow Cu_{(s)}\,]$ $E°_r = +0.34$ V
        $2\,[\,Ga_{(s)} \rightarrow Ga^{3+}_{(aq)} + 3\,e^-\,]$ $E°_r = ?$

       $3\,Cu^{2+}_{(aq)} + 2\,Ga_{(s)} \rightarrow 3\,Cu_{(s)} + 2\,Ga^{3+}_{(aq)}$ $\Delta E° = +0.90$ V

           $\Delta E° = E°_r - E°_r$
               cathode   anode
       $+0.90$ V $= +0.34$ V $- E°_r$
          $E°_r = -0.56$ V

60 | CELL STOICHIOMETRY

1. A student wishes to set up an electrolytic cell to plate copper unto a belt buckle. Predict the length of time it will take to plate out 2.5 g of copper from a copper(II) nitrate solution using a 2.5 A current. At which electrode should the buckle be attached?

2. Determine the mass of chlorine produced when a 200 A current flows for 24.0 h through a cell containing molten sodium chloride (a Downs Cell). At which electrode is the chlorine produced?

3. How long would it take a 500 A current to produce 1.00 kg of aluminum from aluminum oxide dissolved in molten cryolite (Hall-Héroult Cell).

4. A trophy company is setting up a nickel plating cell using an electrolyte containing nickel(II) ions. Predict the current required to produce nickel metal at the rate of 5.00 g/min.

5. In an electrolysis experiment, a student passed 1.57 A of current through an aqueous solution of lead(II) nitrate containing a 7.63 g piece of lead foil as the cathode. What is the final mass of the lead electrode after 17.0 min.

60 CELL STOICHIOMETRY

1. $Cu^{2+}_{(aq)}$ + 2 e⁻ $\rightarrow$ $Cu_{(s)}$ (at the cathode)
   2.5 A, t .... 2.5 g
   9.65×10^4 C/mol ... 63.55 g/mol

   $$n_{Cu} = 2.5 \text{ g} \times \frac{1 \text{ mol}}{63.55 \text{ g}} = 0.039 \text{ mol}$$

   $$n_{e^-} = 0.039 \text{ mol} \times \frac{2}{1} = 0.079 \text{ mol}$$

   $$t = \frac{nF}{I} = \frac{0.079 \text{ mol} \times 9.65 \times 10^4 \text{ C/mol}}{2.5 \text{ C/s}} = 3.0 \text{ ks or 51 min}$$

2. $2 Cl^-_{(l)}$ $\rightarrow$ $Cl_{2(g)}$ + 2 e⁻ (at the anode)
   m .... 200 A, 24.0 h
   70.90 g/mol ... 9.65×10^4 C/mol

   $$n_{e^-} = \frac{It}{F} = \frac{200 \text{ C/s} \times 24.0 \text{ h} \times 3600 \text{ s/h}}{9.65 \times 10^4 \text{ C/mol}} = 179 \text{ mol}$$

   $$n_{Cl_2} = 179 \text{ mol} \times \frac{1}{2} = 89.5 \text{ mol}$$

   $$m_{Cl_2} = 89.5 \text{ mol} \times \frac{70.90 \text{ g}}{1 \text{ mol}} = 6.35 \text{ kg}$$

3. $Al^{3+}_{(cryolite)}$ + 3 e⁻ $\rightarrow$ $Al_{(l)}$
   500 A, t .... 1.00 kg
   9.65×10^4 C/mol ... 26.98 g/mol

   $$n_{Al} = 1.00 \text{ kg} \times \frac{1 \text{ mol}}{26.98 \text{ g}} = 0.0371 \text{ kmol}$$

   $$n_{e^-} = 0.0371 \text{ mol} \times \frac{3}{1} = 0.111 \text{ kmol}$$

   $$t = \frac{nF}{I} = \frac{0.111 \text{ kmol} \times 9.65 \times 10^4 \text{ C/mol}}{500 \text{ C/s}} = 21.5 \text{ ks or 358 min}$$

4. $Ni^{2+}_{(aq)}$ + 2 e⁻ $\rightarrow$ $Ni_{(s)}$
   1.00 min, I .... 5.00 g
   9.65×10^4 C/mol ... 58.69 g/mol

   $$n_{Ni} = 5.00 \text{ g} \times \frac{1 \text{ mol}}{58.69 \text{ g}} = 0.0852 \text{ mol}$$

   $$n_{e^-} = 0.0852 \text{ mol} \times \frac{2}{1} = 0.170 \text{ mol}$$

   $$I = \frac{nF}{t} = \frac{0.170 \text{ mol} \times 9.65 \times 10^4 \text{ C/mol}}{1.00 \text{ min} \times 60 \text{ s/min}} = 274 \text{ A}$$

5. $Pb^{2+}_{(aq)}$ + 2 e⁻ $\rightarrow$ $Pb_{(s)}$
   1.57 A, 17.0 min ... m
   9.65×10^4 C/mol ... 207.20 g/mol

   $$n_{e^-} = \frac{It}{F} = \frac{1.57 \text{ C/s} \times 17.0 \text{ min} \times 60 \text{ s/min}}{9.65 \times 10^4 \text{ C/mol}} = 1.66 \times 10^{-2} \text{ mol}$$

   $$n_{Pb} = 1.66 \times 10^{-2} \text{ mol} \times \frac{1}{2} = 8.30 \times 10^{-3} \text{ mol}$$

   $$m_{Pb} = 8.30 \times 10^{-3} \text{ mol} \times \frac{207.20 \text{ g}}{1 \text{ mol}} = 1.72 \text{ g}$$

   $$m_f = 7.63 \text{ g} + 1.72 \text{ g} = 9.35 \text{ g}$$

61   CHEMICAL EQUILIBRIUM

1. For each of the following, write the chemical reaction equation with appropriate equilibrium arrows.
   (a) pH measurements indicate that acetic acid in vinegar is approximately 1% ionized into hydrogen ions and acetate ions.

   (b) Quantitative analysis of the reaction of sodium sulfate and calcium chloride solutions shows that the products are favored.

   (c) Aluminum sulfate solution reacts quantitatively with a sodium hydroxide solution.

2. Chlorine and carbon monoxide gases are mixed in a 1.00 L container and the following equilibrium is established.

$$CO_{(g)} + Cl_{2(g)} \rightleftharpoons COCl_{2(g)}$$

Initially, 1.5 mol of chlorine was present with excess carbon monoxide. At equilibrium, 0.80 mol of $COCl_{2(g)}$ was found.
   (a) Calculate the percent reaction.

   (b) Write the equilibrium law for this reaction.

   (c) At equilibrium, 1.75 mol of carbon monoxide and 0.70 mol of chlorine were present. Calculate the equilibrium constant.

3. Write the equilibrium law for each of the following chemical reaction equations.
   (a) $2 SO_{2(g)} + O_{2(g)} \rightleftharpoons 2 SO_{3(g)}$

   (b) $2 NO_{2(g)} \rightleftharpoons 2 NO_{(g)} + O_{2(g)}$

   (c) $N_{2(g)} + 3 H_{2(g)} \rightleftharpoons 2 NH_{3(g)}$

4. In an experiment at a high temperature, 0.500 mol/L of hydrogen bromide gas is placed into a sealed container and decomposes into hydrogen and bromine gases.
   (a) Write the equilibrium equation and law for this reaction.

   (b) The equilibrium concentrations in this system are: $[HBr_{(g)}] = 0.240$ mol/L, $[H_{2(g)}] = [Br_{2(g)}] = 0.130$ mol/L. Calculate the equilibrium constant.

61 CHEMICAL EQUILIBRIUM

1. (a)
$$CH_3COOH_{(aq)} \overset{1\%}{\underset{>50\%}{\rightleftharpoons}} H^+_{(aq)} + CH_3COO^-_{(aq)}$$

(b) $Na_2SO_{4(aq)} + CaCl_{2(aq)} \rightleftharpoons CaSO_{4(s)} + 2\ NaCl_{(aq)}$

(c) $Al_2(SO_4)_{3(aq)} + 6\ NaOH_{(aq)} \rightarrow 2\ Al(OH)_{3(s)} + 3\ Na_2SO_{4(aq)}$

2. (a) $n_{COCl_2} = 1.5\ mol \times \dfrac{1}{1} = 1.5\ mol$ (maximum yield)

$\%\ reaction = \dfrac{0.80\ mol}{1.5\ mol} \times 100 = 53\%$

(b) $K = \dfrac{[COCl_{2(g)}]}{[CO_{(g)}][Cl_{2(g)}]}$

(c) $K = \dfrac{0.80\ mol/L}{(1.75\ mol/L)(0.70\ mol/L)}$

$= 0.65\ (mol/L)^{-1}$

3. (a) $K = \dfrac{[SO_{3(g)}]^2}{[SO_{2(g)}]^2[O_{2(g)}]}$

(b) $K = \dfrac{[NO_{(g)}]^2[O_{2(g)}]}{[NO_{2(g)}]^2}$

(c) $K = \dfrac{[NH_{3(g)}]^2}{[N_{2(g)}][H_{2(g)}]^3}$

4. (a) $2\ HBr_{(g)} \rightleftharpoons H_{2(g)} + Br_{2(g)}$

$K = \dfrac{[H_{2(g)}][Br_{2(g)}]}{[HBr_{(g)}]}$

(b) $K = \dfrac{(0.130\ mol/L)^2}{(0.240\ mol/L)^2} = 0.293$

62 LE CHÂTELIER'S PRINCIPLE

1. Nitrogen monoxide, a major air pollutant, is formed in automobile engines from the endothermic reaction of nitrogen and oxygen gases.
   (a) Write the equilibrium reaction equation including the term "energy" in the equation.

   (b) Describe the direction of the equilibrium shift if the concentration of oxygen is increased.

   (c) Describe the direction of the equilibrium shift if the pressure is increased.

   (d) Gasoline burns better at higher temperatures. What are some disadvantages of the operation of automobile engines at higher temperatures?

2. In a sealed container, nitrogen monoxide and oxygen gases are in equilibrium with nitrogen dioxide gas. The reaction of nitrogen monoxide and oxygen is exothermic.

   $$2\,NO_{(g)} \;+\; O_{2(g)} \;\rightleftharpoons\; 2\,NO_{2(g)} \;+\; energy$$

   Predict the equilibrium shift when the following changes are made.
   (a) the temperature is decreased

   (b) the concentration of $NO_{(g)}$ is decreased

   (c) the concentration of $NO_{2(g)}$ is increased

   (d) the volume of the system is decreased

3. The equilibrium of the iron(III)-thiocyanate system (Lab Exercise 14A) is convenient to study. Thiocyanate ions are colorless and, at very low concentrations, iron(III) ions are essentially colorless. However, the $FeSCN^{2+}_{(aq)}$ complex is highly colored even at low concentrations. Predict the color change in the equilibrium mixture when each of the following changes is made.

   $$Fe^{3+}_{(aq)} \;+\; SCN^-_{(aq)} \;\rightleftharpoons\; FeSCN^{2+}_{(aq)}$$
   $\quad$ yellow $\qquad$ colorless $\qquad\quad$ red

   (a) A crystal of $KSCN_{(s)}$ is added to the system.

   (b) A crystal of $FeCl_{3(s)}$ is added to the system.

   (c) A crystal of $NaOH_{(s)}$ is added to the system.

62 LE CHÂTELIER'S PRINCIPLE

1. Nitrogen monoxide, a major air pollutant, is formed in automobile engines from the endothermic reaction of nitrogen and oxygen gases.
   (a) Write the equilibrium reaction equation including the term "energy" in the equation.
   $$N_{2(g)} + O_{2(g)} + energy \rightleftharpoons 2 NO_{(g)}$$

   (b) Describe the direction of the equilibrium shift if the concentration of oxygen is increased.
   The equilibrium shifts to the right.

   (c) Describe the direction of the equilibrium shift if the pressure is increased.
   There will be no shift in the equilibrium because the number of gas molecules is the same on both sides of the equilibrium equation.

   (d) Gasoline burns better at higher temperatures. What are some disadvantages of the operation of automobile engines at higher temperatures?
   At high temperatures, the equilibrium shifts to the right, producing more of the pollutant nitrogen monoxide. There may also be additional technical and safety problems as the temperature is increased.

2. In a sealed container, nitrogen monoxide and oxygen gases are in equilibrium with nitrogen dioxide gas. The reaction of nitrogen monoxide and oxygen is exothermic.

   $$2 NO_{(g)} + O_{2(g)} \rightleftharpoons 2 NO_{2(g)} + energy$$

   Predict the equilibrium shift when the following changes are made.
   (a) the temperature is decreased
   The equilibrium shifts to the right.

   (b) the concentration of $NO_{(g)}$ is decreased
   The equilibrium shifts to the left.

   (c) the concentration of $NO_{2(g)}$ is increased
   The equilibrium shifts to the left.

   (d) the volume of the system is decreased
   The equilibrium shifts to the right.

3. The equilibrium of the iron(III)-thiocyanate system (Lab Exercise 14A) is convenient to study. Thiocyanate ions are colorless and, at very low concentrations, iron(III) ions are essentially colorless. However, the $FeSCN^{2+}_{(aq)}$ complex is highly colored even at low concentrations. Predict the color change in the equilibrium mixture when each of the following changes is made.
   $$Fe^{3+}_{(aq)} + SCN^-_{(aq)} \rightleftharpoons FeSCN^{2+}_{(aq)}$$
   yellow   colorless   red
   (a) A crystal of $KSCN_{(s)}$ is added to the system.
   The red color becomes more intense as the equilibrium shifts to the right.

   (b) A crystal of $FeCl_{3(s)}$ is added to the system.
   The red color becomes more intense as the equilibrium shifts to the right.

   (c) A crystal of $NaOH_{(s)}$ is added to the system.
   The red color becomes less intense as the equilibrium shifts to the left.
   (Iron(III) ions precipitate as iron(III) hydroxide.)

63 K_w, pH, AND pOH CALCULATIONS

1. Calculate the $[OH^-_{(aq)}]$ in limes which have a $[H^+_{(aq)}]$ of 1.3×10^{-2} mol/L.

2. Calculate the $[H^+_{(aq)}]$ in lemons which have a $[OH^-_{(aq)}]$ of 2.0×10^{-12} mol/L.

3. A sodium hydroxide solution is prepared by dissolving 2.50 g to make 2.00 L of solution. Calculate the hydroxide and hydrogen ion concentrations.

4. A 0.728 g sample of hydrogen chloride gas is dissolved in 200 mL of solution. Calculate the hydrogen and hydroxide ion concentrations.

5. A vinegar solution has a hydrogen ion concentration of 1.5×10^{-3} mol/L. Calculate the pH.

6. An ammonia solution has a pOH of 2.92. What is the concentration of hydroxide ions in the solution?

7. Calculate the pOH and pH of a solution made by dissolving 7.50 g of strontium hydroxide to make 500 mL of solution.

Complete the following table.

	Substance	$[H^+_{(aq)}]$ (mol/L)	pH	$[OH^-_{(aq)}]$ (mol/L)	pOH	Acidic, Basic, or Neutral
8.	milk			3.2×10^{-8}		
9.	pure water		7.0			
10.	blood	4.0×10^{-8}				
11.	cleaner				3.20	

63 K_w, pH, AND pOH CALCULATIONS

1. $[OH^-_{(aq)}] = \dfrac{K_w}{[H^+_{(aq)}]} = \dfrac{1.0 \times 10^{-14} \ (mol/L)^2}{1.3 \times 10^{-2} \ mol/L} = 7.7 \times 10^{-13} \ mol/L$

2. $[H^+_{(aq)}] = \dfrac{K_w}{[OH^-_{(aq)}]} = \dfrac{1.0 \times 10^{-14} \ (mol/L)^2}{2.0 \times 10^{-12} \ mol/L} = 5.0 \times 10^{-3} \ mol/L$

3. $n_{NaOH} = 2.50 \ g \times \dfrac{1 \ mol}{40.00 \ g} = 6.25 \times 10^{-2} \ mol$

   $[NaOH_{(aq)}] = \dfrac{6.25 \times 10^{-2} \ mol}{2.00 \ L} = 3.13 \times 10^{-2} \ mol/L$

   $[OH^-_{(aq)}] = [NaOH_{(aq)}] = 3.13 \times 10^{-2} \ mol/L$

   $[H^+_{(aq)}] = \dfrac{K_w}{[OH^-_{(aq)}]} = \dfrac{1.0 \times 10^{-14} \ (mol/L)^2}{3.13 \times 10^{-2} \ mol/L} = 3.2 \times 10^{-13} \ mol/L$

4. $n_{HCl} = 0.728 \ g \times \dfrac{1 \ mol}{36.46 \ g} = 2.00 \times 10^{-2} \ mol$

   $[HCl_{(aq)}] = \dfrac{2.00 \times 10^{-2} \ mol/L}{0.200 \ L} = 9.98 \times 10^{-2} \ mol/L$

   $[H^+_{(aq)}] = [HCl_{(aq)}] = 9.98 \times 10^{-2} \ mol/L$

   $[OH^-_{(aq)}] = \dfrac{K_w}{[H^+_{(aq)}]} = \dfrac{1.0 \times 10^{-14} \ (mol/L)^2}{9.98 \times 10^{-2} \ mol/L} = 1.0 \times 10^{-13} \ mol/L$

5. $pH = -log[H^+_{(aq)}] = -log(1.5 \times 10^{-3}) = 2.82$

6. $[OH^-_{(aq)}] = 10^{-pOH} = 10^{-2.92} \ mol/L = 1.2 \times 10^{-3} \ mol/L$

7. $n_{Sr(OH)_2} = 7.50 \ g \times \dfrac{1 \ mol}{121.64 \ g} = 0.0617 \ mol$

   $[Sr(OH)_{2(aq)}] = \dfrac{0.0617 \ mol}{0.500 \ L} = 0.123 \ mol/L$

   $[OH^-_{(aq)}] = 2 \ [Sr(OH)_{2(aq)}] = 2 \times 0.123 \ mol/L = 0.247 \ mol/L$
   $pOH = -log[OH^-_{(aq)}] = -log(0.247) = 0.608$
   $pH = 14.00 - 0.608 = 13.39$

	Substance	$[H^+_{(aq)}]$ (mol/L)	pH	$[OH^-_{(aq)}]$ (mol/L)	pOH	Acidic, Basic, or Neutral
8.	milk	3.1×10^{-7}	6.51	3.2×10^{-8}	7.49	acidic
9.	pure water	1×10^{-7}	7.0	1×10^{-7}	7.0	neutral
10.	blood	4.0×10^{-8}	7.40	2.5×10^{-7}	6.60	basic
11.	cleaner	1.6×10^{-11}	10.80	6.3×10^{-4}	3.20	basic

64 (ENRICHMENT) THE SOLUBILITY PRODUCT CONSTANT, K_{sp}

A chemical system involving the solubility equilibrium of an ionic compound is a common heterogeneous system. The constant concentration of the solid is incorporated with the equilibrium constant, K_e, to give a special constant known as the **solubility product** constant, K_{sp}. The solubility product constant, K_{sp}, like K_w, is a special case of a condensed constant, K_c.

As an example, the solubility equilibrium equation and solubility product expression for a saturated solution of aluminum hydroxide are presented below.

$$Al(OH)_{3(s)} \rightleftharpoons Al^{3+}_{(aq)} + 3\ OH^-_{(aq)}$$

$$K_e = \frac{[Al^{3+}_{(aq)}][OH^-_{(aq)}]^3}{[Al(OH)_{3(s)}]}$$

$$K_{sp} = K_e[Al(OH)_{3(s)}] = [Al^{3+}_{(aq)}][OH^-_{(aq)}]^3$$

At SATP, the pH indicates that the concentration of hydroxide ions in a saturated solution of aluminum hydroxide is 3.2×10^{-4} mol/L.

According to the balanced equation, the ratio of aluminum ions to hydroxide ions is 1:3.
$$[Al^{3+}_{(aq)}] = 3.2 \times 10^{-4} \text{ mol/L} \times \frac{1}{3} = 1.1 \times 10^{-4} \text{ mol/L}$$

According to the equilibrium law, $K_{sp} = [Al^{3+}_{(aq)}][OH^-_{(aq)}]^3$. Therefore,
$$K_{sp} = (1.1 \times 10^{-4} \text{ mol/L})(3.2 \times 10^{-4} \text{ mol/L})^3$$
$$= 3.5 \times 10^{-15} \text{ (mol/L)}^4.$$

EXAMPLE

Leaching of cations from lake bottoms may be toxic to life in the lake. Leaching increases with the increased acidity of the lake water.
What would be the normal (non-acidic) concentration of calcium ions in a lake lined by limestone?
$$CaCO_{3(s)} \rightleftharpoons Ca^{2+}_{(aq)} + CO_3^{2-}_{(aq)}$$
$$K_{sp} = [Ca^{2+}_{(aq)}][CO_3^{2-}_{(aq)}] = 9.9 \times 10^{-9} \text{ mol}^2/\text{L}^2 \text{ (at 15°C)}$$

$$[Ca^{2+}_{(aq)}] = \sqrt{K_{sp}}$$
$$= \sqrt{9.9 \times 10^{-9} \text{ mol}^2/\text{L}^2}$$
$$= 9.9 \times 10^{-5} \text{ mol/L}$$

At 15°C, the calcium ion concentration due to limestone in the lake should be 9.9×10^{-5} mol/L.

1. Describe, both empirically and theoretically, a saturated table salt solution.
2. Calcium fluoride has been suggested as a reagent for fluoridating drinking water. The concentration of fluoride ion in a saturated solution of calcium fluoride is measured as 4.3×10^{-4} mol/L at 26°C. Predict the K_{sp} for calcium fluoride at this temperature.
3. Schools often concentrate lead compounds in toxic waste solutions by precipitating lead(II) carbonate. The K_{sp} for lead(II) carbonate at 18°C is 3.3×10^{-14} (mol/L)^2.
   Predict the concentration of lead(II) ions in a saturated solution of lead(II) carbonate at 18°C.
4. The solubility product of a saturated magnesium hydroxide solution is 1.2×10^{-11} (mol/L)^3 at 18°C. Predict the pH of a saturated magnesium hydroxide solution at 18°C.

© NELSON CANADA,
A DIVISION OF THOMSON CANADA LIMITED, 1994

64 (ENRICHMENT) THE SOLUBILITY PRODUCT CONSTANT, K_{sp}

1. Empirically, a saturated table salt solution is a solution in which no more table salt will dissolve at that temperature. Theoretically, a saturated table salt solution is a solution in which the rate of dissolving of table salt is equal to the rate of crystallizing.

2. $CaF_{2(s)} \rightleftharpoons Ca^{2+}_{(aq)} + 2\, F^-_{(aq)}$

$[Ca^{2+}_{(aq)}] = 4.3 \times 10^{-4}\, mol/L \times \dfrac{1}{2}$

$\qquad\quad = 2.2 \times 10^{-4}\, mol/L$

$K_{sp} = [Ca^{2+}_{(aq)}][F^-_{(aq)}]$

$\qquad = (2.2 \times 10^{-4}\, mol/L)(4.3 \times 10^{-4}\, mol/L)^2$

$\qquad = 4.0 \times 10^{-11}\, (mol/L)^3$

3. $PbCO_{3(s)} \rightleftharpoons Pb^{2+}_{(aq)} + CO_3^{2-}{}_{(aq)}$

$K_{sp} = [Pb^{2+}_{(aq)}][CO_3^{2-}{}_{(aq)}]$

$\qquad = [Pb^{2+}_{(aq)}]^2$

$[Pb^{2+}_{(aq)}] = \sqrt{K_{sp}}$

$\qquad\quad = \sqrt{3.3 \times 10^{-14}\, (mol.\, L)^2}$

$\qquad\quad = 1.8 \times 10^{-7}\, mol/L$

4. $Mg(OH)_{2(s)} \rightleftharpoons Mg^{2+}_{(aq)} + 2\, OH^-_{(aq)}$

$[Mg^{2+}_{(aq)}] = \dfrac{1}{2}[OH^-_{(aq)}]$

$K_{sp} = [Mg^{2+}_{(aq)}][OH^-_{(aq)}]^2$

$\qquad = \dfrac{1}{2}[OH^-_{(aq)}][OH^-_{(aq)}]^2$

$1.2 \times 10^{-11}\, (mol/L)^3 = \dfrac{1}{2}[OH^-_{(aq)}]^3$ or $[OH^-_{(aq)}] = \sqrt[3]{2K_{sp}} = \sqrt[3]{2 \times 1.2 \times 10^{-11}(mol/l)^3}$

$[OH^-_{(aq)}] = 2.9 \times 10^{-4}\, mol/L$

$[H^+_{(aq)}] = \dfrac{1 \times 10^{-14}(mol/L)^2}{2.9 \times 10^{-4}\, mol/L}$

$\qquad\quad = 3.5 \times 10^{-11}\, mol/L$

$pH = -\log(3.5 \times 10^{-11}) = 10.46$

65 STRENGTHS OF ACIDS

1. List three empirical properties that may be measured to distinguish among acids of different strengths.

2. Calculate the hydrogen ion concentration and pH of a 0.10 mol/L solution of nitrous acid.

3. Calculate the hydrogen ion concentration and pH of a solution prepared by dissolving 10.70 g of ammonium chloride to make 2.00 L of solution.

4. Use the percent reaction value to determine the mass of sodium hydrogen sulfate required to prepare 500 mL of solution with a pH of 1.57.

5. A 0.80 mol/L solution of an unknown acid, $HX_{(aq)}$, has a pH of 3.75.
   (a) Calculate the percent reaction.

   (b) Calculate the acid ionization constant.

6. Calculate the pH of a solution containing 0.25 mol/L of an acid with an acid ionization constant of 3.2×10^{-6} mol/L.

65 STRENGTHS OF ACIDS

1. conductivity, rate of reaction, pH

2. $[H^+_{(aq)}] = \dfrac{8.1}{100} \times 0.10 \text{ mol/L} = 8.1 \times 10^{-3} \text{ mol/L}$

   $pH = -\log[H^+_{(aq)}] = -\log(8.1 \times 10^{-3}) = 2.09$ $\qquad$ **(2.07 from K_a)**

3. $n_{NH_4Cl} = 10.70 \text{ g} \times \dfrac{1 \text{ mol}}{53.50 \text{ g}} = 0.2000 \text{ mol}$

   $[NH_4Cl_{(aq)}] = \dfrac{0.2000 \text{ mol}}{2.00 \text{ L}} = 0.100 \text{ mol/L}$

   $[H^+_{(aq)}] = \dfrac{0.0076}{100} \times 0.100 \text{ mol/L} = 7.6 \times 10^{-6} \text{ mol/L}$

   $pH = -\log[H^+_{(aq)}] = -\log(7.6 \times 10^{-6}) = 5.12$ $\qquad$ **(same answer from K_a)**

4. $[H^+_{(aq)}] = 10^{-pH} = 10^{-1.57} \text{ mol/L} = 2.7 \times 10^{-2} \text{ mol/L}$

   $\dfrac{[H^+_{(aq)}]}{[HSO_4^-{}_{(aq)}]} \times 100 = 27\%$

   $[HSO_4^-{}_{(aq)}] = \dfrac{2.7 \times 10^{-2} \text{ mol/L}}{27} \times 100 = 0.10 \text{ mol/L}$

   $[NaHSO_{4(aq)}] = [HSO_4^-{}_{(aq)}] = 0.10 \text{ mol/L}$

   $n_{NaHSO_4} = 500 \text{ mL} \times \dfrac{0.10 \text{ mol}}{1 \text{ L}} = 50 \text{ mmol}$

   $m_{NaHSO_4} = 50 \text{ mmol} \times \dfrac{120.06 \text{ g}}{1 \text{ mol}} = 6.0 \text{ g}$

5. (a) $[H^+_{(aq)}] = 10^{-pH} = 10^{-3.75} \text{ mol/L} = 1.8 \times 10^{-4} \text{ mol/L}$

   $p = \dfrac{1.8 \times 10^{-7} \text{ mol/L}}{0.80 \text{ mol/L}} \times 100 = 2.2 \times 10^{-2}\%$

   (b) $HX_{(aq)} \rightleftharpoons H^+_{(aq)} + X^-_{(aq)}$

   $K = \dfrac{[H^+_{(aq)}][X^-_{(aq)}]}{[HX_{(aq)}]} = \dfrac{(1.8 \times 10^{-4} \text{ mol/L})^2}{0.80 \text{ mol/L}} = 4.0 \times 10^{-8} \text{ mol/L}$

6. $HX_{(aq)} \rightleftharpoons H^+_{(aq)} + X^-_{(aq)}$

   $K = \dfrac{[H^+_{(aq)}][X^-_{(aq)}]}{[HX_{(aq)}]}$

   $[H^+_{(aq)}] = \sqrt{3.2 \times 10^{-6} \text{ mol/L} \times 0.25 \text{ mol/L}} = 8.9 \times 10^{-4} \text{ mol/L}$

   $pH = -\log[H^+_{(aq)}] = -\log(8.9 \times 10^{-4}) = 3.05$

66 BRØNSTED-LOWRY DEFINITIONS AND INDICATORS

1. According to the Brønsted-Lowry concept, define an acid and a base in terms of an acid-base reaction.

2. (a) What is meant by an "acidic" solution?

   (b) Does a Brønsted-Lowry acid have to form an acidic solution?

3. There are many species that are classified as amphiprotic.
   (a) What does amphiprotic mean?

   (b) What general type of species is amphiprotic?

4. $HOOCCOO^-_{(aq)} + PO_4^{3-}_{(aq)} \rightleftharpoons HPO_4^{2-}_{(aq)} + OOCCOO^{2-}_{(aq)}$
   (a) Label each species as A or B for both forward and reverse reactions.

   (b) Identify all conjugate acid-base pairs.

   (c) According to the Brønsted-Lowry concept, what determines the position of this equilibrium?

5. What is the generalization about the strength of an acid relative to its conjugate base?

6. Use the following pH scale to label the colors for bromothymol blue over the 0–14 pH range. Identify the form of the indicator for each distinct color using conventional symbols.

   | | | | | | | | | | | | | | |
   0   7   14

7. *Problem*   What is the approximate pH of an unknown solution?
   Evidence Separate samples of the unknown solution turned blue litmus to red, congo red to blue, and orange IV to yellow.

   Analysis

8. Three unknown solutions in unlabelled beakers have pH values of 5.8, 7.8 and 9.8. Write two diagnostic tests using indicators to identify the pH of each solution.

66 BRØNSTED-LOWRY DEFINITIONS AND INDICATORS

1. According to the Brønsted-Lowry concept, a proton is transferred from an acid to a base in an acid-base reaction.

2. (a) An acidic solution is an empirical classification of a solution with properties such as turning blue litmus red and having a pH less than 7.

   (b) No. A Brønsted-Lowry acid is defined in terms of proton transfer in a reaction and not in terms of the empirical properties of a solution. For example, the hydrogen corbonate ion forms a basic solution but sometimes, acts as a Brønsted-Lowry acid.

3. (a) An amphiprotic substance is one that appears to act as a Brønsted-Lowry acid in some reactions and as a Brønsted-Lowry base in other reactions.

   (b) Hydrogen-polyatomic species, such as $HCO_3^-_{(aq)}$, are generally amphiprotic.

4. (a) $\underset{A}{HOOCCOO^-_{(aq)}} + \underset{B}{PO_4^{3-}_{(aq)}} \rightleftharpoons \underset{A}{HPO_4^{2-}_{(aq)}} + \underset{B}{OOCCOO^{2-}_{(aq)}}$

   (b) $HOOCCOO^-_{(aq)} / OOCCOO^{2-}_{(aq)}$

   $HPO_4^{2-}_{(aq)} / PO_4^{3-}_{(aq)}$

   (c) According to the Brønsted-Lowry concept, the position of the equilibrium is determined by the competition for proton transfer between the reacting species in the forward reaction and the reacting species in the reverse reaction.

5. In general, the stronger an acid, the weaker its conjugate base; and conversely, the weaker the acid, the stronger its conjugate base.

6.
```
 ├── yellow ──┤gr├── blue ──┤
 │ │ │ │ │ │ │ │ │ │ │ │ │ │ │ │
 0            7             14
```

   yellow: ≤ 6; $HBb_{(aq)}$
   green: in 6–7.6 range; $HBb_{(aq)}$ and $Bb^-_{(aq)}$
   blue: ≥ 7.6; $Bb^-_{(aq)}$

7. If litmus turns red, the pH is 6.0 or lower.
   If congo red turns blue, the pH is 3.0 or lower.
   If orange IV turns yellow, the pH is 2.8 or higher.
   According to the evidence, the approximate pH of the unknown solution is between 2.8 and 3.0.

8. If all three solutions are tested with red and blue litmus paper, the solution that turns the blue litmus to red has a pH of 5.8.
   If the remaining two solutions are tested with thymol blue, the one that turns yellow has a pH of 7.8 and the one that turns blue has a pH of 9.8.
   (Other acceptable diagnostic tests are possible.)

67 PREDICTING ACID-BASE EQUILIBRIA

1. What is the generalization used to predict the position of an acid-base equilibrium?

Predict the acid-base reaction for each of the following questions. Communicate your answer using the five-step method.

2. A nitrous acid spill is neutralized with sodium hydrogen carbonate (baking soda).

3. In a chemical analysis, a sample of methanoic acid is titrated in a quantitative reaction with standardized sodium hydroxide.

4. Could hypochlorous acid in a bleach solution be neutralized with baking soda?

5. Small quantities of the poisonous hydrocyanic acid can be produced in a research laboratory by quantitatively reacting sodium cyanide and hydrochloric acid.

6. A student mixes solutions of sodium acetate and ammonium chloride to test the predictive power of the Brønsted-Lowry concept.

7. Some excess hydrofluoric acid remained after its use in glass etching. To dispose of this hazardous acid, a technician decided to neutralize the solution.
   (a) What readily available base would she choose to ensure a complete neutralization?

   (b) Predict the neutralization reaction assuming a quantitative reaction.

67 PREDICTING ACID-BASE EQUILIBRIA

1. **If the strongest acid present is listed higher in the acid-base table than the strongest base, then the products are favored (the extent of the reaction is greater than fifty percent).**

2.

$$\text{HNO}_{2(aq)}, \quad \text{Na}^+_{(aq)}, \quad \overset{\text{A}}{\text{HCO}_3^-}_{(aq)}, \quad \overset{\text{A}}{\text{H}_2\text{O}}_{(l)}$$

SA ⟶ SB, B

$$\text{HNO}_{2(aq)} + \text{HCO}_3^-_{(aq)} \overset{>50\%}{\rightleftharpoons} \text{H}_2\text{CO}_{3(aq)} + \text{NO}_2^-_{(aq)}$$

3. SA

$$\text{HCOOH}_{(aq)}, \quad \text{Na}^+_{(aq)}, \quad \overset{}{\text{OH}^-}_{(aq)}, \quad \overset{\text{A}}{\text{H}_2\text{O}}_{(l)}$$

SB, B

$$\text{HCOOH}_{(aq)} + \text{OH}^-_{(aq)} \rightarrow \text{HCOO}^-_{(aq)} + \text{HOH}_{(l)}$$

4. SA

$$\text{HClO}_{(aq)}, \quad \text{Na}^+_{(aq)}, \quad \overset{\text{A}}{\text{HCO}_3^-}_{(aq)}, \quad \overset{\text{A}}{\text{H}_2\text{O}}_{(l)}$$

SB, B

$$\text{HClO}_{(aq)} + \text{HCO}_3^-_{(aq)} \overset{<50\%}{\rightleftharpoons} \text{ClO}^-_{(aq)} + \text{H}_2\text{CO}_{3(aq)}$$

No, only a partial neutralization, if any, will occur.

5.

$$\text{Na}^+_{(aq)}, \quad \text{CN}^-_{(aq)}, \quad \overset{\text{SA}}{\text{H}_3\text{O}^+}_{(aq)}, \quad \text{Cl}^-_{(aq)}, \quad \overset{\text{A}}{\text{H}_2\text{O}}_{(l)}$$

SB, B, B

$$\text{CN}^-_{(aq)} + \text{H}_3\text{O}^+_{(aq)} \rightarrow \text{HCN}_{(aq)} + \text{H}_2\text{O}_{(l)}$$

6.

$$\text{Na}^+, \quad \text{CH}_3\text{COO}^-_{(aq)}, \quad \overset{\text{SA}}{\text{NH}_4^+}_{(aq)}, \quad \text{Cl}^-_{(aq)}, \quad \overset{\text{A}}{\text{H}_2\text{O}}_{(l)}$$

SB, B, B

$$\text{NH}_4^+_{(aq)} + \text{CH}_3\text{COO}^-_{(aq)} \overset{<50\%}{\rightleftharpoons} \text{NH}_{3(aq)} + \text{CH}_3\text{COOH}_{(aq)}$$

7. (a) **The strongest readily-available base is sodium hydroxide.**

(b) SA

$$\text{HF}_{(aq)}, \quad \text{Na}^+_{(aq)}, \quad \overset{}{\text{OH}^-}_{(aq)}, \quad \overset{\text{A}}{\text{H}_2\text{O}}_{(l)}$$

SB, B

$$\text{HF}_{(aq)} + \text{OH}^-_{(aq)} \rightarrow \text{H}_2\text{O}_{(l)} + \text{F}^-_{(aq)}$$

68   pH CURVES

1. Use the accompanying sketch of a pH curve for a titration to answer the following questions.

   (a) Does the buret contain the acid or the base?

   (b) Is the sample reacted an acid or a base?

   (c) How many endpoints are present? Estimate each pH endpoint.

   (d) How many quantitative reactions have occurred?

   (e) Choose the best indicator for each endpoint.

   (f) What part of the curve represents a possible buffering region?

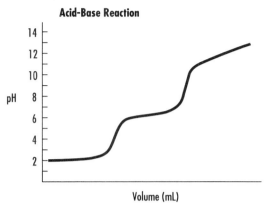

Acid-Base Reaction

2. Sketch a pH curve for the reaction of sulfuric acid with sodium hydroxide solution. All reaction steps are known to be quantitative. Include reaction equations.

3. A sodium hydrogen phosphate solution is to be titrated with hydrochloric acid. Only one quantitative reaction is observed. Sketch the pH curve and write equilibrium equations.

4. What are two advantages of a color endpoint using an indicator as opposed to a pH endpoint using a pH meter?

68 pH CURVES

1. (a) The buret contains the base.

   (b) The sample is the acid.

   (c) Two endpoints are apparent at about pH 4.0 and pH 8.5.

   (d) Two quantitative reactions have occurred.

   (e) Methyl orange or congo red are suitable for the first endpoint, and metacresol purple or thymol blue are suitable for the second endpoint.

   (f) The nearly horizontal sections of the graph represent buffering regions.

2. $H_3O^+_{(aq)} + OH^-_{(aq)} \rightarrow 2\,H_2O_{(l)}$
   $HSO_4^-_{(aq)} + OH^-_{(aq)} \rightarrow SO_4^{2-}_{(aq)} + HOH_{(l)}$

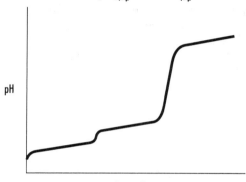

Reaction of $H_2SO_{4(aq)}$ with $NaOH_{(aq)}$

3. $HPO_4^{2-}_{(aq)} + H_3O^+_{(aq)} \rightarrow H_2PO_4^-_{(aq)} + H_2O_{(l)}$

   $H_2PO_4^-_{(aq)} + H_3O^+_{(aq)} \overset{>50\%}{\rightleftharpoons} H_3PO_{4(aq)} + H_2O_{(l)}$

Reaction of $Na_2HPO_{4(aq)}$ with $HCl_{(aq)}$

4. Two advantages are that the method of analysis does not require any expensive equipment, and the endpoint is more visible and can be seen during the titration process (it does not require that you watch a meter while performing the titration).

69 (ENRICHMENT) HYDROLYSIS

The Brønsted-Lowry definitions provide adequate explanations for most situations once the empirical results are known. However, their ability to predict is limited, especially when certain ions (e.g. amphiprotic ions and highly charged metal ions), are dissolved in water. Part of the difficulty is believed to be the result of a process called *hydrolysis*.

Empirically, hydrolysis is the reaction of a chemical with water to form an acidic or a basic solution. A theoretical definition of hydrolysis includes the concept of entities reacting with water to increase the hydronium or hydroxide ion concentration. Research indicates that not all ions undergo detectable hydrolysis. Further research shows that metal cations of Groups 1 and 2 elements and the conjugate bases of the six strong acids do not appear to hydrolyze to any appreciable extent. This generalization is based on the evidence that solutions of compounds such as sodium chloride and calcium nitrate are neutral.

The evidence from the laboratory shows that copper(II) sulfate, aluminum sulfate, and iron(III) nitrate form acidic solutions. Many highly charged (especially >2+) metal ions are believed to have such high charge densities that they bond to specific numbers of water molecules. Studies of these compounds show that our previous representations were inadequate for describing the acidic properties of these ions. A revised representation is presented below.

Previous Formula	Revised Formula
$Fe^{3+}_{(aq)}$	$Fe(H_2O)_6^{3+}_{(aq)}$
$Al^{3+}_{(aq)}$	$Al(H_2O)_6^{3+}_{(aq)}$

Within the complex ions, the high positive charge on the central metal ion is believed to repel protons from the complex ion, making the protons of the water molecules more acidic and therefore able to act as Brønsted-Lowry acids. These ions produce a relatively high concentration of hydronium ions compared with some well-known acids.

$$Fe(H_2O)_6^{3+}_{(aq)} + H_2O_{(l)} \overset{24\%}{\rightleftharpoons} H_3O^+_{(aq)} + [Fe(H_2O)_5OH]^{2+}_{(aq)}$$

$$Al(H_2O)_6^{3+}_{(aq)} + H_2O_{(l)} \overset{1.2\%}{\rightleftharpoons} H_3O^+_{(aq)} + [Al(H_2O)_5OH]^{2+}_{(aq)}$$

1. Our present knowledge and information allow only limited predictions of which solutions are acidic and basic. What previously mentioned classes of substances creates problems for predicting acidity?

2. Write an empirical definition of hydrolysis.

3. Which common cations are considered spectator ions and which produce acidic solutions?

4. Predict the chemical reaction for the hydrolysis of each of the following chemicals. Indicate whether the solution will be acidic or basic.
   (a) copper(II) nitrate
   (b) sodium acetate
   (c) ammonium chloride

5. What is the pH of a 0.10 mol/L iron(III) chloride solution? (See equation above)

6. If the pH of a 0.10 mol/L copper(II) nitrate solution is 4.5, what is the percent reaction of the copper(II) complex ion with water?

69 (ENRICHMENT) HYDROLYSIS

1. Highly charged metal ions and amphiprotic substances both create problems in predicting acidity.

2. In the context of acids and bases, hydrolysis is the reaction of a chemical with water to form an acidic or a basic solution.

3. Cations of Group 1 and 2 elements are considered spectator ions. Highly charged metal cations generally behave as acids.

4. (a) SA   A
$Cu(H_2O)_6{}^{2+}{}_{(aq)}$, $NO_3{}^-{}_{(aq)}$, $H_2O_{(l)}$
                                          SB

$$Cu(H_2O)_6{}^{2+}{}_{(aq)} + H_2O_{(l)} \overset{<50\%}{\rightleftharpoons} H_3O^+ + Cu(H_2O)_5OH^+{}_{(aq)} \quad \text{(acidic)}$$

(b)   SA
$Na^+{}_{(aq)}$, $CH_3COO^-{}_{(aq)}$, $H_2O_{(l)}$
              SB   B

$$CH_3COO^-{}_{(aq)} + H_2O_{(l)} \overset{<50\%}{\rightleftharpoons} OH^-{}_{(aq)} + CH_3COOH_{(aq)} \quad \text{(basic)}$$

(c) SA   A
$NH_4{}^+{}_{(aq)}$, $Cl^-{}_{(aq)}$, $H_2O_{(l)}$
              B   SB

$$NH_4{}^+{}_{(aq)} + H_2O_{(l)} \overset{<50\%}{\rightleftharpoons} H_3O^+{}_{(aq)} + NH_{3(aq)} \quad \text{(acidic)}$$

5. $[H_3O^+{}_{(aq)}] = \dfrac{p}{100} [Fe^{3+}{}_{(aq)}]$

$\qquad\quad = \dfrac{24}{100} \times 0.10 \text{ mol/L}$

$\qquad\quad = 2.4 \times 10^{-3} \text{ mol/L}$

$pH = -\log[H_3O^+{}_{(aq)}]$

$\qquad = -\log(2.4 \times 10^{-3})$

$\qquad = 2.62$

6. $[H_3O^+{}_{(aq)}] = 10^{-pH}$

$\qquad\qquad = 10^{-4.5} \text{ mol/L}$

$\qquad\qquad = 3 \times 10^{-5} \text{ mol/L}$

$p = \dfrac{[H_3O^+{}_{(aq)}]}{[Cu^{2+}{}_{(aq)}]} \times 100$

$\quad = \dfrac{3 \times 10^{-5} \text{ mol/L}}{0.10 \text{ mol/L}} \times 100$

$\quad = 3 \times 10^{-2}\% \text{ or } 0.03\%$

70 ACID-BASE STOICHIOMETRY

1. In a chemical analysis, 25.0 mL of sulfuric acid solution was titrated to the second endpoint with 0.358 mol/L KOH$_{(aq)}$. In the titration, an average volume of 18.2 mL was required. Calculate the molar concentration of the sulfuric acid.

2. Several 10.0 mL vinegar samples were titrated with a standardized 0.582 mol/L solution of sodium hydroxide. An average volume of 13.8 mL of sodium hydroxide was required to reach the phenolphthalein endpoint. What is the concentration of the vinegar solution?

3. A sodium borate solution was titrated to the second endpoint with 0.241 mol/L hydrobromic acid. An average volume of 15.2 mL of hydrobromic acid was required to react with 20.0 mL samples of sodium borate. Calculate the molar concentration of sodium borate.

4. *Problem*   What is the molar concentration of a hydrochloric acid solution?
   Experimental Design   100.0 mL of a standard solution of sodium oxalate was prepared using 1.85 g of the dry solid. Using the second endpoint, 10.0 mL samples were titrated with hydrochloric acid.

   Evidence

TITRATION OF 10.0 mL SAMPLES OF Na$_2$OOCCOO$_{(aq)}$ WITH HCl$_{(aq)}$				
Trial	**1**	**2**	**3**	**4**
Final buret reading (mL)	16.1	31.5	46.9	16.9
Initial buret reading (mL)	0.3	16.1	31.5	1.5
Comment on endpoint	poor	good	good	good

   Analysis

70 ACID-BASE STOICHIOMETRY

1. $H_2SO_{4(aq)}$ + 2 $KOH_{(aq)}$ → 2 $HOH_{(l)}$ + $K_2SO_{4(aq)}$
    25.0 mL   18.2 mL
    C 0.358 mol/L

    $$n_{KOH} = 18.2 \text{ mL} \times \frac{0.358 \text{ mol}}{1 \text{ L}} = 6.52 \text{ mmol}$$

    $$n_{H_2SO_4} = 6.52 \text{ mmol} \times \frac{1}{2} = 3.26 \text{ mmol}$$

    $$C_{H_2SO_4} = \frac{3.26 \text{ mmol}}{25.0 \text{ mL}} = 0.130 \text{ mol/L}$$

2. $NaOH_{(aq)}$ + $CH_3COOH_{(aq)}$ → $NaCH_3COO_{(aq)}$ + $HOH_{(l)}$
    13.8 mL   10.0 mL
    0.582 mol/L C

    $$n_{NaOH} = 13.8 \text{ mL} \times \frac{0.582 \text{ mol}}{1 \text{ L}} = 8.03 \text{ mmol}$$

    $$n_{CH_3COOH} = 8.03 \text{ mmol} \times \frac{1}{1} = 8.03 \text{ mmol}$$

    $$C_{CH_3COOH} = \frac{8.03 \text{ mmol}}{10.0 \text{ mL}} = 0.803 \text{ mol/L}$$

3. $Na_3BO_{3(aq)}$ + 2 $HBr_{(aq)}$ → $NaH_2BO_{3(aq)}$ + 2 $NaBr_{(aq)}$
    20.0 mL   15.2 mL
    C 0.241 mol/L

    $$n_{HBr} = 15.2 \text{ mL} \times \frac{0.241 \text{ mol}}{1 \text{ L}} = 3.66 \text{ mmol}$$

    $$n_{Na_3BO_3} = 3.66 \text{ mmol} \times \frac{1}{2} = 1.83 \text{ mmol}$$

    $$C_{Na_3BO_3} = \frac{1.83 \text{ mmol}}{20.0 \text{ mL}} = 91.6 \text{ mmol/L}$$

4. $$n_{Na_2OOCCOO} = 1.85 \text{ g} \times \frac{1 \text{ mol}}{134.00 \text{ g/mol}} = 0.0138 \text{ mol}$$

    $$C_{Na_2OOCCOO} = \frac{0.0138 \text{ mol}}{0.1000 \text{ L}} = 0.138 \text{ mol/L}$$

    2 $HCl_{(aq)}$ + $Na_2OOCCOO_{(aq)}$ → 2 $NaCl_{(aq)}$ + $HOOCCOOH_{(aq)}$
    15.4 mL   10.0 mL
    C 0.138 mol/L

    $$n_{Na_2OOCCOO} = 10.0 \text{ mL} \times \frac{0.138 \text{ mol}}{1 \text{ L}} = 1.38 \text{ mmol}$$

    $$n_{HCl} = 1.38 \text{ mmol} \times \frac{2}{1} = 2.76 \text{ mmol}$$

    $$C_{HCl} = \frac{2.76 \text{ mmol}}{15.4 \text{ mL}} = 0.179 \text{ mol/L}$$

(*The average volume of hydrochloric acid was calculated ignoring the evidence from Trial 1, since an endpoint with low accuracy was indicated.*)

71 (ENRICHMENT) EXCESS PROBLEMS

Titrations involve a procedure where the reaction is stopped when it is quantitative and complete. A pH meter or indicator is used for the purpose of knowing when to stop an acid-base titration. Other contexts, such as plotting pH curves, involve adding an excess of one chemical or not adding enough of one chemical. In both cases the reaction may be stoichiometric and quantitative, but is *not complete*.

Excess problems are recognized in texts and tests by too much information being provided. For reactions in solution, *an excess problem is recognized when the concentration and volume of two chemicals in a reaction are provided*. Questions usually ask for the final concentration of an acid, base, hydronium ion, or hydroxide ion. Other questions may ask for the pH or pOH of the final mixture. The general procedure for solving excess problems is outlined below.

1. Convert both of the concentration and volume values into amounts in moles.

2. Use the mole ratio, and then subtract the amounts, to determine the identity and extent of the excess amount.

3. Add the volumes of the two solutions to determine the final volume of the mixture.

4. Calculate the concentration of the excess chemical from the excess amount and the total volume of solution.

5. Calculate the excess hydronium or hydroxide ion concentration from the percentage ionization or K_a.

6. Convert the hydronium or hydroxide ion concentration to a pH or pOH if required.

EXAMPLE

What is the pH of a solution produced by mixing 50 mL of 0.1 mol/L sodium hydroxide with 20 mL of 0.1 mol/L aqueous hydrogen sulfide?

$$2\,NaOH_{(aq)} \;+\; H_2S_{(aq)} \;\rightarrow\; 2\,HOH_{(l)} \;+\; Na_2S_{(aq)}$$
50 mL   20 mL
0.1 mol/L   0.1 mol/L

Step 1: n_{NaOH} = 50 mL × 0.1 mol/L n_{H_2S} = 20 mL × 0.1 mol/L
                          = 5 mmol   (excess reagent)   = 2 mmol   (limiting reagent)

Step 2: n_{NaOH} = 2 mmol × $\dfrac{2}{1}$ = 4 mmol   (required)

              n_{NaOH} = 5 mmol − 4 mmol = 1 mmol   (in excess)

Steps 3/4: C_{NaOH} = $\dfrac{1\ mmol}{(50 + 20)\ mL}$ = 0.014 mol/L

Step 5: $[OH^-]$ = C_{NaOH} = 0.014 mol/L

            $[H_3O^+_{(aq)}]$ = $\dfrac{1.0 \times 10^{-14}\ mol^2/L^2}{0.014\ mol/L}$ = 7.0 × 10^{-13} mol/L

Step 6:   pH = −log (7.0 × 10^{-13}) = 12.15

1. What is the pH of a solution produced by mixing 100 mL of 0.25 mol/L nitric acid with 50 mL of 0.40 mol/L potassium hydroxide solution?

2. Calculate the hydrogen and hydroxide ion concentrations after 60 mL of 0.1 mol/L hydrochloric acid is reacted with 25 mL of 0.1 mol/L barium hydroxide solution.

3. Calculate the hydrogen and hydroxide ion concentrations after 100 mL of 0.20 mol/L sodium hydroxide solution is reacted with 40 mL of a 0.20 mol/L sulfuric acid solution.

4. A 10 mL sample of a hydrochloric acid solution of unknown concentration was titrated with a standardized 0.0759 mol/L sodium hydroxide solution. The endpoint of 11.5 mL was *overshot* by 1.5 mL of $NaOH_{(aq)}$. According to the evidence, what is the pH of the Erlenmeyer solution after overshooting the endpoint?

71 (ENRICHMENT) EXCESS PROBLEMS

1. $HNO_{3(aq)} + KOH_{(aq)} \rightarrow HOH_{(l)} + KNO_{3(aq)}$
   100 mL   50 mL
   0.25 mol/L 0.4 mol/L

   n_{HNO_3} = 100 mL $\times$ 0.25 mol/L = 25 mmol (excess reagent)
   n_{KOH} = 50 mL $\times$ 0.40 mol/L = 20 mmol (limiting reagent)

   n_{HNO_3} = 20 mmol $\times \dfrac{1}{1}$ = 20 mmol (required)

   n_{HNO_3} = 25 mmol – 20 mmol = 5 mmol (in excess)

   C_{HNO_3} = $\dfrac{5 \text{ mmol}}{100 \text{ mL} + 50 \text{ mL}}$ = 0.03 mol/L (in final mixture)

   $[H^+_{(aq)}] = [HNO_{3(aq)}]$ = 0.03 mol/L
   pH = –log 0.03 = 1.5

2. $2 HCl_{(aq)} + Ba(OH)_{2(aq)} \rightarrow BaCl_{2(aq)} + 2 HOH_{(l)}$
   60 mL   25 mL
   0.1 mol/L   0.1 mol/L

   n_{HCl} = 60 mL $\times$ 0.1 mol/L = 6 mmol (excess reagent)
   $n_{Ba(OH)_2}$ = 25 mL $\times$ 0.1 mol/L = 3 mmol (limiting reagent)

   n_{HCl} = 3 mmol $\times \dfrac{2}{1}$ = 5 mmol (required)

   n_{HCl} = 6 mmol – 5 mmol = 1 mmol (in excess)

   C_{HCl} = $\dfrac{1 \text{ mmol}}{60 \text{ mL} + 25 \text{ mL}}$ = 0.01 mol/L (in final mixutre)

   $[H^+_{(aq)}] = [HCl_{(aq)}]$ = 0.01 mol/L

   $[OH^-_{(aq)}]$ = $\dfrac{1 \times 10^{-14} \text{ (mol/L)}^2}{0.01 \text{ mol/L}}$ = 9×10^{-13} mol/L

3. $H_2SO_{4(aq)} + 2 NaOH_{(aq)} \rightarrow 2 HOH_{(l)} + Na_2SO_{4(aq)}$
   40 mL   100 mL
   0.20 mol/L 0.20 mol/L

   $n_{H_2SO_4}$ = 40 mL $\times$ 0.20 mol/L = 8.0 mmol (limiting reagent)
   n_{NaOH} = 100 mL $\times$ 0.20 mol/L = 20 mmol (excess reagent)

   n_{NaOH} = 8.0 mmol $\times \dfrac{2}{1}$ = 16 mmol (required)

   n_{NaOH} = 20 mmol – 16 mmol = 4 mmol (in excess)

   C_{NaOH} = $\dfrac{4 \text{ mmol}}{100 \text{ mL} + 40 \text{ mL}}$ = 0.03 mol/L

   $[OH^-_{(aq)}] = [NaOH_{(aq)}]$ = 0.03 mol/L

   $[H^+_{(aq)}]$ = $\dfrac{1 \times 10^{-14} \text{ (mol/L)}^2}{0.03 \text{ mol/L}}$ = 4×10^{-13} mol/L

4. $HCl_{(aq)} + NaOH_{(aq)} \rightarrow HOH_{(l)} + NaCl_{(aq)}$
   10.0 mL   (11.5 + 1.5) mL
   C   0.0759 mol/L

   n_{NaOH} = 1.5 mL $\times$ 0.0759 mol/L = 0.11 mmol (in excess)

   C_{NaOH} = $\dfrac{0.11 \text{ mmol}}{10.0 \text{ mL} + 13.0 \text{ mL}}$ = 0.0050 mol/L

   $[H^+_{(aq)}]$ = $\dfrac{1.0 \times 10^{-14} \text{ (mol/L)}^2}{0.0050 \text{ mol/L}}$ = 2.0×10^{-12} mol/L

   pH = –log (2×10^{-12}) = 11.69

VIDEO 1
COLOR

(15 min)   *World of Chemistry Series*

Chemistry is closely related to other branches of science, such as biology and physics, and is an important part of the interaction among science, technology, and society. Although the subject of this video is *color*, the main purpose is to illustrate some aspects of the history and nature of chemistry. You should be able to answer the following questions while viewing the video.

1. State two examples of a connection between

    (a) science and society

    (b) science and technology

    (c) scientific research and color

2. What characteristics are important or necessary for scientists?

3. What is chemistry?

VIDEO 1
COLOR

(15 min)   *World of Chemistry Series*

Chemistry is closely related to other branches of science, such as biology and physics, and is an important part of the interaction among science, technology, and society. Although the subject of this video is *color*, the main purpose is to illustrate some aspects of the history and nature of chemistry. You should be able to answer the following questions while viewing the video.

1. State two examples of a connection between

   (a) science and society

   (Students could also be asked to provide two examples not mentioned in the video for each of (a), (b), and (c).)
   - **the development of new paints, dyes, and drugs as required by society**
   - **Paul Ehrlich's research to find ways to stain bacteria, which led to chemotherapy**

   (b) science and technology
   - **William Perkin's study of coal tar led to the development of commercial dyes.**
   - **The technology of staining tissue samples facilitates scientific research.**
   - **Current technology provides simple means of testing the pH of water in swimming pools, as well as testing for pregnancy and diabetes.**

   (c) scientific research and color
   - **Issac Newton's study of light with prisms (separating white light into a color spectrum)**
   - **Professor Barton's research using dyes to study the structure of DNA**

2. What characteristics are important or necessary for scientists?
   - **curiosity about nature (a scientific attitude)**
   - **keen powers of observation (a scientific skill)**

3. What is chemistry?
   Chemistry is the study of the composition, properties, and changes in matter.

VIDEO 2
PERIODIC TABLE

(15 min)   *World of Chemistry Series*

Understanding the organization of the periodic table is essential for the study of chemistry. This video illustrates some of the evidence that led to the development of the periodic law and the periodic table. Try to answer as many questions as you can while viewing the video. The answers to some of the questions can be found in your textbook.

1. List some properties that are unique to each element.

2. Differentiate between a period and a group in the periodic table.

3. When alkali metals are stored, they must be kept from contacting air and water. In what ways are the following alkali metals stored?
   (a) lithium, sodium and potassium

   (b) rubidium and cesium

4. (a) Name the gas produced when alkali metals come into contact with water.

   (b) Describe the diagnostic test for the gas produced when alkali metals react with water.

5. What is the name for the systematic variation in the properties of the elements in the periodic table?

6. What is the effect on glass when sodium atoms are replaced with potassium atoms?

7. Name the three elements that Mendeleyev predicted would be discovered.

8. Which elements did Glenn Seaborg identify as a series?

9. Provide a theoretical definition of atomic number, mass number, and isotope.

10. How does the size of atoms change as you consider elements going down a group in the periodic table?

11. Which electrons in an atom determine the chemical properties of an element?

VIDEO 2
PERIODIC TABLE

(15 min) *World of Chemistry Series*

Understanding the organization of the periodic table is essential for the study of chemistry. This video illustrates some of the evidence that led to the development of the periodic law and the periodic table. Try to answer as many questions as you can while viewing the video. The answers to some of the questions can be found in your textbook.

1. List some properties that are unique to each element.
   - **melting point**
   - **boiling point**
   - **conductivity**
   - **reactivity**

2. Differentiate between a period and a group in the periodic table.
   A period is a horizontal row of elements in the periodic table.
   A group is a vertical column of elements in the periodic table.

3. When alkali metals are stored, they must be kept from contacting air and water. In what ways are the following alkali metals stored?
   (a) lithium, sodium and potassium

   stored in oil

   (b) rubidium and cesium

   sealed in glass

4. (a) Name the gas produced when alkali metals come into contact with water.

   hydrogen

   (b) Describe the diagnostic test for the gas produced when alkali metals react with water.

   If a flame is inserted into the gas, and a squeal or pop is heard, then hydrogen is likely present.

5. What is the name for the systematic variation in the properties of the elements in the periodic table?

   periodicity

6. What is the effect on glass when sodium atoms are replaced with potassium atoms?
   The glass becomes stronger and resists shattering.

7. Name the three elements that Mendeleyev predicted would be discovered.
   gallium, germanium, and scandium

8. Which elements did Glenn Seaborg identify as a series?

   the actinide series

9. Provide a theoretical definition of atomic number, mass number, and isotope.
   Atomic number is the number of protons in an atom of an element.
   Mass number is the sum of the numbers of protons and neutrons in an atom of an element.
   Isotopes are atoms of the same element that have the same number of protons but different numbers of neutrons — same atomic number but different mass number.

10. How does the size of atoms change as you consider elements going down a group in the periodic table?
    Atomic size increases.

11. Which electrons in an atom determine the chemical properties of an element?
    the valance (outermost) electrons

VIDEO 3
ATOM

(15 min)   *World of Chemistry Series*

This video illustrates some of the experiments that led to the development of modern atomic theory. These experiments provide evidence to support the model of the atom used in the study of high school chemistry. You should be able to answer most of the questions while viewing the video. The answers to some of the questions can be found in your textbook.

1. What kind of electric charges exist and how do they affect one another?

2. List the three main particles believed to be in an atom: name, symbol, location.

3. Consult your text to provide a theoretical definition for atomic number, mass number, and isotope.

4. Draw a simple diagram showing the design and general result of Rutherford's experiment.

5. What part of the atom did Rutherford discover?

6. Different electron cloud arrangements have different _____ .

7. How do atoms get "excited"? What happens after they are excited?

8. The most recent evidence for atoms comes from the scanning tunnelling microscope. (See *Nelson Chemistry* page 64.)

   (a) What do the images show?

   (b) Is it possible to see parts of the atom?

VIDEO 3
ATOM

(15 min)   *World of Chemistry Series*

This video illustrates some of the experiments that led to the development of modern atomic theory. These experiments provide evidence to support the model of the atom used in the study of high school chemistry. You should be able to answer most of the questions while viewing the video. The answers to some of the questions can be found in your textbook.

1. What kind of electric charges exist and how do they affect one another?

   Positive and negative charges exist.

   Like charges repel and unlike charges attract.

2. List the three main particles believed to be in an atom: name, symbol, location.

electron	**e−**	**outside the nucleus**
proton	**p+**	**inside the nucleus**
neutron	**n**	**inside the nucleus**

3. Consult your text to provide a theoretical definition for atomic number, mass number, and isotope.

   atomic number: the number of protons in an atom of an element

   mass number: the sum of the number of protons and neutrons in an atom of an element

   isotopes: atoms of the same element with the same number of protons but different numbers of neutrons

4. Draw a simple diagram showing the design and general result of Rutherford's experiment.

   (See *Nelson Chemistry* page 58.)

5. What part of the atom did Rutherford discover?

   the nucleus

6. Different electron cloud arrangements have different **energy states**.

7. How do atoms get "excited"? What happens after they are excited?

   Atoms are excited by absorbing energy. After they are excited, they lose energy by radiating electromagnetic energy such as light.

8. The most recent evidence for atoms comes from the scanning tunnelling microscope. (See *Nelson Chemistry* page 64.)

   (a) What do the images show?

   The images show the contours of the electron clouds of the atom.

   (b) Is it possible to see parts of the atom?

   Not at present.

VIDEO 4
REACTIONS AND THEIR DRIVING FORCES

(15 min)   *World of Chemistry Series*

This video illustrates a variety of exothermic and endothermic chemical reactions to illustrate the driving forces of reactions. Try to answer as many questions as you can while viewing the video.

1. The following questions refer to the reaction between potassium permanganate and glycerine.

   (a) Is the reaction exothermic or endothermic?

   (b) How does the reaction illustrate the drive toward decreasing energy?

   (c) Sketch a graph of the energy change in this reaction.

2. The entropy of a system is a measure of the degree of disorder in the system. The diffusion of a colored dye in water is an example of increasing entropy. List four general cases where entropy increases.

   (a)   (c)
   (b)   (d)

3. The following questions refer to the reaction between barium hydroxide and ammonium thiocyanate.

   (a) Is the reaction exothermic or endothermic?

   (b) How does the reaction illustrate the drive toward increasing entropy?

   (c) Sketch a graph of the energy change in this reaction.

4. What are the two driving forces of chemical reactions?

VIDEO 4
REACTIONS AND THEIR DRIVING FORCES

(15 min) *World of Chemistry Series*

This video illustrates a variety of exothermic and endothermic chemical reactions to illustrate the driving forces of reactions. Try to answer as many questions as you can while viewing the video. *The content and questions concerning entropy go beyond the scope of this course and the textbook.*

1. The following questions refer to the reaction between potassium permanganate and glycerine.

   (a) Is the reaction exothermic or endothermic?

   exothermic

   (b) How does the reaction illustrate the drive toward decreasing energy?

   When potassium permanganate and glycerine are brought together, a spontaneous reaction occurs that releases energy from the system to the surroundings.

   (c) Sketch a graph of the energy change in this reaction.

   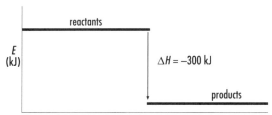

2. The entropy of a system is a measure of the degree of disorder in the system. The diffusion of a colored dye in water is an example of increasing entropy. List four general cases where entropy increases.

   (a) **more molecules form**   (c) **gas forms from liquid or solid**
   (b) **liquid forms from solid**   (d) **a mixture is formed**

3. The following questions refer to the reaction between barium hydroxide and ammonium thiocyanate.

   (a) Is the reaction exothermic or endothermic?

   endothermic

   (b) How does the reaction illustrate the drive toward increasing entropy?

   When barium hydroxide and ammonium thiocyanate are mixed, a spontaneous reaction occurs, producing a more disordered state even though the reaction requires that energy be absorbed from the surroundings.

   (c) Sketch a graph of the energy change in this reaction.

   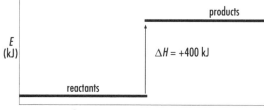

4. What are the two driving forces of chemical reactions?

   • **decreasing the energy of the system**

   • **increasing the entropy of the system**

VIDEO 5
THE CHEMISTRY OF SOLUTIONS

(12 min) *Senior High Science Series*

Understanding the properties of solutions is essential for the study of chemistry. This video illustrates several solutions and explains the interaction of the solutes and solvents. Try to answer as many questions as you can while viewing the video. The answers to some of the questions can be found in your textbook.

1. What is a solution?

2. Give an example of each of the following.

   (a) a solid dissolved in a liquid

   (b) a liquid dissolved in a liquid

   (c) a solid dissolved in a solid

   (d) a gas dissolved in a gas

   (e) a gas dissolved in liquid

3. What is an amalgam?

4. How can the components of a chlorophyll solution be separated by paper chromatography?

5. What is a polar molecule?

6. What is a non-polar molecule?

7. What is the general rule about the solubility of polar and non-polar substances?

8. Why do solutions of ionic compounds conduct electricity while pure ionic compounds do not conduct?

VIDEO 5
THE CHEMISTRY OF SOLUTIONS

(12 min)   *Senior High Science Series*

Understanding the properties of solutions is essential for the study of chemistry. This video illustrates several solutions and explains the interaction of the solutes and solvents. Try to answer as many questions as you can while viewing the video. The answers to some of the questions can be found in your textbook.

1. What is a solution?

   A solution is a homogeneous mixture of dissolved substances containing at least one solute and one solvent.

2. Give an example of each of the following.

   (a) a solid dissolved in a liquid

   sugar in coffee

   (b) a liquid dissolved in a liquid

   soap in water

   (c) a solid dissolved in a solid

   Bronze is a solution (alloy) of copper and tin.

   (d) a gas dissolved in a gas

   Air is a solution of nitrogen and oxygen.

   (e) a gas dissolved in liquid

   Carbonated beverages are solutions of carbon dioxide in water.

3. What is an amalgam?

   an alloy of mercury and another metal

4. How can the components of a chlorophyll solution be separated by paper chromatography?

   Place a drop of chlorophyll solution on a strip of chromatography paper. Then dip the end of the paper into a solvent. As the solvent moves up the paper by capillary action, it carries the chlorophyll with it. The different components of chlorophyll travel at different rates, causing the components to separate.

5. What is a polar molecule?

   a molecule with a slightly uneven charge distribution, with oppositely charged ends

6. What is a non-polar molecule?

   a molecule with an even charge distribution that is not affected by nearby charged objects

7. What is the general rule about the solubility of polar and non-polar substances?

   Like dissolves like.

8. Why do solutions of ionic compounds conduct electricity while pure ionic compounds do not conduct?

   In order to conduct electricity, ions must be free to move. Ions in a solution can move but those in a solid cannot.

VIDEO 6
CHEMICAL BONDS

(15 min)   *World of Chemistry Series*

Chemical bonds are the interatomic glue that hold together the building blocks of matter. This video illustrates the properties of ionic, network covalent, and molecular compounds, and explains their properties in terms of bonding theory. Watch for the answers to the following questions while viewing the video. The answers to some of the questions can be found in your textbook.

1. Write a theoretical description of each of the following.

    (a) ionic bonds

    (b) covalent bonds

2. List the two basic rules of bond formation.

    (a)

    (b)

3. Sodium chloride is an important ionic compound.

    (a) Describe the formation of an ionic bond between sodium and chlorine.

    (b) What does the formula NaCl indicate about the composition of sodium chloride?

4. Mineral crystals, such as quartz, involve covalent bonds.

    (a) Describe the properties of quartz.

    (b) Describe the chemical bonds in quartz.

5. Molecular compounds also involve covalent bonds.

    (a) Describe the properties of molecular compounds.

    (b) How does the bonding in molecular compounds differ from that in network covalent compounds?

6. What is the energy change that accompanies the formation of chemical bonds?

VIDEO 6
CHEMICAL BONDS

(15 min)   *World of Chemistry Series*

Chemical bonds are the interatomic glue that hold together the building blocks of matter. This video illustrates the properties of ionic, network covalent, and molecular compounds, and explains their properties in terms of bonding theory. Watch for the answers to the following questions while viewing the video. The answers to some of the questions can be found in your textbook.

1. Write a theoretical description of each of the following.
   (a) ionic bonds

   the simultaneous attraction of positive and negative ions

   (b) covalent bonds

   the simultaneous attraction of two nuclei for a shared pair of electrons

2. List the two basic rules of bond formation.
   (a) **Atoms tend to lose, gain, or share electrons.**
   (b) **Electrons tend to exist in pairs.**

3. Sodium chloride is an important ionic compound.
   (a) Describe the formation of an ionic bond between sodium and chlorine.

   A sodium atom gives up a valence electron which is accepted by a chlorine atom. As a result of this electron transfer, the sodium atom acquires a positive charge and the chlorine atom acquires a negative charge. The sodium and chloride ions formed are bonded together by the attraction between their opposite charges.

   (b) What does the formula NaCl indicate about the composition of sodium chloride?

   The ratio of sodium ions to chloride ions is 1 to 1.

4. Mineral crystals, such as quartz, involve covalent bonds.
   (a) Describe the properties of quartz.

   Quartz has a very high melting point, is hard, and is generally non-reactive.

   (b) Describe the chemical bonds in quartz.

   Silicon and oxygen form covalent bonds, which form a three-dimensional network. The ratio of silicon to oxygen atoms in the network is 1 to 2.

5. Molecular compounds also involve covalent bonds.
   (a) Describe the properties of molecular compounds.

   Molecular compounds have relatively low melting points and boiling points.

   (b) How does the bonding in molecular compounds differ from that in network covalent compounds?

   The covalent bonding within molecular compounds involves relatively small numbers of atoms. There are no covalent bonds between molecules.

   The bonding in covalent compounds involves all of the atoms in the crystal.

6. What is the energy change that accompanies the formation of chemical bonds?

   Energy is released by the formation of chemical bonds.

VIDEO 7
WATER

(15 min) *World of Chemistry Series*

Understanding the properties of water is important for the study of chemistry. This video illustrates some of the unique properties of water and explains them in terms of hydrogen bonding. Try to answer as many questions as you can while viewing the video. The answers to some of the questions can be found in your textbook.

1. Draw a diagram of a water molecule, showing its shape and polarity.

2. Draw a diagram to show how hydrogen bonds between water molecules result in a regular hexagonal crystal structure in ice.

3. Why is ice less dense than liquid water?

4. How can snow be produced on ski slopes at temperatures above 0°C?

5. Why are ionic and polar compounds generally soluble in water?

6. Draw a diagram to show the arrangement of water molecules around
    (a) a sodium ion   (b) a chloride ion

7. Why are non-polar compounds insoluble in water?

8. List the major uses of water in industry.

263

VIDEO 7
WATER

(15 min)   *World of Chemistry Series*

Understanding the properties of water is important for the study of chemistry. This video illustrates some of the unique properties of water and explains them in terms of hydrogen bonding. Try to answer as many questions as you can while viewing the video. The answers to some of the questions can be found in your textbook.

1. Draw a diagram of a water molecule, showing its shape and polarity.

$\delta^+ H \underset{}{\overset{O \, \delta^-}{\diagup}} H \, \delta^+$

2. Draw a diagram to show how hydrogen bonds between water molecules result in a regular hexagonal crystal structure in ice. (**See Nelson Chemistry page 229.**)

3. Why is ice less dense than liquid water?

    Water molecules in ice are arranged in a six-sided open pattern that is more closely packed than the more random arrangement of water molecules in the liquid phase.

4. How can snow be produced on ski slopes at temperatures above 0°C?

    Adding bacteria to the water aids the formation of ice crystals, and thereby raises the freezing point of water.

5. Why are ionic and polar compounds generally soluble in water?

    Water molecules are polar and they attract the positive and negative ions of an ionic compound and pull them out of the crystal into the solution. Water molecules are also attracted to other polar molecules and draw them into solution.

6. Draw a diagram to show the arrangement of water molecules around

    (a) a sodium ion   (b) a chloride ion (**See Nelson Chemistry page 115.**)

    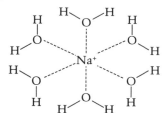   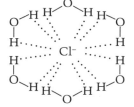

7. Why are non-polar compounds insoluble in water?

    Water molecules are polar and strongly attracted to each other, but are not attracted to non-polar molecules.

8. List the major uses of water in industry.
    *   **heat transfer**   • **as a chemical reactant**   • **as a solvent**

                                        CHAPTER 8   PAGE 229

VIDEO 8
CARBON THE COMPROMISER

(10 min)   *Organic Chemistry Series* TV Ontario

Organic chemistry refers to the study of the molecular compounds of carbon. This video describes the origins of organic chemistry in the work of the German chemists Wöhler and Kekulé. Watch for the answers to the following questions while viewing the video. The answers to some of the questions can be found in your textbook.

1. Describe the theory of *vitalism*.

2. In 1828, Friedrich Wöhler conducted an experiment that falsified the theory of vitalism. Describe Wöhler's experiment.

3. In the mid-19th century, August Kekulé proposed a theory to explain the molecular formulas of organic compounds.

   (a) Describe Kekulé's theory.

   (b) How did Kekulé's theory account for the large number of compounds that carbon forms?

   (c) What structural formula did Kekulé propose for benzene?

4. What general rule is used to predict the gain or loss of electrons when atoms form ions?

5. *Electronegativity* is a number that describes the relative ability of an atom to attract a pair of electrons in its valence level.

   (a) What is the range of electronegativity values?

   (b) What is the electronegativity of carbon?

   (c) How does carbon achieve a noble gas electronic structure?

VIDEO 8
CARBON THE COMPROMISER

(10 min)   *Organic Chemistry Series* TV Ontario

Organic chemistry refers to the study of the molecular compounds of carbon. This video describes the origins of organic chemistry in the work of the German chemists Wöhler and Kekulé. Watch for the answers to the following questions while viewing the video. The answers to some of the questions can be found in your textbook.

1. Describe the theory of *vitalism*.

   The theory of vitalism stated that organic chemicals could only be produced by living organisms.

2. In 1828, Friedrich Wöhler conducted an experiment that falsified the theory of vitalism. Describe Wöhler's experiment.

   Wöhler heated the inorganic compound ammonium cyanate, NH_4OCN, to synthesize the organic compound urea, H_2NCONH_2. It was previously believed that urea could only be produced by an animal.

3. In the mid-19th century, August Kekulé proposed a theory to explain the molecular formulas of organic compounds.

   (a) Describe Kekulé's theory.

      A carbon atom always forms four bonds.

   (b) How did Kekulé's theory account for the large number of compounds that carbon forms?

      Kekulé's theory could account for straight chain and branched chain molecules.

   (c) What structural formula did Kekulé propose for benzene?

      Kekulé proposed a ring structure for benzene.

4. What general rule is used to predict the gain or loss of electrons when atoms form ions?

   When atoms form ions, they gain or lose electrons to acquire the same number of electrons as atoms of the nearest noble gas.

5. *Electronegativity* is a number that describes the relative ability of an atom to attract a pair of electrons in its valence level.

   (a) What is the range of electronegativity values?

      Electronegativity values range from 0.7 to 4.0.

   (b) What is the electronegativity of carbon?

      The electronegativity of carbon is 2.5.

   (c) How does carbon achieve a noble gas electronic structure?

      Carbon shares electrons with other atoms to achieve a noble gas electronic structure.

VIDEO 9
FIXING FUELS

(10 min) *Organic Chemistry Series* TV Ontario

Most of the fuels in use today are compounds of carbon. This video discusses the structure of fuels extracted from natural gas and crude oil. Watch for the answers to the following questions while viewing the video. The answers to some of the questions can be found in your textbook.

1. Compounds with the same molecular formula but with different structures are called *isomers*.

   (a) Draw a structural diagram of a straight chain isomer of C_4H_{10}.

   (b) Draw a structural diagram of a branched chain isomer of C_4H_{10}.

   (c) Draw a structural diagram of a ring structure of C_6H_{12}.

   (d) As the number of carbon atoms in a compound increases, how does the number of possible isomers change?

2. List the five main components of natural gas, and indicate which ones are removed before consumer use.

3. Crude oil is composed of a wide variety of organic compounds.

   (a) What property of matter is used to separate the components of crude oil?

   (b) Describe the process of fractional distillation.

   (c) Describe the process of catalytic cracking.

   (d) Why is catalytic cracking necessary?

VIDEO 9
FIXING FUELS

(10 min)   *Organic Chemistry Series* TV Ontario

Most of the fuels in use today are compounds of carbon. This video discusses the structure of fuels extracted from natural gas and crude oil. Watch for the answers to the following questions while viewing the video. The answers to some of the questions can be found in your textbook.

1. Compounds with the same molecular formula but with different structures are called *isomers*.

   (a) Draw a structural diagram of a straight chain isomer of C_4H_{10}.

   (b) Draw a structural diagram of a branched chain isomer of C_4H_{10}.

   (c) Draw a structural diagram of a ring structure of C_6H_{12}.

    or

   (d) As the number of carbon atoms in a compound increases, how does the number of possible isomers change?

   The number of possible isomers increases rapidly.

2. List the five main components of natural gas, and indicate which ones are removed before consumer use.

   - **methane**
   - **ethane**
   - **water vapor (removed)**
   - **propane**
   - **hydrogen sulfide (removed)**

3. Crude oil is composed of a wide variety of organic compounds.

   (a) What property of matter is used to separate the components of crude oil?

   boiling point

   (b) Describe the process of fractional distillation.

   Crude oil is heated until it begins to boil, and is then piped into the bottom of a fractionating column. As the vapors from the boiling crude pass up the fractionating column, they cool and condense on trays. The compounds composed of heavier molecules condense first, followed by progressively lighter fractions as they move up the tower.

   (c) Describe the process of catalytic cracking.

   Heavier hydrocarbons, C_{30} to C_{40}, are broken down into smaller hydrocarbons by passing them through a reaction vessel containing a catalyst.

   (d) Why is catalytic cracking necessary?

   Not enough straight-run gasoline is obtained by fractional distillation of crude oil to satisfy consumer demand for gasoline.

VIDEO 10
ATMOSPHERE

(15 min)   *World of Chemistry Series*

This video describes the origin and composition of the atmosphere of Earth, and the impact of recent human activities on the atmosphere. The content of the video is closely related to the features *Global Warming* (page 294) and *Ozone Depletion* (page 257) in your textbook.

1. List the gases that scientists believe were part of Earth's primitive atmosphere.

2. What event do scientists believe that caused oxygen to become a major component of Earth's atmosphere?

3. List, in order of abundance, the gases that make up Earth's present atmosphere.

4. Name three methods that are used to gather evidence on the atmosphere of ancient Earth.

5. (a) Draw a diagram of the carbon cycle.

   (b) Why has the level of carbon dioxide increased during the last few centuries?

6. What is the *Greenhouse Effect* and why is it a concern?

7. (a) What is the *ozone layer* and how does it benefit humans?

   (b) What is the apparent cause of ozone depletion?

   (c) Write a series of chemical equations to represent the reactions between chlorine and ozone in the upper atmosphere.

VIDEO 10
ATMOSPHERE

(15 min)   *World of Chemistry Series*

This video describes the origin and composition of the atmosphere of Earth, and the impact of recent human activities on the atmosphere. The content of the video is closely related to the features *Global Warming* (page 294) and *Ozone Depletion* (page 257) in your textbook.

1. List the gases that scientists believe were part of Earth's primitive atmosphere.

   $N_{2(g)}$, $CO_{(g)}$, $H_2O_{(g)}$, $CH_{4(g)}$, $NH_{3(g)}$

2. What event do scientists believe that caused oxygen to become a major component of Earth's atmosphere?

   the evolution of photosynthetic plants

3. List, in order of abundance, the gases that make up Earth's present atmosphere.

   $N_{2(g)}$, $O_{2(g)}$, $Ar_{(g)}$, $CO_{2(g)}$, $Ne_{(g)}$, $He_{(g)}$, $Kr_{(g)}$, $Xe_{(g)}$, $H_{2(g)}$, $O_{3(g)}$

4. Name three methods that are used to gather evidence on the atmosphere of ancient Earth.

   analyzing samples taken from tree rings, glacial ice, and ocean floor sediments

5. (a) Draw a diagram of the carbon cycle.

   (See *Nelson Chemistry* page 275.)

   (b) Why has the level of carbon dioxide increased during the last few centuries?

   increased burning of fossil fuels as a result of world wide industrialization

6. What is the *Greenhouse Effect* and why is it a concern?

   Sunlight passes through the atmosphere and warms up the surface of Earth, which then emits long wavelength infrared radiation. Some of the infrared radiation is trapped by greenhouse gases such as carbon dioxide, CFCs, and methane, which causes an overall warming of the atmosphere. Even small increases in the average temperature of Earth could have a profound effect on world food production and ocean levels.

7. (a) What is the *ozone layer* and how does it benefit humans?

   The layer of ozone in the upper atmosphere filters out certain frequencies of ultraviolet radiation that is harmful to life on Earth, e.g. causing skin cancer in humans.

   (b) What is the apparent cause of ozone depletion?

   Chlorine from CFCs reacts with ozone in the upper atmosphere.

   (c) Write a series of chemical equations to represent the reactions between chlorine and ozone in the upper atmosphere.

   $Cl_{(g)} + O_{3(g)} \rightarrow ClO_{(g)} + O_{2(g)}$

   $ClO_{(g)} + O_{(g)} \rightarrow Cl_{(g)} + O_{2(g)}$

   The chlorine is recycled to react with more ozone, so that for every chlorine atom released into the atmosphere, thousands of ozone molecules are destroyed.

VIDEO 11
CONSERVATION OF ENERGY

(29 min)   *Mechanical Universe Series*

The laws of thermodynamics are fundamental to understanding the energy changes in chemical reactions. This video illustrates several energy transformations and explains them in terms of the First Law of Thermodynamics.

1. State the First Law of Thermodynamics.

2. List several different forms of energy.

3. Write a mathematical expression for gravitational potential energy (E_p).

4. Write a mathematical expression for kinetic energy (E_k).

5. (a) Write a mathematical expression to show the total energy of a system as the sum of its potential and kinetic energies.

    (b) Describe the energy changes involved when a pole vaulter makes a successful jump.

6. Describe the experimental work done by James Prescott Joule that helped establish the law of conservation of energy.

7. What becomes of the mechanical energy of an object when it comes to rest on the ground?

8. Complete the statement, "Energy is conserved, but . . ."

VIDEO 11
CONSERVATION OF ENERGY

(29 min) *Mechanical Universe Series*

The laws of thermodynamics are fundamental to understanding the energy changes in chemical reactions. This video illustrates several energy transformations and explains them in terms of the First Law of Thermodynamics.

1. State the First Law of Thermodynamics.

   Energy cannot be created or destroyed, but is converted from one form into another. (Conservation of Energy)

2. List several different forms of energy.

   solar, nuclear, chemical, mechanical

3. Write a mathematical expression for gravitational potential energy (E_p).

   $E_p = mgh$

4. Write a mathematical expression for kinetic energy (E_k).

   $E_k = \frac{1}{2}mv^2$

5. (a) Write a mathematical expression to show the total energy of a system as the sum of its potential and kinetic energies.

   $E_{total} = E_p + E_k$

   (b) Describe the energy changes involved when a pole vaulter makes a successful jump.

   The pole changes the kinetic energy of the pole vaulter into potential energy as the vaulter gains height and passes over the bar. As the vaulter falls potential energy changes back into kinetic energy and then thermal energy.

6. Describe the experimental work done by James Prescott Joule that helped establish the law of conservation of energy.

   Joule used a system of weights and pulleys to stir water with a paddle. From the evidence collected he was able to establish that a given quantity of mechanical energy is always converted into the same quantity of thermal energy.

7. What becomes of the mechanical energy of an object when it comes to rest on the ground?

   The mechanical energy is converted into thermal energy and sound.

8. Complete the statement, "Energy is conserved, but ..."

   ...we render it useless."

VIDEO 12
THE BUILDING BLOCKS OF ELECTROCHEMISTRY

(10 min)   *Electrochemistry Series*

Electron transfer reactions are common to all voltaic and electrolytic cells. This video uses the context of voltaic cells to define the basic terms and conventions used in electrochemistry.

1. Write a half-reaction equation to represent each of the following chemical processes.

    (a) the reduction of chlorine

    (b) the oxidation of sodium

2. Define the following terms theoretically.

    (a) oxidation
    (b) reduction
    (c) oxidizing agent
    (d) reducing agent

3. Why do oxidation and reduction always occur together?

4. Identify the following parts of a voltaic cell.

    (a) the electrode where oxidation occurs
    (b) the electrode where reduction occurs
    (c) the positive electrode
    (d) the negative electrode

5. (a) What direction do the ions move in a voltaic cell?

    (b) What is the apparent discrepancy between the direction of ion flow and the charges on the electrodes?

    (c) How is this discrepancy resolved?

(10 min)   *Electrochemistry Series*

Electron transfer reactions are common to all voltaic and electrolytic cells. This video uses the context of voltaic cells to define the basic terms and conventions used in electrochemistry.

1. Write a half-reaction equation to represent each of the following chemical processes.

    (a) the reduction of chlorine

    $$Cl + e^- \rightarrow Cl^-$$

    (b) the oxidation of sodium

    $$Na \rightarrow Na^+ + e^-$$

2. Define the following terms theoretically.

    (a) oxidation   **loss of electrons**

    (b) reduction   **gain of electrons**

    (c) oxidizing agent **the chemical species that causes oxidation (by gaining electrons)**

    (d) reducing agent   **the chemical species that causes reduction (by losing electrons)**

3. Why do oxidation and reduction always occur together?

    Electrons are transferred from one chemical species to another in a redox reaction, so one species gains electrons (is reduced) while the other species loses electrons (is oxidized).

4. Identify the following parts of a voltaic cell.

    (a) the electrode where oxidation occurs   **anode**

    (b) the electrode where reduction occurs   **cathode**

    (c) the positive electrode   **cathode**

    (d) the negative electrode   **anode**

5. (a) What direction do the ions move in a voltaic cell?

    Anions move toward the anode (in the same circuit direction as the electrons).

    Cations move toward the cathode.

    (b) What is the apparent discrepancy between the direction of ion flow and the charges on the electrodes?

    It seems as if positive ions are being attracted to the positive electrode, and as if negative ions are being attracted to the negative electrode.

    (c) How is this discrepancy resolved?

    The reduction reaction at the cathode reduces cations, which removes positive charge from the solution surrounding the cathode. Other cations move into this region. Similarly, the oxidation reaction at the anode removes negative ions from the solution surrounding the anode, and other anions move into the region to replace them.

VIDEO 13
REVERSIBLE REACTIONS AND DYNAMIC EQUILIBRIUM

(11 min)   *Senior High Science Series*

This video illustrates several chemical systems that are in a state of dynamic equilibrium. Le Châtelier's principle is used to explain the shifts in equilibrium that occur when the systems are disturbed.

1. The effect of changing temperature on an equilibrium is illustrated by the dinitrogen tetraoxide-nitrogen dioxide equilibrium.

    $N_2O_{4(g)}$ + energy $\rightleftharpoons$ $2NO_{2(g)}$
    colorless   reddish brown

    (a) Predict and explain the change that occurs when the system is placed in hot water.

    (b) Predict and explain the change that occurs when the system is placed in ice water.

2. The effect of changing concentration on an equilibrium is illustrated by the chromate-dichromate equilibrium.

    $2CrO_4^{2-}{}_{(aq)}$ + $2H^+{}_{(aq)}$ $\rightleftharpoons$ $Cr_2O_7^{2-}{}_{(aq)}$ + $H_2O_{(l)}$
    yellow   orange

    (a) Predict and explain the change that occurs when $HCl_{(aq)}$ is added to the system.

    (b) Predict and explain the change that occurs when $KOH_{(aq)}$ is added to the system.

3. The effect of changing pressure on an equilibrium is illustrated by the carbon dioxide-carbonic acid equilibrium.

    $CO_{2(g)}$ + $H_2O_{(l)}$ $\rightleftharpoons$ $H_2CO_{3(aq)}$

    (a) Predict and explain the change caused by increasing the pressure on the system.

    (b) Predict and explain the change caused by decreasing the pressure on the system.

4. Additional evidence of dynamic equilibrium is obtained from a saturated solution of aqueous silver iodide in contact with solid silver iodide.

    $AgI_{(s)}$ $\rightleftharpoons$ $Ag^+{}_{(aq)}$ + $I^-{}_{(aq)}$

    Describe and explain what happens when solid radioactive silver iodide is added.

5. Stalactites and stalagmites in limestone caverns are formed by the following equilibrium system.

    $CaCO_{3(s)}$ + $CO_{2(g)}$ + $H_2O_{(l)}$ $\rightleftharpoons$ $Ca^{2+}{}_{(aq)}$ + $2HCO_3^-{}_{(aq)}$

    Predict the effect on this equilibrium of large numbers of people visiting the limestone caverns.

NELSON CHEMISTRY
TEACHER'S RESOURCE MASTERS

VIDEO 13
REVERSIBLE REACTIONS AND DYNAMIC EQUILIBRIUM

(11 min) *Senior High Science Series*

This video illustrates several chemical systems that are in a state of dynamic equilibrium. Le Châtelier's principle is used to explain the shifts in equilibrium that occur when the systems are disturbed.

1. The effect of changing temperature on an equilibrium is illustrated by the dinitrogen tetraoxide-nitrogen dioxide equilibrium.

   $N_2O_{4(g)}$ + energy $\rightleftharpoons$ $2NO_{2(g)}$
   colorless   reddish brown

   (a) Predict and explain the change that occurs when the system is placed in hot water.

   The reddish brown color becomes darker because the equilibrium shifts to the right, using up some of the thermal energy that is added and producing more $NO_{2(g)}$.

   (b) Predict and explain the change that occurs when the system is placed in ice water.

   The reddish brown color becomes lighter because the equilibrium shifts to the left, producing thermal energy and consuming $NO_{2(g)}$.

2. The effect of changing concentration on an equilibrium is illustrated by the chromate-dichromate equilibrium.

   $2CrO_4^{2-}{}_{(aq)}$ + $2H^+{}_{(aq)}$ $\rightleftharpoons$ $Cr_2O_7^{2-}{}_{(aq)}$ + $H_2O_{(l)}$
   yellow   orange

   (a) Predict and explain the change that occurs when $HCl_{(aq)}$ is added to the system.

   The solution turns a dark orange color because the equilibrium shifts to the right, consuming some of the added $H^+{}_{(aq)}$ and producing more $Cr_2O_7^{2-}{}_{(aq)}$.

   (b) Predict and explain the change that occurs when $KOH_{(aq)}$ is added to the system.

   The solution turns yellow because the equilibrium shifts to the left, producing more $H^+{}_{(aq)}$ and $CrO_4^{2-}{}_{(aq)}$ — replacing the $H^+{}_{(aq)}$ removed by the $OH^-{}_{(aq)}$.

3. The effect of changing pressure on an equilibrium is illustrated by the carbon dioxide-carbonic acid equilibrium.

   $CO_{2(g)}$ + $H_2O_{(l)}$ $\rightleftharpoons$ $H_2CO_{3(aq)}$

   (a) Predict and explain the change caused by increasing the pressure on the system.

   The pH decreases because the equilibrium shifts to the right, producing more acid.

   (b) Predict and explain the change caused by decreasing the pressure on the system.

   The pH increases because the equilibrium shifts to the left, reducing the concentration of the acid.

4. Additional evidence of dynamic equilibrium is obtained from a saturated solution of aqueous silver iodide in contact with solid silver iodide.

   $AgI_{(s)}$ $\rightleftharpoons$ $Ag^+{}_{(aq)}$ + $I^-{}_{(aq)}$

   Describe and explain what happens when solid radioactive silver iodide is added.

   The solution becomes radioactive because some of the solid dissolves.

5. Stalactites and stalagmites in limestone caverns are formed by the following equilibrium system.

   $CaCO_{3(s)}$ + $CO_{2(g)}$ + $H_2O_{(l)}$ $\rightleftharpoons$ $Ca^{2+}{}_{(aq)}$ + $2HCO_3^-{}_{(aq)}$

   Predict the effect on this equilibrium of large numbers of people visiting the limestone caverns.

   Increasing the number of people visiting the caverns will raise the $CO_{2(g)}$ levels and cause the equilibrium to shift to the right, allowing more $CaCO_{3(s)}$ to dissolve.

VIDEO 14
THE HABER PROCESS

(10 min) *Chemical Equilibrium Series*

This video describes the industrial method for producing ammonia developed by Fritz Haber in Germany during World War I. The Haber process illustrates Le Châtelier's principle as well as factors affecting reaction rates. The content of the video is closely related to the feature *The Haber Process: A Case Study in Technology* (page 443) in your textbook.

1. What is "fixed nitrogen" and how is it produced in nature?

2. The formation of ammonia is represented by the following chemical equation.

$$N_{2(g)} + 3H_{2(g)} \rightleftharpoons 2NH_{3(g)} + 92 \text{ kJ}$$

   (a) According to Le Châtelier's principle, what conditions are required to give a high yield of $NH_{3(g)}$?

   (b) What conditions did Haber use to produce commercial quantities of $NH_{3(g)}$?

   (c) How did Haber deal with the slowness of the reaction?

3. What is the main use of the Haber process today?

VIDEO 14
THE HABER PROCESS

(10 min)   *Chemical Equilibrium Series*

This video describes the industrial method for producing ammonia developed by Fritz Haber in Germany during World War I. The Haber process illustrates Le Châtelier's principle as well as factors affecting reaction rates. The content of the video is closely related to the feature *The Haber Process: A Case Study in Technology* (page 443) in your textbook.

1. What is "fixed nitrogen" and how is it produced in nature?

    Fixed nitrogen is nitrogen in compounds such as nitrogen monoxide, nitrogen dioxide, ammonia, and nitrates.

2. The formation of ammonia is represented by the following chemical equation.
    $$N_{2(g)} + 3H_{2(g)} \rightleftharpoons 2NH_{3(g)} + 92 \text{ kJ}$$

    (a) According to Le Châtelier's principle, what conditions are required to give a high yield of $NH_{3(g)}$?
    - **high pressure**
    - **low temperature**
    - **high concentration of nitrogen and hydrogen**
    - **continuous removal of ammonia**

    (b) What conditions did Haber use to produce commercial quantities of $NH_{3(g)}$?
    - **high pressure**
    - **high temperature**
    - **remove ammonia as it is produced**

    (c) How did Haber deal with the slowness of the reaction?
    He introduced a catalyst into the reaction vessel.

3. What is the main use of the Haber process today?

    The Haber process is used to produce large quantities of ammonia for use directly as a chemical fertilizer and indirectly as a reactant for the production of fertilizers such as ammonia sulfate, nitrate, and phosphate.

VIDEO 15
ACIDS AND BASES

(15 min)   *World of Chemistry Series*

This video illustrates the properties of acids and bases, and explains acid-base reactions in terms of proton transfer.

1. Classify the following compounds as acids or bases and provide a name and use for each one.
    (a) $NH_{3(aq)}$
    (b) $CaO_{(s)}$

    (c) $H_2SO_{4(aq)}$
    (d) $C_3H_4OH(COOH)_{3(aq)}$
    (e) $CH_3COOH_{(aq)}$

2. List four empirical properties of acids.

3. Write non-ionic equations to represent the following acid-base reactions.
    (a) hydrogen chloride and ammonia

    (b) nitric acid and aqueous sodium hydroxide

    (c) acetic acid and aqueous sodium bicarbonate

    (d) stomach acid and aqueous sodium bicarbonate

4. Write an equation to represent each of the following reactions of water.
    (a) accepting a proton from aqueous hydrogen chloride

    (b) donating a proton to aqueous ammonia

5. By how much does a pH change of one unit change the concentration of hydronium ions?

6. What are the factors that determine the pH of an acid?

7. Acid rain is a serious concern in many parts of the world.
    (a) Name two natural and two man-made causes of acid rain.

    (b) Name two compounds that are precursors of acid rain.

    (c) List two effects of acid rain on the environment.

VIDEO 15
ACIDS AND BASES

(15 min)   *World of Chemistry Series*

This video illustrates the properties of acids and bases, and explains acid-base reactions in terms of proton transfer.

1. Classify the following compounds as acids or bases and provide a name and use for each one.

   (a) $NH_{3(aq)}$ **base, ammonia, household cleaner**

   (b) $CaO_{(s)}$ **base, calcium oxide (lime), neutralize acid soil and lakes and is used for water and sewage treatment**

   (c) $H_2SO_{4(aq)}$ **acid, sulfuric acid, lead-acid car battery**

   (d) $C_3H_4OH(COOH)_{3(aq)}$ **acid, citric acid, beverages**

   (e) $CH_3COOH_{(aq)}$ **acid, acetic acid, vinegar**

2. List four empirical properties of acids.

   - **turn blue litmus red**
   - **produce hydrogen gas when reacted with active metals like magnesium**
   - **produce carbon dioxide gas when reacted with limestone**
   - **neutralize bases**

3. Write non-ionic equations to represent the following acid-base reactions.

   (a) hydrogen chloride and ammonia

   $$HCl_{(g)} + NH_{3(g)} \rightarrow NH_4Cl_{(s)}$$

   (b) nitric acid and aqueous sodium hydroxide

   $$HNO_{3(aq)} + NaOH_{(aq)} \rightarrow NaNO_{3(aq)} + HOH_{(l)}$$

   (c) acetic acid and aqueous sodium bicarbonate

   $$CH_3COOH_{(aq)} + NaHCO_{3(aq)} \rightarrow NaCH_3COO + H_2O_{(l)} + CO_{2(g)}$$

   (d) stomach acid and aqueous sodium bicarbonate

   $$HCl_{(aq)} + NaHCO_{3(aq)} \rightarrow NaCl_{(aq)} + H_2CO_{3(aq)} \qquad (H_2O_{(l)} + CO_{2(g)})$$

4. Write an equation to represent each of the following reactions of water.

   (a) accepting a proton from aqueous hydrogen chloride

   $$H_2O_{(l)} + HCl_{(aq)} \rightarrow H_3O^+_{(aq)} + Cl^-_{(aq)}$$

   (b) donating a proton to aqueous ammonia

   $$H_2O_{(l)} + NH_{3(aq)} \rightarrow OH^-_{(aq)} + NH_4^+_{(aq)}$$

5. By how much does a pH change of one unit change the concentration of hydronium ions?
   by a factor of 10

6. What are the factors that determine the pH of an acid?
   the strength and concentration of the acid

7. Acid rain is a serious concern in many parts of the world.

   (a) Name two natural and two man-made causes of acid rain.

   natural causes of acid rain — forest fires and volcanos

   man-made causes of acid rain — automobiles and coal burning power plants

   (b) Name two compounds that are precursors of acid rain.

   nitrogen dioxide and sulfur dioxide

   (c) List two effects of acid rain on the environment.

   - **damage to forests and other plant life**
   - **aquatic life threatened by changing pH of lakes**

Problem
Prediction
Design
Materials
Procedure
✔ **Evidence**
✔ **Analysis**
Evaluation
Synthesis

CAUTION

You must never look directly at burning magnesium. The bright flame emits ultraviolet radiation that could harm your eyes.

Because of its hazardous nature, Investigation 1.1 should never be carried out by students. The first three steps of the Procedure are written to the teacher.

INVESTIGATION

1.1 Demonstration of Combustion of Magnesium

Classification systems are useful for organizing information. In Investigation 1.1, you will observe the combustion of magnesium and classify what you learn from this activity. While observing magnesium burn, record your observations in a table of evidence. Classify your observations in as many ways as you can, and then classify what you have learned as observation or interpretation. (See "Introduction to the Student," page 17, and Appendix B, page 525, for an outline of the parts of a laboratory report.)

This investigation, the first in your study of chemistry, provides you with an opportunity to follow strict safety procedures. The burning of magnesium as demonstrated by your teacher is, like many investigations in chemistry, potentially dangerous. The bright flame present during the burning of magnesium emits ultraviolet radiation that can permanently damage one's eyes. Thus, it is imperative that no one look directly at the burning magnesium. Note that your teacher has taken precautions to ensure that as the magnesium is lighted, eyes are protected from the radiation. The glass beaker reduces the ultraviolet light transmitted to a level that is safe to observe. **Only observe the burning magnesium when it is within the glass beaker**.

Problem

What changes occur when magnesium burns?

Experimental Design

Magnesium is observed before, during, and after being burned in air. All observations are recorded and then classified.

Materials

lab apron	steel wool
safety glasses	laboratory burner and striker
rubber gloves	crucible tongs
magnesium ribbon (approx. 5 cm)	large glass beaker

Procedure

1. Take safety precautions, then light the laboratory burner. (Instructions for lighting a laboratory burner are in Appendix C, page 529.)

2. Clean the magnesium ribbon with steel wool; record any observations.

3. Light the magnesium ribbon in the burner flame (Figure 1.2) and hold the burning magnesium inside the glass beaker to observe.

4. Record any observations while the magnesium burns.

5. Record any observations after the magnesium has burned.

6. Classify the observations you have made in as many ways as you can.

| Problem |
| Prediction |
| Design |
| Materials |
| Procedure |
| ✔ **Evidence** |
| ✔ **Analysis** |
| ✔ **Evaluation** |
| Synthesis |

INVESTIGATION

1.2 Classifying Pure Substances

Before 1800, scientists distinguished elements from compounds by heating them to find out if they decomposed. This experimental design was the only one known at that time. The purpose of Investigation 1.2 is to evaluate this experimental design.

Problem

Are water, bluestone, malachite, table salt, and sugar empirically classified as elements or compounds?

Prediction

According to the theoretical definitions of element and compound and the given chemical formulas, these substances are all compounds. The reasoning behind this prediction is that the chemical formulas for water, bluestone, malachite, table salt, and sugar include more than one kind of atom.

water	$H_2O_{(l)}$
bluestone	$CuSO_4 \cdot 5\,H_2O_{(s)}$
malachite	$Cu(OH)_2 \cdot CuCO_{3(s)}$
table salt	$NaCl_{(s)}$
sugar	$C_{12}H_{22}O_{11(s)}$

Experimental Design

A sample of each substance is heated using a laboratory burner and any evidence of chemical decomposition is recorded.

Materials

lab apron	laboratory burner and striker
safety glasses	ring stand and wire gauze
distilled water	crucible
bluestone	clay triangle
malachite	hot plate
table salt	large test tube (18×150 mm)
sugar	utility clamp and stirring rod
cobalt chloride paper	medicine dropper
250 mL Erlenmeyer flask	piece of aluminum foil
laboratory scoop	

Procedure

1. (a) Test some distilled water with cobalt chloride paper and notice the change in color (Figure 1.9).
   (b) Pour distilled water into an Erlenmeyer flask until the water is about 1 cm deep. Set up the apparatus as shown in Figure 1.10 (a).
   (c) Dry the inside of the top of the Erlenmeyer flask. Place a piece of cobalt chloride paper across the mouth of the flask.
   (d) Boil the water. Record any evidence of decomposition of the water.

CAUTION

Some of the materials are toxic and irritant. Avoid contact with skin and eyes.

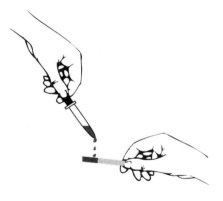

Figure 1.9
A strip of cobalt chloride paper is blue when it is dry, but turns a pale pink when wet with water.

◇ CAUTION

When a laboratory burner is used, long hair should be tied back. If you must leave the laboratory station temporarily and cannot turn off the burner, make sure that it has a visible, yellow flame.

While heating the test tube, be careful not to point it at anyone. There may be some splattering, due to trapped air, as a dry chemical is heated. A face shield may be used, along with safety glasses. Be careful when you pick up the crucible with the tongs. After the investigation, place the contents of the crucible in the waste container provided.

2. (a) Place some bluestone to a depth of about 0.5 cm in a clean, dry test tube. Set up the apparatus as shown in Figure 1.10 (b).
   (b) Record any evidence of the decomposition of the sample.

3. (a) Set a crucible in the clay triangle on the iron ring as shown in Figure 1.10 (c). Add only enough malachite to cover the bottom of the dish with a thin layer.
   (b) Heat the sample slowly at first, with a uniform, almost invisible flame; then heat it strongly with a two-part flame (Appendix C, page 529).
   (c) Record any evidence of decomposition of the sample.

4. (a) Place a few grains of table salt and a few grains of sugar in two separate locations on a piece of aluminum foil. Place the foil on a hot plate.
   (b) Set the hot plate to maximum heat and record any evidence of decomposition.

(a)

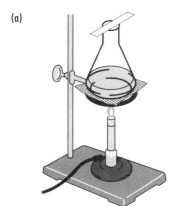

(b)

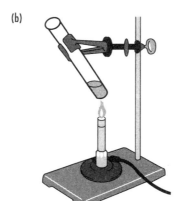

(c)

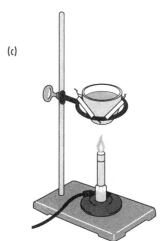

Figure 1.10
Heating substances. (a) An Erlenmeyer flask is used to funnel vapors. (b) A test tube is used when heating small quantities of a chemical. (c) A crucible is required when a substance must be heated strongly.

Problem
✔ **Prediction**
Design
Materials
Procedure
✔ **Evidence**
✔ **Analysis**
✔ **Evaluation**
Synthesis

CAUTION

Ⓣ Ⓗ Because of the many corro-sive substances used in this investigation, wear eye protection at all times. Some of the substances are also toxic.

INVESTIGATION

4.1 Testing Replacement Reaction Generalizations

The purpose of this investigation is to evaluate the single and dou-ble replacement generalizations by testing predictions for a number of reactions. For each combination of reactants given, assume that the reaction is *spontaneous*; that is, it occurs when reactants are mixed.

Problem

What reaction products result when the following substances are mixed?

1. A piece of aluminum is put in a copper(II) chloride solution.
2. Aqueous solutions of barium hydroxide and sulfuric acid are mixed.
3. Aqueous chlorine is added to a sodium bromide solution.
4. A clean zinc strip is placed in a copper(II) sulfate solution.
5. Hydrochloric acid is added to a magnesium hydroxide suspen-sion (a mixture of a solid in a liquid).
6. Solutions of calcium chloride and sodium carbonate are mixed.
7. Solutions of cobalt(II) chloride and sodium hydroxide are mixed.
8. Calcium metal is placed in water.
9. Hydrochloric acid is added to a sodium acetate solution.
10. A magnesium strip is placed in hydrochloric acid.

Experimental Design

Predictions of possible products are made, including any diagnostic test information such as evidence of chemical reactions (Table 4.2, page 94) and specific tests for products (Appendix C, page 537). The general plan is to observe the reactants before and after mixing, and to note any evidence supporting or contradicting the predictions.

Prediction

Procedure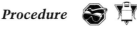

1. Record observations of the reactant chemicals at one of the lab-oratory stations.
2. Perform the reaction according to the directions given at the station (for example, Figure 4.15) and record observations before, during, and after the reaction.
3. Clean all apparatus and the laboratory bench before proceeding to the next station.

- Problem
- Prediction
- ✔ **Design**
- ✔ **Materials**
- ✔ **Procedure**
- ✔ **Evidence**
- ✔ **Analysis**
- Evaluation
- Synthesis

INVESTIGATION

5.1 Chemical Analysis

The purpose of this investigation is to perform an introductory chemical analysis of several unknown white solid solutes, using the diagnostic tests discussed so far.

Problem

Which of the white solids labelled 1, 2, 3, and 4 is calcium chloride, citric acid, glucose, calcium hydroxide — $CaCl_{2(s)}$, $C_3H_4OH(COOH)_{3(s)}$, $C_6H_{12}O_{6(s)}$, and $Ca(OH)_{2(s)}$?

CAUTION

Calcium hydroxide is corrosive. Do not touch any of the solids.

Problem

Prediction

✔ **Design**

Materials

Procedure

✔ **Evidence**

✔ **Analysis**

Evaluation

Synthesis

CAUTION

Solutions used may be toxic and irritant; avoid eye and skin contact.

INVESTIGATION

5.2 The Iodine Clock Reaction

Technological problem solving often involves a systematic trial and error approach that is guided by knowledge and experience. Usually one variable at a time is manipulated while all other variables are controlled. Variables that may be manipulated include concentration, volume, and temperature. The purpose of this investigation is to find a method for getting a reaction to occur in a specified time period.

Problem

What technological process can be employed to have solution A react with solution B in a time of 20 ± 1 s?

- Problem
- Prediction
- Design
- ✔ **Materials**
- ✔ **Procedure**
- ✔ **Evidence**
- ✔ **Analysis**
- Evaluation
- Synthesis

CAUTION

Lead and strontium ions are toxic. Dispose of all materials as directed by your teacher.

INVESTIGATION

5.3 Sequential Qualitative Analysis

The purpose of this investigation is to do a qualitative chemical analysis of an unknown solution using precipitation reactions.

Problem

Are there any lead(II) ions and/or strontium ions present in a sample solution?

Experimental Design

A sodium chloride solution is used as a diagnostic test for the presence of lead(II) ions. The solution is known to contain only lead and/or strontium ions. If a precipitate forms, the solution is filtered to remove the precipitate from the test solution. A sodium sulfate solution is then added to the filtrate as a diagnostic test for strontium ions.

▨ Problem
▨ Prediction
▨ Design
▨ Materials
✔ **Procedure**
▨ Evidence
▨ Analysis
▨ Evaluation
▨ Synthesis

INVESTIGATION

5.4 Solution Preparation

The purpose of this investigation is to prepare, by dilution, a solution which could not be prepared directly from a mass of solute (see Appendix C, page 535). Write a procedure that uses materials from the following list and obtain the approval of your teacher. Do any necessary calculations before starting the laboratory work.

Problem

Any practical problem that requires a 500 ppm solution.

Experimental Design

100 mL of a 500 ppm solution of sodium chloride will be prepared by diluting a 5.0% W/V solution prepared from pure sodium chloride. The solution will be used to ...

Materials

lab apron	(2) 50 mL beakers or (2) 250 mL beakers
safety glasses	distilled water
sodium chloride	stirring rod
centigram balance	10 mL graduated cylinder
scoop	100 mL graduated cylinder
weighing boat	medicine dropper

- Problem
- ✔ **Prediction**
- Design
- Materials
- Procedure
- ✔ **Evidence**
- ✔ **Analysis**
- ✔ **Evaluation**
- Synthesis

CAUTION

Always wear safety glasses when handling or heating chemicals. A face shield is advisable, in case the solution splatters. Be careful when handling hot objects.

INVESTIGATION

5.5 The Solubility of Sodium Chloride in Water

The purpose of this investigation is to test graphing of data (Table 5.5) as a method for predicting the solubility of sodium chloride in water at a particular temperature. The predicted value is then compared with the value determined experimentally by crystallization from a saturated solution.

Problem

What is the solubility of sodium chloride at room temperature?

Prediction

Experimental Design

A precisely measured volume of a saturated $NaCl_{(aq)}$ solution at room temperature is heated to crystallize the salt.

Materials

lab apron	pipet bulb
safety glasses	distilled water
saturated $NaCl_{(aq)}$ solution	alcohol or gas burner
centigram balance	matches or striker
thermometer	laboratory stand
100 mL beaker	iron ring
evaporating dish	wire gauze
10 mL pipet	

Procedure

1. Measure and record the mass of an evaporating dish to a precision of 0.01 g.

2. Obtain 40 mL to 50 mL of saturated $NaCl_{(aq)}$ in a 100 mL beaker.

3. Measure and record the temperature of the solution to a precision of 0.2°C.

4. Pipet a 10.00 mL sample of the saturated solution into the evaporating dish. (See Appendix C on page 532 for instructions on using a pipet.)

5. Using a burner, heat the solution until the water boils away, and dry crystalline $NaCl_{(s)}$ remains (Figure 5.13).

6. Allow the evaporating dish to cool.

7. Measure and record the mass of the evaporating dish and its contents.

8. Reheat the evaporating dish and the residue and repeat steps 6 and 7. If the mass remains constant, this confirms that the sample is dry.

Problem
✔ **Prediction**
Design
Materials
Procedure
✔ **Evidence**
✔ **Analysis**
✔ **Evaluation**
Synthesis

CAUTION

🔥 Butane is flammable. Do not conduct this experiment near an open flame.

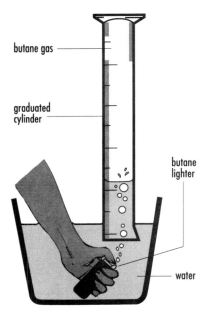

butane gas

graduated cylinder

butane lighter

water

Figure 6.13
The gas from a butane lighter can be collected by downward displacement of water. This apparatus can be used to determine the molar mass of butane.

INVESTIGATION

6.1 Determining the Molar Mass of a Gas

In this investigation you will determine the molar mass of a sample of gas with the purpose of evaluating the experimental design used to obtain the result. Butane is suggested but you may substitute another gas.

Problem

What is the molar mass of butane?

Prediction

Experimental Design

A sample of butane gas from a lighter is collected in a graduated cylinder by downward displacement of water. The volume, temperature, and pressure of the gas are measured, along with the change in mass of the butane lighter. The design is evaluated on the basis of the accuracy of the experimental value for the molar mass of butane, which is compared with the accepted value.

Materials

lab apron	500 mL graduated cylinder or
safety glasses	600 mL graduated beaker
butane lighter	thermometer
plastic bucket	barometer
balance	

Procedure

1. Determine the initial mass of the butane lighter.
2. Pour water into the bucket until it is two-thirds full and then completely fill the cylinder with water and invert it in the bucket (Figure 6.13). Ensure that no air has been trapped in the cylinder.
3. Hold the butane lighter in the water and under the cylinder and release the gas until you have collected 400 mL to 500 mL of gas.
4. Equalize the pressures inside and outside the cylinder by adjusting the position of the cylinder until the water levels inside and outside the cylinder are the same.
5. Read the measurement on the cylinder and record the volume of gas collected.
6. Record the ambient (room) temperature and pressure.
7. Dry the butane lighter and determine its final mass.
8. Release the butane gas from the cylinder in a fume hood or outdoors.

- Problem
- Prediction
- Design
- ✔ **Materials**
- ✔ **Procedure**
- ✔ **Evidence**
- ✔ **Analysis**
- ✔ **Evaluation**
- Synthesis

CAUTION

 Sodium carbonate is toxic.

Table 7.2

REFERENCE DATA: REACTION OF SODIUM CARBONATE AND CALCIUM CHLORIDE	
Mass of CaCO₃ Produced (g)	Mass of Na₂CO₃ Reacting (g)
0.47	0.50
0.94	1.00
1.42	1.50
1.89	2.00
2.36	2.50

INVESTIGATION

7.1 Analysis of Sodium Carbonate Solution

In your procedure, specify the quantities of chemicals to be used. For a description of the method for filtering a precipitate, see Appendix C, page 534.

Problem

What mass of sodium carbonate is present in a 50.0 mL sample of a solution?

Experimental Design

The mass of sodium carbonate present in the sample solution is determined by having it react with an excess quantity of a calcium chloride solution. The mass of calcium carbonate precipitate produced is used to determine the mass of sodium carbonate that reacted; the figure is read from a graph of the reference data in Table 7.2.

- ▨ Problem
- ✔ **Prediction**
- ▨ Design
- ▨ Materials
- ✔ **Procedure**
- ✔ **Evidence**
- ✔ **Analysis**
- ✔ **Evaluation**
- ▨ Synthesis

INVESTIGATION

7.2 Decomposing Malachite

The chemical in this investigation is known by several names. The systematic or IUPAC name for it is *copper(II) hydroxide carbonate*. From this name the formula of the substance can be determined. Geologists, however, refer to this substance as *malachite*. It also has a common name that appears in chemical supply catalogues — *basic copper carbonate*. Copper(II) hydroxide carbonate is a green *double salt* with the chemical formula $Cu(OH)_2 \cdot CuCO_{3(s)}$. This double salt decomposes completely when heated to 200°C, forming copper(II) oxide, carbon dioxide, and water vapor. The purpose of this investigation is to provide direct experience in making quantitative predictions about chemical reactions. You should develop a hypothesis on which to base a prediction.

Problem

How is the mass of copper(II) oxide formed from the decomposition of malachite related to the mass of malachite reacted?

Prediction

CAUTION

☠ **Malachite is toxic.**

Figure 7.4
Use the glass stirring rod to break up lumps of malachite and to mix the contents of the dish while they are being heated. Large lumps may decompose on the outside but not on the inside.

Experimental Design

The mass of a sample of malachite is determined, then the sample is heated strongly until the color changes completely from green to black (Figure 7.4). The mass of black product is determined. The results from several laboratory groups are combined in a graph to answer the Problem question and to provide reference data to predict future decompositions of malachite.

Materials

lab apron
safety glasses
porcelain dish (or crucible and clay triangle)
small ring stand
laboratory burner (Figure 1.10 (c), page 38) or hot plate
glass stirring rod
sample of malachite (not more than 3 g)
centigram balance
laboratory scoop or plastic spoon

Problem
- ✔ **Prediction**
- ✔ **Design**
- ✔ **Materials**
- ✔ **Procedure**
- ✔ **Evidence**
- ✔ **Analysis**
- ✔ **Evaluation**

Synthesis

INVESTIGATION

7.3 Testing the Stoichiometric Method

The purpose of this investigation is to test the stoichiometric method. As scientific knowledge is continually tested, it becomes increasingly trustworthy.

Problem

What is the mass of precipitate produced by the reaction of 2.00 g of strontium nitrate in solution with excess (2.56 g) copper(II) sulfate in solution?

CAUTION

Strontium nitrate is moderately toxic; there is risk of fire when it is in contact with organic chemicals and it may explode when bumped or heated. Copper(II) sulfate is a strong irritant and is toxic if ingested.

Problem
✔ **Prediction**
Design
Materials
Procedure
✔ **Evidence**
✔ **Analysis**
✔ **Evaluation**
Synthesis

CAUTION

Hydrochloric acid in 6 mol/L concentration is very corrosive. If acid is splashed into your eyes, immediately rinse them with water for 15 to 20 min. Acid splashed onto the skin should be rinsed immediately with plenty of water. If acid is splashed onto clothes, neutralize with baking soda, then wash thoroughly with plenty of water. Notify your teacher.

INVESTIGATION

7.4 The Universal Gas Constant

Most scientific constants, such as molar mass and molar volume, are determined empirically. The purpose of this investigation is to test a simple experimental design for determining the value of the universal gas constant R. The calculations in the Analysis of this experiment are similar to those in the example involving ammonia gas on page 177, except that you solve for R in the last step. In your Evaluation, focus on judging the experimental design.

Problem

What is the experimental value of the universal gas constant?

Prediction

Experimental Design

A known mass of magnesium ribbon reacts with excess hydrochloric acid. The temperature, pressure, and volume of the hydrogen gas that is produced are measured. The experimental value of the gas constant determined with this design (illustrated in Figure 7.11) is judged by comparing it with the accepted value.

Materials

lab apron
safety glasses
100 mL graduated cylinder
2-hole stopper to fit cylinder
thermometer
barometer
large beaker (600 mL or 1000 mL)
magnesium ribbon (60 mm to 80 mm)
6 mol/L hydrochloric acid
piece of fine copper wire (100 mm to 150 mm)

Procedure

1. Obtain a strip of magnesium ribbon about 60 mm to 80 mm long.
2. Measure and record the mass of the magnesium.
3. Fold the magnesium ribbon to make a small compact bundle, no larger than a pencil eraser.

© NELSON CANADA,
A DIVISION OF THOMSON CANADA LIMITED, 1994

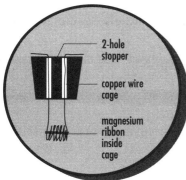

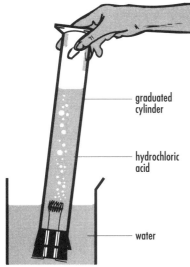

Figure 7.11
While holding the cylinder so it does not tip, rest it on the bottom of the beaker. The acid, which is more dense than water, will flow down toward the stopper and react with the magnesium. The hydrogen produced should remain trapped in the graduated cylinder.

CAUTION

Rinse your hands well after step 9.

4. Wrap the fine copper wire all around the magnesium, making a cage to hold it but leaving 30 mm to 50 mm of the wire free for a handle.

5. Carefully pour 10 mL to 15 mL of 6 mol/L hydrochloric acid into the graduated cylinder.

6. Slowly fill the graduated cylinder to the brim with water from a beaker. As you fill the cylinder, pour slowly down the side of the cylinder to minimize mixing of the water with the acid at the bottom. In this way, the liquid at the top of the cylinder is relatively pure water and the acid remains at the bottom.

7. Half-fill the large beaker with water.

8. Bend the copper-wire handle through the holes in the stopper so that the cage holding the magnesium is positioned about 10 mm below the bottom of the stopper.

9. Insert the stopper into the graduated cylinder — the liquid in the cylinder will overflow a little. Cover the holes in the stopper with your finger. Working quickly, invert the cylinder, and immediately lower it so that the stopper is below the surface of the water in the beaker before you remove your finger from the stopper holes (Figure 7.11).

10. Observe the reaction, then wait about 5 min after the bubbling stops to allow the contents of the graduated cylinder to reach room temperature.

11. Raise or lower the graduated cylinder so that the level of liquid inside the beaker is the same as the level of liquid in the graduated cylinder. (This equalizes the gas pressure in the cylinder with the pressure of the air in the room.)

12. Measure and record the volume of gas in the graduated cylinder.

13. Record the laboratory (ambient) temperature and pressure.

14. The liquids in this investigation may be poured down the sink, but rinse the sink with lots of water.

- Problem
- Prediction
- Design
- Materials
- Procedure
- Evidence
- Analysis
- Evaluation
- Synthesis

INVESTIGATION

7.5 Preparation of a Standard Solution from a Solid

The purpose of this demonstration is to illustrate the skills required to prepare a standard solution from a pure solid. You will need these skills in many investigations in this book. No problem is stated, since no specific analysis is being done — your task is to learn the procedure.

Materials

lab apron
safety glasses
$CuSO_4 \cdot 5H_2O_{(s)}$
150 mL beaker
centigram balance
laboratory scoop
stirring rod
wash bottle of pure water (deionized or distilled)
100 mL volumetric flask with stopper
small funnel
medicine dropper
meniscus finder

Procedure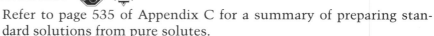

Refer to page 535 of Appendix C for a summary of preparing standard solutions from pure solutes.

1. (Pre-lab) Calculate the mass of solid copper(II) sulfate-5-water needed to prepare 100.0 mL of a 0.5000 mol/L solution.

2. Obtain the calculated mass of copper(II) sulfate-5-water in a clean, dry 150 mL beaker.

3. Dissolve the solid in 40 mL to 50 mL of pure water. (The term *pure water* does not mean that the water is absolutely pure. It means that the water has been treated in a laboratory in an ion exchange column or in a distillation apparatus.)

4. Transfer the solution into a 100 mL volumetric flask.

5. Add pure water until the volume is 100.0 mL.

6. Stopper the flask and mix the contents thoroughly by repeatedly inverting the flask.

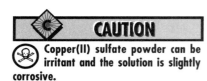

CAUTION

Copper(II) sulfate powder can be irritant and the solution is slightly corrosive.

- Problem
- Prediction
- Design
- Materials
- Procedure
- Evidence
- Analysis
- Evaluation
- Synthesis

CAUTION

Copper(II) sulfate solution is slightly corrosive.

INVESTIGATION

7.6 Preparing a Standard Solution by Dilution

The purpose of this demonstration is to illustrate the procedure and skills for precisely diluting a stock solution in order to prepare a standard solution. As in Investigation 7.5, no problem is stated since no specific chemical analysis is being done. A typical problem statement would be in the context of a chemical analysis and the preparation of the standard solution would be part of an overall procedure.

Materials

lab apron
safety glasses
0.5000 mol/L $CuSO_{4(aq)}$
150 mL beaker
10 mL volumetric pipet
pipet bulb
wash bottle of pure water
100 mL volumetric flask with stopper
small funnel
medicine dropper
meniscus finder

Procedure

Refer to pages 532 and 536 of Appendix C for information on using a pipet and on preparing standard solutions by dilution.

1. (Pre-lab) Calculate the volume of 0.5000 mol/L stock solution of $CuSO_{4(aq)}$ required to prepare 100.0 mL of 0.050 00 mol/L solution.

2. Add 40 mL to 50 mL of pure water to a clean 100 mL volumetric flask.

3. Measure 10.00 mL of the stock solution using a 10 mL volumetric pipet.

4. Transfer the 10.00 mL of solution into the 100 mL volumetric flask.

5. Add pure water until the final volume is reached.

6. Stopper the flask and mix the solution thoroughly.

Problem
Prediction
Design
Materials
Procedure
✔ **Evidence**
✔ **Analysis**
✔ **Evaluation**
Synthesis

CAUTION

Hydrochloric acid is corrosive; avoid skin and eye contact.

INVESTIGATION

7.7 Titration Analysis of Hydrochloric Acid

The indicator used in this investigation, methyl orange, changes from yellow to pink as the endpoint of the reaction. The indicator changes in color just after enough $HCl_{(aq)}$ has been added to react completely with the $Na_2CO_{3(aq)}$ sample.

Problem

What is the molar concentration of a given hydrochloric acid solution?

Experimental Design

A 0.100 mol/L standard solution of sodium carbonate is prepared. Measured samples of this solution are titrated with a sample of the hydrochloric acid solution. The color change of methyl orange is the endpoint of the titration. The titration is repeated until three consistent results are obtained; that is, until reacting volumes agree within 0.1 mL to 0.2 mL.

Materials

lab apron	400 mL waste beaker
safety glasses	(2) 150 mL beakers
methyl orange indicator	(2) 250 mL Erlenmeyer flasks
$Na_2CO_{3(s)}$	small funnel
$HCl_{(aq)}$ of unknown concentration	100 mL volumetric flask and stopper
pure water	50 mL buret
medicine dropper	buret clamp
laboratory scoop	10 mL volumetric pipet
meniscus finder	pipet bulb
centigram balance	ring stand

Procedure

1. (Pre-lab) Calculate the mass of sodium carbonate required to prepare 100.0 mL of a 0.100 mol/L solution.

2. Prepare the standard solution of $Na_2CO_{3(aq)}$ and transfer it to a clean, dry, labelled 150 mL beaker.

3. Place 70 mL to 80 mL of $HCl_{(aq)}$ in a clean, dry, labelled 150 mL beaker.

4. Set up the buret to contain the $HCl_{(aq)}$.

5. Pipet a 10.00 mL sample of $Na_2CO_{3(aq)}$ into a clean Erlenmeyer flask and add 1 to 3 drops of methyl orange indicator.

6. Record the initial buret reading (to a precision of 0.1 mL).

7. Titrate the sodium carbonate solution with $HCl_{(aq)}$ until a single drop produces a permanent change from pale yellow to pink.

8. Record the final buret reading (to 0.1 mL).

9. Repeat steps 5 to 8 until three consistent results are obtained. (You should have enough standard sodium carbonate solution to do about 8 trials, if necessary.)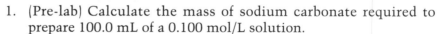

Problem
Prediction
✔ **Design**
✔ **Materials**
✔ **Procedure**
✔ **Evidence**
✔ **Analysis**
✔ **Evaluation**
Synthesis

CAUTION

Before beginning your experiment, ask your teacher to approve your procedure.

INVESTIGATION

8.1 Activity Series

The purpose of this investigation is to develop an activity series of a number of common metals with dilute acid (Figure 8.1). In this investigation, assume that the rate of reaction (as opposed to the completeness of reaction) is an appropriate indicator of chemical reactivity. In your procedure, specify safety precautions against possible "spitting" when the dilute acid reacts with the most reactive metals.

Hint: Develop a procedure to be sure that you observe the reaction of the metal, and not the reaction of its oxide coating.

Problem

What is the order of reactivity (activity series) of calcium, copper, iron, lead, magnesium, and zinc metals with dilute hydrochloric acid?

Problem
Prediction
Design
Materials
Procedure
✔ **Evidence**
✔ **Analysis**
✔ **Evaluation**
Synthesis

INVESTIGATION

8.2 Molecular Models

Chemists use molecular models to explain and predict molecular structure, relating structure to the properties and reactions of substances. The purpose of this investigation is to explain some known chemical reactions by using molecular models. Evaluate your explanation by assessing whether it is logical, consistent, and as simple as possible.

Problem

How can theory, represented by molecular models, explain the following series of chemical reactions that have occurred in a laboratory?

(a) $CH_4 + Cl_2 \rightarrow CH_3Cl + HCl$
(b) $C_2H_4 + Cl_2 \rightarrow C_2H_4Cl_2$
(c) $N_2H_4 + O_2 \rightarrow N_2 + 2\,H_2O$
(d) $CH_3CH_2CH_2OH \rightarrow CH_3CHCH_2 + H_2O$
(e) $HCOOH + CH_3OH \rightarrow HCOOCH_3 + H_2O$

Experimental Design

Molecular model kits, designed to show the bonding capacities of atoms, test the ability of theory to explain the specified chemical reactions.

Materials

molecular model kit (Figure 8.20)

Procedure

1. For the first reaction, construct a structural model for each reactant and record the structures in your notebook.

2. Study your models to determine the minimum rearrangement that would produce structural models of the products.

3. Rearrange the models to form the products, and record the structures in your notebook.

4. Repeat steps 1 to 3 for each chemical equation.

INVESTIGATION

8.3 Evidence for Double Bonds

The purpose of this investigation is to obtain empirical evidence for the existence of double bonds. Two compounds, cyclohexane and cyclohexene (Figure 8.21), are believed to be almost identical, except for the presence of a double bond between two carbon atoms in cyclohexene. These compounds illustrate a relationship between structure and reactivity — cyclohexene reacts rapidly with bromine but cyclohexane does not. The reaction is indicated by the disappearance of the color of the bromine.

Problem

Which of the common substances supplied contain molecules with double covalent bonds between carbon atoms?

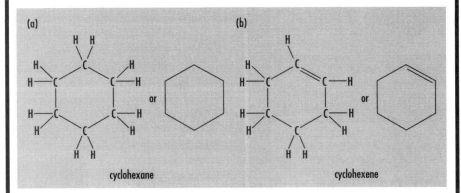

Experimental Design

The unknown samples and two controls (cyclohexane and cyclohexene) are tested by adding a few drops of a solution containing bromine in trichlorotrifluoroethane (or bromine in water). After each sample is mixed with the bromine solution, any evidence of a chemical reaction is noted.

Materials

lab apron
safety glasses
small test tubes with stoppers
test tube rack
waste container, with lid, for organic substances
bromine solution in a dropper bottle
cyclohexane in dropper bottle
cyclohexene in dropper bottle
common substances, such as mineral oil, paint thinner, liquid paraffin, soybean oil, corn oil, peanut oil

---

Problem
Prediction
Design
Materials
Procedure
✔ **Evidence**
✔ **Analysis**
Evaluation
Synthesis

Figure 8.21
In (a), the structural diagrams of cyclohexane show that all bonds are single bonds. In (b), the cyclohexene structures indicate one carbon-carbon double bond. The second structure in diagrams (a) and (b) represents the same molecule with a line diagram where each corner of the structure represents a carbon atom, and any bonds not shown are assumed to join hydrogen atoms to carbon atoms.

Procedure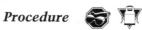

1. Add 10 drops of cyclohexane to a clean test tube.

2. Add 1 drop of bromine solution to the test tube. Shake the test tube gently. Repeat this procedure with up to 4 drops of bromine solution.

3. Dispose of all materials into the labelled waste container.

4. Repeat steps 1 to 3 using cyclohexene. Use a clean test tube.

5. Repeat steps 1 to 3 using the samples provided. Use a clean test tube each time.

Problem
Prediction
Design
Materials
Procedure
✔ **Evidence**
✔ **Analysis**
Evaluation
Synthesis

CAUTION

All unknown substances are potentially hazardous. Avoid inhaling vapors and avoid skin contact.

INVESTIGATION

8.4 Evidence for Polar Molecules

The purpose of this investigation is to test for the presence of polar molecules in a variety of pure chemical substances.

Problem

Which of various molecular substances contain polar molecules?

Experimental Design

A thin stream of each liquid is tested by holding a positively charged acetate strip or a negatively charged vinyl strip near the liquid (Figure 8.27, page 228).

Materials

lab apron   buret funnel
safety glasses   400 mL beaker or pan
50 mL samples of various liquids   acetate strip
50 mL buret or 10 mL pipet   vinyl strip
buret clamp and stand   paper towel

Procedure

1. Fill the buret with one of the liquids provided.

2. Rub the acetate strip back and forth several times in a piece of paper towel.

3. Open the stopcock of the buret so that a thin stream of the liquid pours into the beaker about 15 cm below it.

4. Hold the charged acetate strip close to the liquid stream and observe the stream of liquid.

5. Repeat steps 1 to 4 with the charged vinyl strip.

6. Clean the buret thoroughly. Repeat steps 1 to 5 with the other liquids provided.

Figure 8.27
(a) Testing a liquid with a charged strip provides evidence for the existence of polar molecules in a substance.
(b) Polar molecules in a liquid become oriented so that their positive poles are closer to a negatively charged material. Near a positively charged material they become oriented in the opposite direction. Polar molecules are thus attracted by either kind of charge.

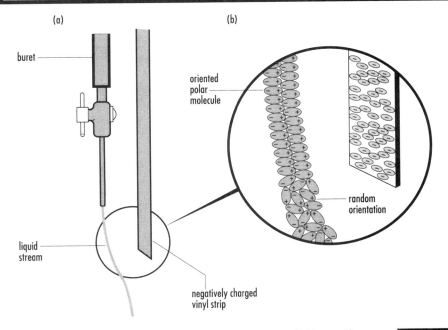

Problem
✔ **Prediction**
Design
Materials
✔ **Procedure**
✔ **Evidence**
✔ **Analysis**
✔ **Evaluation**
Synthesis

INVESTIGATION

8.5 Hydrogen Bond Formation

You have learned that energy is required to break chemical bonds and that energy is released when new bonds are formed. The purpose of this investigation is to test these ideas in relation to hydrogen bonding.

Problem

Are additional hydrogen bonds formed when water and glycerol are mixed?

Prediction

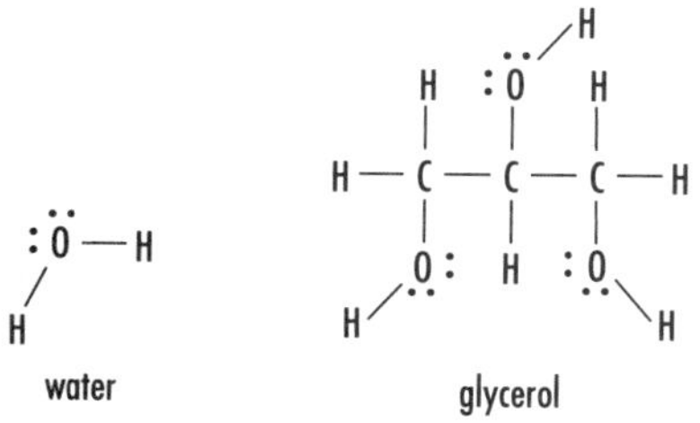

Figure 8.29
The structures of water and glycerol suggest the possibility of more hydrogen bonds after these two substances form a solution.

Experimental Design

Water, $H_2O_{(l)}$, and glycerol, $C_3H_5(OH)_{3(l)}$, liquids that contain O—H bonds (Figure 8.29) are mixed, and any change in energy is measured using a thermometer. If a significant temperature increase is noted, then additional hydrogen bonds are probably present.

Materials

lab apron
safety glasses
30 mL of water
30 mL of glycerol
(2) 50 mL graduated cylinders

nested pair of polystyrene cups
cup lid with center hole
thermometer
250 mL beaker (for support)

- Problem
- Prediction
- Design
- Materials
- Procedure
- Evidence
- ✔ **Analysis**
- Evaluation
- Synthesis

INVESTIGATION

9.1 Models of Organic Compounds

The purpose of this investigation is to examine the structure of some isomers of organic compounds and to practice drawing structural diagrams.

Problem

What are the structures of the isomers of C_4H_{10}, $C_2H_3Cl_3$, C_2H_6O, and C_2H_7N?

Materials

molecular model kit (Figure 9.5)

Procedure

1. Assemble two different isomeric models of C_4H_{10} and record three different structural diagrams for each model.
2. Assemble two different models for each of the other molecular formulas and record their complete and condensed structural diagrams.

Problem
Prediction
Design
Materials
Procedure
✔ **Evidence**
✔ **Analysis**
Evaluation
Synthesis

INVESTIGATION

9.2 Classifying Organic Compounds

The purpose of this investigation is to provide practice in classification. You will also learn about some organic compounds found in various commercial and consumer products.

Problem

How can selected organic compounds be classified according to the functional groups in their molecular structures?

Experimental Design

Information from chemical references and empty containers of commercial and consumer products containing one or more organic substances are investigated to determine the name, toxicity, and structure of selected organic compounds. The compounds are then classified according to similar functional groups. (Some compounds may contain more than one functional group and thus fit more than one classification.)

Procedure

1. Observe one of the samples provided and read the names of compounds listed on the product label.

2. Using the information sheet provided, record the product's commercial name and its use. For the selected organic compound contained in the product, record the toxicity rating, the IUPAC name, and a structural diagram.

3. Repeat steps 1 and 2 for the other samples provided.

Problem
Prediction
Design
Materials
Procedure
✔ **Evidence**
✔ **Analysis**
Evaluation
Synthesis

INVESTIGATION

9.3 Structures and Properties of Isomers

The purpose of this investigation is to examine the structures and physical properties of some isomers of unsaturated hydrocarbons.

Problem

What are the structures and physical properties of the isomers of C_4H_8 and C_4H_6?

Experimental Design

Structures of possible isomers are determined by means of a molecular model kit. Once each structure is named, the boiling and melting points are obtained from a reference such as *The CRC Handbook of Chemistry and Physics* or *The Merck Index*.

Materials

molecular model kits
chemical reference

Procedure

1. Use the required "atoms" to make a model of C_4H_8.
2. Draw a structural diagram of the model and write the IUPAC name for the structure.
3. By rearranging bonds, produce models for all other isomers of C_4H_8, including cyclic structures. Draw structural diagrams and write the IUPAC name for each structure before disassembling the models.
4. Repeat steps 1 to 3 for C_4H_6.
5. In a reference, find the melting point and the boiling point of each of the compounds identified.

▒	Problem
▒	Prediction
▒	Design
▒	Materials
▒	Procedure
✔	**Evidence**
✔	**Analysis**
▒	Evaluation
▒	Synthesis

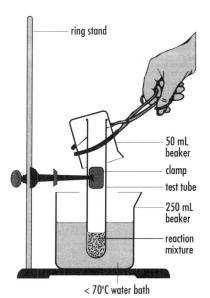

Figure 9.28
Set-up for Investigation 9.4.

 **CAUTION**

Both ethanoic and sulfuric acids are dangerously corrosive. Protect your eyes, and do not allow the acids to come into contact with skin, clothes, or lab desks. Both methanol and ethanol are flammable; do not use near an open flame.

 **CAUTION**

Excessive inhalation of the products may cause headaches or dizziness. Use your hand to waft the odor from the beaker towards your nose.

INVESTIGATION

9.4 Synthesizing an Ester

The purpose of this investigation is to synthesize and observe the properties of two esters.

Problem

What are some physical properties of ethyl ethanoate (ethyl acetate) and methyl salicylate?

Experimental Design

The esters are produced by the reaction of appropriate alcohols and acids, using sulfuric acid as a catalyst. The solubility and the odor of the esters are observed.

Materials

lab apron
safety glasses
dropper bottles of ethanol, methanol, glacial ethanoic (acetic) acid,
    and concentrated sulfuric acid
vial of salicylic acid
(2) 25×250 mm test tubes
250 mL beaker
(2) 50 mL beakers
(2) 10 mL graduated cylinders
laboratory scoop
balance
hot plate
thermometer
ring stand with test tube clamp

Procedure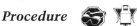

1. Add about 5 mL of ethanol and 6 mL of ethanoic acid to one of the test tubes.

2. Have your teacher add 8 to 10 drops of concentrated sulfuric acid to the mixture.

3. Set up a hot water bath using the 250 mL beaker. (The temperature of the water should not exceed 70°C.)

4. Clamp the test tube so that it is immersed in hot water to the depth of the mixture.

5. As a safety precaution to block any eruption of the volatile mixture, invert a 50 mL beaker above the end of the test tube and heat the mixture for about 10 min (Figure 9.28).

6. After heating the mixture, rinse the 50 mL beaker with cold tap water and add about 30 mL of cold water to the beaker.

7. Cool the test tube with cold tap water and pour the contents of the test tube into the cold water in the beaker. Observe and smell the mixture carefully, using the proper technique for smelling chemicals.

8. Repeat steps 1 to 7, using 3.0 g of salicylic acid, 10 mL of methanol, and 20 drops of sulfuric acid.

| Problem |
| Prediction |
| Design |
| Materials |
| Procedure |
| ✔ **Evidence** |
| ✔ **Analysis** |
| ✔ **Evaluation** |
| Synthesis |

INVESTIGATION

9.5 Some Properties and Reactions of Organic Compounds

The purpose of this investigation is to observe some properties and reactions of organic compounds.

Problem

What observations can be made about the properties and reactions of some organic compounds?

Experimental Design

Part I

Physical and chemical properties of isomers of $C_4H_{10}O$ are tabulated and compared. Solubilities, melting points, and boiling points of the isomers are found in a reference, such as *The CRC Handbook of Chemistry and Physics* or *The Merck Index*. Evidence for the reaction of the alcohol isomers with a potassium permanganate solution, and with concentrated hydrochloric acid solution, are obtained. A teacher demonstration provides evidence for the reaction of the three alcohol compounds with metallic sodium.

Part II

The reactions of cyclohexane and cyclohexene with a basic solution of potassium permanganate are investigated. Carbon compounds containing double and triple bonds react rapidly with bromine (page 248), so it seems reasonable that they should react quickly with other reactive substances such as potassium permanganate.

Part III

The chemical and physical properties of benzoic acid are examined, using sodium hydroxide and hydrochloric acid. Carboxylic acids are acidic because a hydrogen ion is easily removed from the –COOH group. Like inorganic acids, organic acids should undergo simple double replacement reactions. The solubility in water of organic acids should depend on the size of the non-polar part of the molecule; small molecules such as ethanoic (acetic) acid, $CH_3COOH_{(l)}$, are known to be soluble.

Materials

lab apron
safety glasses
safety screen (for teacher demonstration)
overhead projector (optional for teacher demonstration)
sodium metal (for teacher demonstration only)
3 small beakers (for teacher demonstration)
knife (for teacher demonstration)
forceps (for teacher demonstration)

dropper bottle of 1-butanol
dropper bottle of 2-butanol
dropper bottle of 2-methyl-2-propanol
dropper bottle of concentrated (12 mol/L) $HCl_{(aq)}$
dropper bottle of 6.0 mol/L $NaOH_{(aq)}$
dropper bottle of 0.010 mol/L $KMnO_{4(aq)}$
dropper bottle of cyclohexane, $C_6H_{12(l)}$
dropper bottle of cyclohexene, $C_6H_{10(l)}$
vial of benzoic acid, $C_6H_5COOH_{(s)}$
litmus paper
laboratory scoop
stirring rod
6 small test tubes with stoppers
test tube rack
(3) 50 mL beakers

Procedure

Part I Teacher Demonstration

1. Prepare the safety screen. Pour approximately 20 mL of 1-butanol, 2-butanol, and 2-methyl-2-propanol into labelled 50 mL beakers.

2. Add a *small* piece of sodium metal to each beaker. Observe and record evidence of reaction.

3. Add water to each beaker *slowly, with stirring,* so that any remaining sodium metal reacts completely.

Part I Student Investigation

1. (a) Pour about 2 mL of 1-butanol, 2-butanol, and 2-methyl-2-propanol into labelled test tubes and place them in a test tube rack.
   (b) Add approximately 2 mL of potassium permanganate solution to each of the test tubes.
   (c) Stopper and shake each test tube using the accepted technique for this procedure. Remove the stoppers after shaking the test tubes and replace each test tube in the rack.
   (d) Observe and record any evidence of change that becomes apparent over the next 5 min.

2. (a) Add approximately 2 mL of 1-butanol, 2-butanol, and 2-methyl-2-propanol to labelled 50 mL beakers.
   (b) Add approximately 10 mL of concentrated hydrochloric acid to each beaker.
   (c) Stir *carefully* to mix the contents of the beaker. After 1 min, look for cloudiness in the water layer — an indication of the formation of an alkyl halide with low solubility.

Part II

1. Pour about 1 mL of cyclohexane and cyclohexene into separate labelled test tubes, and place the test tubes in a test tube rack.

2. Add about 2 mL of potassium permanganate solution to each test tube.

3. Add about 2 mL of sodium hydroxide solution to each test tube.

4. Stopper and shake the tubes, and observe any changes. Immediately remove the stoppers. After 5 min, observe any further changes.

Part III

1. Add benzoic acid into a test tube to a depth of about 1 cm.

2. Add about 10 mL of water to the test tube; stopper and shake it. Immediately remove the stopper.

3. Test the liquid with litmus paper.

4. Add sodium hydroxide drop by drop to the test tube, mixing after each drop, until no further change occurs.

5. Add hydrochloric acid drop by drop to the test tube, mixing after each drop, until no further change occurs.

DISPOSAL TIP

Pour the contents of the test tubes and the beakers into waste containers marked with the names of the chemicals.

- Problem
- ✔ **Prediction**
- Design
- ✔ **Materials**
- Procedure
- ✔ **Evidence**
- ✔ **Analysis**
- ✔ **Evaluation**
- Synthesis

⬥ CAUTION

Do not use flammable materials to construct your water heater.

INVESTIGATION

10.1 Designing and Evaluating a Water Heater

A water heater is an insulated container with an energy source to heat the water. The insulation is not perfect, so the water tends to cool as heat flows from the container to the surroundings. In this investigation, two criteria, *specific energy* and *cooling rate* are used to evaluate a water heater. Specific energy is the heat that flows into the water (for the temperature change from room temperature to 60°C) per unit mass of the total container. A well-designed water heater has a high specific energy in joules per kilogram. Cooling rate is the heat flowing out of the water (q) per minute. Based on this criterion, a well-designed water heater has a low cooling rate in joules per minute.

Problem

What is the best design for a simple water heater?

Prediction

Materials

Procedure

1. Obtain your teacher's approval of the safety of your design, then construct your water heater, including a measured volume of water at room temperature.

2. Measure the total mass of the water heater (without the energy source).

3. Measure the initial temperature of the water.

4. Heat the water to 60°C while stirring it constantly.

5. Remove the heat source and let the water heater sit for 10 min.

6. Stir well and measure the final temperature of the water.

7. Repeat steps 1 to 6 using either the same design or a modified design.

■ Problem
■ Prediction
■ Design
✔ **Materials**
■ Procedure
✔ **Evidence**
✔ **Analysis**
■ Evaluation
■ Synthesis

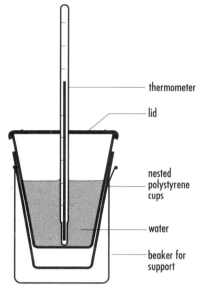

thermometer

lid

nested
polystyrene
cups

water

beaker for
support

Figure 10.23
A simple laboratory calorimeter consists of nested polystyrene cups as the insulated container, a measured quantity of water, and a thermometer. The chemical system that will undergo an enthalpy change is placed in or dissolved in the water of the calorimeter. Energy transfers between the chemical system and the surrounding water are monitored by measuring changes in the temperature of the water.

INVESTIGATION

10.2 Molar Enthalpy of Solution

The purpose of this investigation is to practice the scientific and technological skills associated with calorimetry. Before you do the investigation, decide how precise each measurement should be to provide maximum certainty for your experimental result.

Problem

What is the molar enthalpy of solution of an ionic compound?

Experimental Design

Once an ionic compound is chosen, the approximate mass of the compound required to make 100 mL of a 1.00 mol/L solution is calculated and then is measured precisely. Use the MSDS to determine the hazards associated with the compound and take necessary precautions. The temperature change is measured as the compound dissolves in the water in a calorimeter (Figure 10.23).

Materials

Procedure

1. Measure 100 mL of water in a graduated cylinder and place it in the calorimeter.
2. Obtain the required mass of the compound in a suitable container.
3. Record the initial temperature of the water.
4. Add the compound to the water.
5. Cover the calorimeter and stir until a maximum temperature change is obtained.
6. Record the final temperature of the water.
7. Dispose of the contents of the calorimeter by an acceptable method.
8. If time permits, repeat the experiment with the same chemical or with one that produces the opposite temperature change.

Problem
Prediction
✔ **Design**
✔ **Materials**
✔ **Procedure**
✔ **Evidence**
✔ **Analysis**
✔ **Evaluation**
Synthesis

CAUTION

Wear safety glasses. Both sodium hydroxide and sulfuric acid are corrosive chemicals. Rinse with lots of cold water if these chemicals contact your skin.

INVESTIGATION

10.3 Molar Enthalpy of Reaction

Evaluating experimental designs and estimating the certainty of empirically determined values are important skills in interpreting scientific statements. The purpose of this investigation is to test the calorimeter design and calorimetry procedure by verifying a widely accepted value for the molar enthalpy of a neutralization reaction. The accuracy (percent difference) obtained in this investigation is used to evaluate the calorimeter and the assumptions made in the analysis, not to evaluate the prediction and its authority, *The CRC Handbook of Chemistry and Physics*. The ultimate authority in this experiment is considered to be the reference value used in the prediction.

Problem

What is the molar enthalpy of neutralization for sodium hydroxide when 50 mL of aqueous 1.0 mol/L sodium hydroxide reacts with an excess quantity of 1.0 mol/L sulfuric acid?

Prediction

According to *The CRC Handbook of Chemistry and Physics*, the molar enthalpy of neutralization for sodium hydroxide with sulfuric acid is –57 kJ/mol.

- ▨ Problem
- ✔ **Prediction**
- ▨ Design
- ✔ **Materials**
- ✔ **Procedure**
- ✔ **Evidence**
- ✔ **Analysis**
- ✔ **Evaluation**
- ▨ Synthesis

CAUTION

Avoid using flammable materials in your calorimeter.

INVESTIGATION

10.4 Designing a Calorimeter for Combustion Reactions

The aim of technological problem solving is to develop and test a product or a process. This requires scientific knowledge of substances and their changes, as well as the invention of a suitable design. Successful problem solving requires open-mindedness, flexibility, and persistence, because preliminary designs are often tested by trial and error. The purpose of this investigation is to design, test, and evaluate a metal can calorimeter.

Problem

Which design for a metal can calorimeter gives the largest molar enthalpy of combustion for paraffin wax?

Prediction

Experimental Design

A metal can containing water is altered to receive as much energy as possible from the burning of a simple paraffin candle, $C_{25}H_{52(s)}$, and to minimize the energy lost to the surroundings. The molar enthalpy of combustion for paraffin H_c is determined by the temperature change of the water in the can and the measurements of the mass of the candle before and after heating. The most efficient designs, as determined by the measured molar enthalpies of combustion, are evaluated based on the criteria of reliability, economy, and simplicity.

The following assumptions are made: the candle consists entirely of paraffin wax, $C_{25}H_{52(s)}$; the incomplete combustion of the paraffin yields carbon dioxide and carbon according to some constant stoichiometric ratio; and the heat flowing into the calorimeter materials is insignificant compared with the heat flowing into the water.

Problem
Prediction
Design
Materials
✔ **Procedure**
✔ **Evidence**
✔ **Analysis**
✔ **Evaluation**
Synthesis

CAUTION

Hydrochloric acid is corrosive. Avoid contact with skin, eyes, clothing, or the desk. If you spill this acid on your skin, wash immediately with lots of cool water.

Hydrogen gas, produced in the reaction of hydrochloric acid and magnesium, is flammable. Ensure that there is adequate ventilation and that there are no open flames in the classroom.

Magnesium oxide is an extremely fine dust. Do not inhale magnesium oxide powder, as it is irritating.

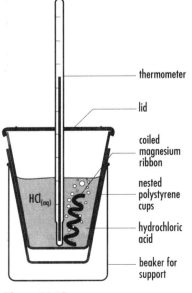

— thermometer

— lid

— coiled magnesium ribbon

— nested polystyrene cups

$HCl_{(aq)}$ — hydrochloric acid

— beaker for support

Figure 11.12
Magnesium ribbon reacts rapidly in dilute hydrochloric acid. With nested polystyrene cups, the enthalpy change can be determined by measuring the temperature change of the HCl solution.

INVESTIGATION

11.1 Applying Hess's Law

Magnesium burns rapidly, releasing heat and light (Figure 1.2, page 27).

$$Mg_{(s)} + \frac{1}{2} O_{2(g)} \rightarrow MgO_{(s)}$$

The enthalpy change of this reaction can be measured using a bomb calorimeter, but not a polystyrene cup calorimeter. The enthalpy change for the combustion of magnesium can be determined by applying Hess's law to the following three chemical equations.

$$MgO_{(s)} + 2\,HCl_{(aq)} \rightarrow MgCl_{2(aq)} + H_2O_{(l)}$$

$$Mg_{(s)} + 2\,HCl_{(aq)} \rightarrow MgCl_{2(aq)} + H_{2(g)}$$

$$H_{2(g)} + \frac{1}{2} O_{2(g)} \rightarrow H_2O_{(l)} \qquad \Delta H_f^\circ = -285.8 \text{ kJ}$$

Problem

What is the molar enthalpy of combustion for magnesium?

Prediction

According to the table of standard molar enthalpies of formation (Appendix F, page 551), the standard molar enthalpy of combustion for magnesium is –601.6 kJ/mol. The molar enthalpy of combustion of magnesium is the same as the molar enthalpy of formation of magnesium oxide.

$$H_c^\circ = H_f^\circ = -601.6 \text{ kJ/mol}$$
$$_{Mg} \quad _{MgO}$$

$$Mg_{(g)} + \frac{1}{2} O_{2(g)} \rightarrow MgO_{(s)} \qquad \Delta H_c^\circ = \Delta H_f^\circ = -601.6 \text{ kJ}$$

Experimental Design

The enthalpy changes for the first two reactions with hydrochloric acid are determined empirically using a polystyrene calorimeter (Figure 11.12). The three ΔH° values are used, along with Hess's law, to obtain the molar enthalpy of combustion for magnesium.

Materials

lab apron
safety glasses
magnesium ribbon (maximum 15 cm strip)
magnesium oxide powder (maximum 1.00 g sample)
1.00 mol/L hydrochloric acid (use 50 mL each time)
polystyrene calorimeter with lid
50 mL or 100 mL graduated cylinder
laboratory scoop or plastic spoon
steel wool
weighing boat or paper
centigram balance
ruler

- Problem
- ✔ **Prediction**
- ✔ **Design**
- Materials
- Procedure
- ✔ **Evidence**
- ✔ **Analysis**
- ✔ **Evaluation**
- Synthesis

CAUTION

Toxic, corrosive, and irritant chemicals are used in this investigation. Avoid skin contact. Wash any splashes on the skin or clothing with plenty of water. If any chemical is splashed in the eye, rinse for at least 15 min and inform your teacher.

DISPOSAL TIP

Keep the trichlorotrifluoroethane (a chlorofluorocarbon, or CFC) sealed to avoid evaporation and inhalation of the vapors. Dispose of the CFC mixtures as directed by your teacher.

INVESTIGATION

12.1 Single Replacement Reactions

The purpose of this investigation is to explain some single replacement reactions (page 104) in terms of oxidation and reduction. As part of the Experimental Design, include diagnostic tests (as in Appendix C, page 537) for the predicted products.

Problem

What are the products of the single replacement reactions for the following sets of reactants?

- copper and aqueous silver nitrate
- aqueous chlorine and aqueous sodium bromide
- magnesium and hydrochloric acid
- zinc and aqueous copper(II) sulfate
- aqueous chlorine and aqueous potassium iodide

Pediction

Experimental Design

Materials

lab apron	magnesium ribbon
safety glasses	zinc strip
five small test tubes	aqueous silver nitrate
one test tube stopper	aqueous sodium bromide
test tube rack	aqueous copper(II) sulfate
steel wool	aqueous potassium iodide
wash bottle	hydrochloric acid
matches	chlorine water
copper strip	trichlorotrifluoroethane

Procedure

1. Set up five test tubes, each half-filled with one of the five aqueous solutions.

2. Add the element indicated to each test tube.

3. Perform diagnostic tests on each of the five mixtures. Record your evidence.

▓	Problem
✔	**Prediction**
✔	**Design**
▓	Materials
✔	**Procedure**
✔	**Evidence**
✔	**Analysis**
✔	**Evaluation**
▓	Synthesis

INVESTIGATION

12.2 Spontaneity of Redox Reactions

Until now in this textbook, it has been assumed that all chemical reactions are **spontaneous**; that is, they occur without a continuous addition of energy to the system. Spontaneous redox reactions in solution generally provide visible evidence of a reaction within a few minutes. The scientific purpose of this investigation is to test the assumption that all single replacement reactions are spontaneous.

Problem

Which combinations of copper, lead, silver, and zinc metals and their aqueous metal ion solutions produce spontaneous reactions?

Prediction

Experimental Design

Materials

lab apron

safety glasses

reusable strips of copper, lead, silver, and zinc metals
(*Note that the lead strips bend much more easily than the zinc strips, which look similar.*)

0.10 mol/L solutions of copper(II) nitrate, lead(II) nitrate, silver nitrate, and zinc nitrate

test tube rack

supply of small test tubes

(4) 50 mL or 100 mL beakers

steel wool

grease pencil or labels

pure water (deionized or distilled)

waste container for lead and silver solutions

- Problem
- ✔ **Prediction**
- ✔ **Design**
- Materials
- Procedure
- ✔ **Evidence**
- ✔ **Analysis**
- ✔ **Evaluation**
- Synthesis

CAUTION

This reaction of sodium metal must be demonstrated with great care, because a great deal of heat is produced. Use only a piece the size of a small pea, use a safety screen, wear a lab apron and safety glasses, and keep observers at least two metres away.

INVESTIGATION

12.3 Demonstration with Sodium Metal

The purpose of this demonstration is to test the five-step method for predicting redox reactions. As part of the Experimental Design, include a list of diagnostic tests using the "If (procedure), and (evidence), then (analysis)" format for every product predicted. (This format is described in Appendix C on page 537.)

Problem

What are the products of the reaction of sodium metal with water?

- Problem
- ✔ **Prediction**
- Design
- Materials
- Procedure
- ✔ **Evidence**
- ✔ **Analysis**
- ✔ **Evaluation**
- Synthesis

INVESTIGATION

12.4 Analysis of a Hydrogen Peroxide Solution

Titration is an efficient, reliable, and precise experimental design used by laboratory technicians for testing the concentration of oxidizing and reducing agents. In this investigation you assume the role of a laboratory technician working in a consumer advocacy laboratory, testing the concentration of a hydrogen peroxide solution (Figure 12.15). The technological purpose of this investigation is to test and evaluate the concentration of the consumer solution of hydrogen peroxide.

Problem

What is the percent concentration of hydrogen peroxide in a consumer product?

Prediction

Experimental Design

A solution of the primary standard, iron(II) ammonium sulfate-6-water, is prepared and the potassium permanganate solution is standardized by a titration with this primary standard. A 25.0 mL sample of a consumer solution of hydrogen peroxide is diluted to 1.00 L with water (that is, it is diluted by a factor of 40). The standardized potassium permanganate solution is used to titrate the diluted hydrogen peroxide. The molar concentration of the original hydrogen peroxide is obtained by analysis of the titration evidence, and by using a graph of the information in Table 12.3.

Table 12.3

PERCENT AND MOLAR CONCENTRATION OF $H_2O_{2(aq)}$	
Percent Concentration (%)	Molar Concentration (mol/L)
2.5	0.73
2.6	0.76
2.7	0.79
2.8	0.82
2.9	0.85
3.0	0.88
3.1	0.91
3.2	0.94
3.3	0.97
3.4	1.0

Materials

lab apron
safety glasses
$FeSO_4 \cdot (NH_4)_2SO_4 \cdot 6\,H_2O_{(s)}$
2 mol/L $H_2SO_{4(aq)}$
diluted $H_2O_{2(aq)}$
$KMnO_{4(aq)}$
wash bottle
50 mL buret and clamp
10 mL graduated cylinder
(2) 100 mL beakers
(2) 250 mL beakers
(2) 250 mL Erlenmeyer flasks
100 mL volumetric flask and stopper
10 mL volumetric pipet and bulb
medicine dropper
stirring rod
centigram balance

© NELSON CANADA,
A DIVISION OF THOMSON CANADA LIMITED, 1994

small funnel
laboratory stand
laboratory scoop

Procedure

1. (Pre-lab) Calculate the mass of $FeSO_4 \cdot (NH_4)_2SO_4 \cdot 6\,H_2O_{(s)}$ required to prepare 100.0 mL of a 0.0500 mol/L solution.

2. Dissolve the iron(II) compound in about 40 mL of $H_2SO_{4(aq)}$ before preparing the standard solution in the 100 mL volumetric flask.

3. Transfer 10.00 mL of the standard iron(II) solution by pipet into a clean 250 mL Erlenmeyer flask.

4. Titrate the acidic iron(II) sample with $KMnO_{4(aq)}$.

5. Repeat steps 3 and 4 until three consistent volumes (within 0.1 mL) are obtained.

6. Transfer 10.00 mL of the diluted hydrogen peroxide solution by pipet into a clean 250 mL Erlenmeyer flask.

7. Using a 10 mL graduated cylinder, add 5 mL of $H_2SO_{4(aq)}$ to the hydrogen peroxide solution.

8. Titrate the acidic hydrogen peroxide solution with $KMnO_{4(aq)}$.

9. Repeat steps 6 to 8 until three consistent volumes (within 0.1 mL) are obtained.

INVESTIGATION

- ▨ Problem
- ▨ Prediction
- ▨ Design
- ▨ Materials
- ▨ Procedure
- ☑ **Evidence**
- ☑ **Analysis**
- ☑ **Evaluation**
- ▨ Synthesis

13.1 Demonstration of a Simple Electric Cell

The purpose of this investigation is to demonstrate an electric cell.

Problem

What electrical properties are observed when two metals come in contact with a conducting solution?

Prediction

According to the hypothesis of Luigi Galvani (1737 – 1798), electricity will only be produced if metals are in contact with animal tissue. Galvani was the Italian scientist who discovered that an electric current flows when two different metals are in contact with a muscle in a frog's leg.

Experimental Design

Different pairs of metal strips are placed in contact with fruits, vegetables, and inorganic solutions to test Galvani's hypothesis.

Materials

lab apron
safety glasses
paper towel
salt water
strips of metals such as $Zn_{(s)}$, $Cu_{(s)}$, $Pb_{(s)}$, and $Ag_{(s)}$
potato, orange, apple
clothespin
ammeter (sensitive current meter)
voltmeter and connecting wires

Procedure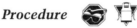

1. Place a paper towel soaked in salt water between strips of two different metals. Hold this "sandwich" together with a clothespin.

2. Connect the two metals to the terminals of an ammeter and observe the reading.

3. Connect the two metals to the terminals of a voltmeter and observe the reading.

4. Remove the paper towel and insert the two metals into one of the fruits or vegetables (Figure 13.3, page 388) and repeat steps 2 and 3. (The metals should not touch each other.)

5. Repeat steps 1 to 4 using different combinations of metals.

- Problem
- ✔ **Prediction**
- Design
- Materials
- ✔ **Procedure**
- ✔ **Evidence**
- ✔ **Analysis**
- ✔ **Evaluation**
- Synthesis

CAUTION

Be careful when handling acidic and basic solutions used for electrolytes, as they are corrosive. Wear eye protection and work near a source of water. Some electrolytes may be toxic or irritant; follow all safety precautions. Avoid eye and skin contact.

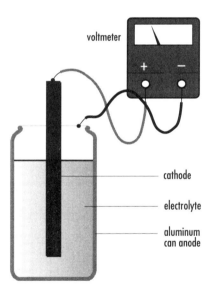

Figure 13.7
An aluminum-can cell is an efficient design since one of the electrodes also serves as the container.

INVESTIGATION

13.2 Designing an Electric Cell

The purpose of this investigation is to use everyday materials to simulate the technological development of an electric cell in which an aluminum soft-drink can is one of the electrodes (Figure 13.7). The other electrode is a solid conductor such as graphite from a pencil, an iron nail, or a piece of copper wire or pipe. The electrolyte may be a salt solution or an acidic or basic solution. Although many characteristics of a cell are important for an overall evaluation of performance, only one characteristic, voltage, is investigated here. Check with your teacher if you wish to evaluate other designs and materials.

Problem

What combination of electrodes and electrolyte gives the largest voltage for an aluminum-can cell?

Prediction

Experimental Design

(a) Using the same electrolyte and aluminum can as the controlled variables, two or three different materials are employed as the second electrode. The voltage of each cell is measured.

(b) Using the same two electrodes as the controlled variables, two or three possible electrolytes are tested. The voltage of each cell is measured.

(c) Additional combinations are tested, based on the analysis of the initial trials.

■ Problem
■ Prediction
■ Design
■ Materials
■ Procedure
✔ **Evidence**
✔ **Analysis**
■ Evaluation
■ Synthesis

CAUTION

Solutions used are toxic and irritant. Avoid contact with skin and eyes.

INVESTIGATION

13.3 Demonstration of a Voltaic Cell

The purpose of this investigation is to demonstrate the design and operation of a voltaic cell used in scientific research.

Problem

What is the design and operation of a voltaic cell?

Experimental Design

An electric cell with only one electrolyte is compared with similar voltaic cells containing the same electrodes but two electrolytes.

Procedure

1. Construct the three cells shown in Figure 13.13.

2. For each design, use a voltmeter to determine which electrode is positive and which is negative (see Appendix C on page 532), and measure the electric potential difference of each cell.

3. With the voltmeter connected, remove and then replace the various parts of the cell.

4. For each cell, connect the two electrodes with a wire. Record any evidence of a reaction after several minutes, and after one or two days.

(a)

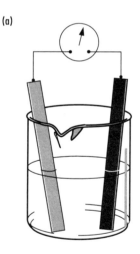

No porous boundary: $Ag_{(s)}$ | $NaNO_{3(aq)}$ | $Cu_{(s)}$

(b)

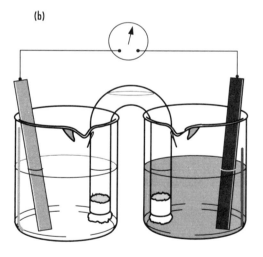

Salt bridge: $Ag_{(s)}$ | $AgNO_{3(aq)}$ || $Cu(NO_3)_{2(aq)}$ | $Cu_{(s)}$

(c)

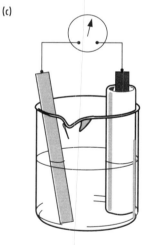

Porous cup: $Ag_{(s)}$ | $AgNO_{3(aq)}$ || $Cu(NO_3)_{2(aq)}$ | $Cu_{(s)}$

Figure 13.13
Investigation 13.3 compares three different cell designs.

Problem
✔ **Prediction**
✔ **Design**
  Materials
✔ **Procedure**
✔ **Evidence**
✔ **Analysis**
✔ **Evaluation**
  Synthesis

CAUTION

The materials used are toxic and irritant. Avoid skin and eye contact.

INVESTIGATION

13.4 Testing Voltaic Cells

The purpose of this investigation is to test the predictions of cell potentials and the charge on the electrodes of various cells.

Problem

In cells constructed from various combinations of copper, lead, silver, and zinc half-cells, what are the standard cell potentials, and which is the anode and cathode in each case?

Prediction

(The prediction is presented on an attached sheet.)

Experimental Design

Materials

lab apron
safety glasses
voltmeter and connecting wires
U-tube and/or porous cups
(4) 100 mL or 150 mL beakers
distilled water
steel wool
cotton
$Cu_{(s)}$, $Pb_{(s)}$, $Ag_{(s)}$, and $Zn_{(s)}$ strips
0.10 mol/L $CuSO_{4(aq)}$, $Pb(NO_3)_{2(aq)}$, $AgNO_{3(aq)}$, $NaNO_{3(aq)}$, and $ZnSO_{4(aq)}$

- ▓ Problem
- ▓ Prediction
- ▓ Design
- ▓ Materials
- ▓ Procedure
- ✔ **Evidence**
- ✔ **Analysis**
- ▓ Evaluation
- ▓ Synthesis

◆ **CAUTION**

Avoid skin or eye contact with the solutions. Avoid inhaling fumes of trichlorotrifluoroethane.

INVESTIGATION

13.5 A Potassium Iodide Electrolytic Cell

The purpose of this investigation is to observe the operation of an electrolytic cell and to determine its reaction products.

Problem

What are the products of the reaction during the operation of an aqueous potassium iodide electrolytic cell?

Experimental Design

Inert electrodes are placed in a 0.50 mol/L solution of potassium iodide and a battery or power supply provides a direct current of electricity to the cell. The litmus and halogen diagnostic tests (Appendix C, page 537) are conducted to test the solution near each electrode before and after the reaction.

Materials

lab apron	safety glasses
petri dish	two carbon electrodes
two connecting wires	3 V to 9 V battery or power
red and blue litmus paper	supply
ring stand and two utility clamps	small test tube with stopper
dropper bottle of trichlorotri-	0.50 mol/L $KI_{(aq)}$
fluoroethane	

Procedure

1. Set up the $KI_{(aq)}$ cell as shown in Figure 13.23 (or as shown but with one ring stand) but without connecting the power supply.
2. Observe the cell and test the solution with litmus paper and trichlorotrifluoroethane.
3. Use a wire to join the two electrodes and observe the cell.
4. Connect and turn on the power supply.
5. Record all observations at each electrode.
6. Perform both diagnostic tests at each electrode.

◆ **DISPOSAL TIP**

Dispose of the solutions as directed by your teacher.

Figure 13.23
A petri dish is a convenient container for the aqueous potassium iodide solution of this electrolytic cell. Carbon rods serve as inert electrodes.

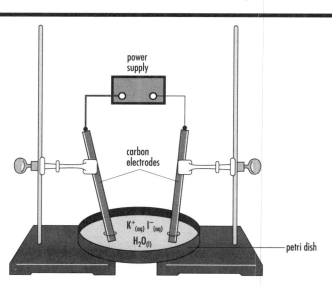

Problem
✔ **Prediction**
Design
Materials
Procedure
✔ **Evidence**
✔ **Analysis**
✔ **Evaluation**
Synthesis

CAUTION

Copper(II) sulfate is toxic and irritant. Avoid skin and eye contact. If you spill copper(II) sulfate solution on your skin, wash the affected area with lots of cool water. Remember to wash your hands before leaving the laboratory.

INVESTIGATION

13.6 Demonstration of Electrolysis

The purpose of this investigation is to test the method of predicting the products of electrolytic cells.

Problems

What are the products of electrolytic cells containing

- aqueous copper(II) sulfate?
- aqueous sodium sulfate?
- aqueous sodium chloride?

Prediction

Experimental Design

The electrolysis of the aqueous copper(II) sulfate is carried out in a U-tube, and the electrolysis of aqueous sodium sulfate and sodium chloride is carried out in a Hoffman apparatus (Figure 8.18, page 219) so that any gases produced can be collected. Diagnostic tests with necessary control tests are conducted to determine the presence of the predicted products.

- ▢ Problem
- ✔ **Prediction**
- ▢ Design
- ✔ **Materials**
- ✔ **Procedure**
- ✔ **Evidence**
- ✔ **Analysis**
- ✔ **Evaluation**
- ▢ Synthesis

CAUTION

Copper(II) sulfate is toxic and irritant. Avoid skin and eye contact. If you spill copper sulfate solution on your skin, wash the affected area with lots of cool water. Remember to wash your hands before leaving the laboratory.

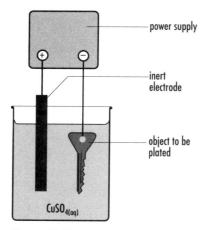

Figure 13.29
An electrolytic cell for copper-plating small objects.

(labels in figure: power supply; inert electrode; object to be plated; CuSO$_{4(aq)}$)

INVESTIGATION

13.7 Copper Plating

The purpose of this investigation is to determine the best procedures for plating copper onto various objects. Evaluate both the process and the product.

Problem

Which procedure causes a smooth layer of copper metal to adhere to a conducting object?

Prediction

Experimental Design

A small metal object, such as a spoon, a key, or a piece of metal, is carefully cleaned. A 0.50 mol/L copper(II) sulfate electrolytic cell that uses an inert electrode as the anode is constructed. The object to be plated is used as the cathode (Figure 13.29). Potentially relevant variables are identified and systematically manipulated. The success of the plating is evaluated by the appearance of the object and by polishing the plated object with steel wool to test adherence of the copper coating.

- ▦ Problem
- ✔ **Prediction**
- ▦ Design
- ▦ Materials
- ✔ **Procedure**
- ✔ **Evidence**
- ✔ **Analysis**
- ✔ **Evaluation**
- ▦ Synthesis

INVESTIGATION

14.1 Extent of a Chemical Reaction

Evidence supporting the assumption that reactions are quantitative was obtained in Chapter 7 with stoichiometry experiments that produced precipitates. In a quantitative reaction, the *limiting reagent* is completely consumed. To identify the limiting reagent you can test the final reaction mixture for the presence of the original reactants. For example, in a diagnostic test you might try to precipitate ions from the final reaction mixture that were present in the original reactants.

The purpose of this investigation is to test the validity of the assumption that chemical reactions are quantitative.

Problem

What are the limiting and excess reagents in the chemical reaction of selected quantities of aqueous sodium sulfate and aqueous calcium chloride?

Prediction

Experimental Design

Samples of sodium sulfate solution and calcium chloride solution are mixed in different proportions and the final mixture is filtered. Samples of the filtrate are tested for the presence of excess reagents, using the following diagnostic tests.

- If a few drops of $Ba(NO_3)_{2(aq)}$ are added to the filtrate and a precipitate forms, then excess sulfate ions are present.

  $$Ba^{2+}_{(aq)} + SO_4^{2-}_{(aq)} \rightarrow BaSO_{4(s)}$$

- If a few drops of $Na_2CO_{3(aq)}$ are added to the filtrate and a precipitate forms, then excess calcium ions are present.

  $$Ca^{2+}_{(aq)} + CO_3^{2-}_{(aq)} \rightarrow CaCO_{3(s)}$$

Materials

lab apron	safety glasses
25 mL of 0.50 mol/L $CaCl_{2(aq)}$	25 mL of 0.50 mol/L $Na_2SO_{4(aq)}$
1.0 mol/L $Na_2CO_{3(aq)}$ in dropper bottle	saturated $Ba(NO_3)_{2(aq)}$ in dropper bottle
(2) 50 mL or 100 mL beakers	two small test tubes
10 mL or 25 mL graduated cylinder	filtration apparatus
	wash bottle
filter paper	stirring rod

⚠ CAUTION

☠ **Barium compounds are toxic. Remember to wash your hands before leaving the laboratory.**

▩ Problem
✔ **Prediction**
▩ Design
▩ Materials
▩ Procedure
✔ **Evidence**
✔ **Analysis**
✔ **Evaluation**
▩ Synthesis

INVESTIGATION

14.2 Demonstration of Equilibrium Shifts

The purpose of this demonstration is to test Le Châtelier's principle by studying two chemical equilibrium systems: the equilibrium between two oxides of nitrogen (Figure 14.9), and the equilibrium of carbon dioxide gas and carbonic acid.

$$N_2O_{4(g)} + energy \rightleftharpoons 2\,NO_{2(g)}$$
$$\text{colorless} \qquad\qquad \text{reddish brown}$$
$$CO_{2(g)} + H_2O_{(l)} \rightleftharpoons H^+_{(aq)} + HCO_3^-_{(aq)}$$

The second equilibrium system, produced by the reaction of carbon dioxide gas and water, is commonly found in the human body and in carbonated drinks. A diagnostic test is necessary to detect shifts in this equilibrium. Methyl red, an acid-base indicator, can detect an increase or decrease in the hydrogen ion concentration in this system. Methyl red turns yellow when the hydrogen ion concentration decreases and it turns red when the hydrogen ion concentration increases.

Problem

How does a change in temperature affect the nitrogen dioxide-dinitrogen tetraoxide equilibrium system? How does a change in pressure affect the carbon dioxide–carbonic acid equilibrium system?

Prediction

Materials

lab apron
(2) $NO_{2(g)}/N_2O_{4(g)}$ sealed flasks
methyl red in dropper bottle
small syringe with needle
    removed (5 to 50 mL)
beaker of ice-water mixture

safety glasses
25 mL cold carbonated water
    (soda water)
solid rubber stopper to seal
    end of syringe
beaker of hot water

CAUTION

☠ Be careful with the flasks containing nitrogen dioxide: this gas is highly toxic.

Procedure

1. Place the sealed $NO_{2(g)}/N_2O_{4(g)}$ flasks in hot and cold water baths and record your observations.

2. Place two or three drops of methyl red indicator in the carbonated water.

3. Draw some carbonated water into the syringe, then block the end with a rubber stopper.

4. Slowly move the syringe plunger and record your observations.

Problem
✔ **Prediction**
Design
Materials
✔ **Procedure**
✔ **Evidence**
✔ **Analysis**
✔ **Evaluation**
Synthesis

CAUTION

Silver nitrate is toxic and irritant; avoid skin and eye contact. Cobalt(II) chloride is toxic. Ethanol is flammable. Make sure there are no flames in the laboratory before using the ethanol solution of cobalt(II) chloride. Remember to wash your hands before leaving the laboratory.

INVESTIGATION

14.3 Testing Le Châtelier's Principle

The purpose of this investigation is to test Le Châtelier's principle by studying the equilibrium between two complex ions containing the cobalt(II) ion dissolved in ethyl alcohol (al).

$$CoCl_4{}^{2-}{}_{(al)} + 6\,H_2O_{(al)} \rightleftharpoons Co(H_2O)_6{}^{2+}{}_{(al)} + 4\,Cl^-{}_{(al)} + energy$$
blue — pink

Problem

How does changing the temperature affect this chemical equilibrium system? How does changing the concentration affect this chemical equilibrium system?

Prediction

Experimental Design

Heat is added or removed by immersing samples of the equilibrium mixture in hot or cold water. In separate samples the concentration of chemicals in the system is changed by adding water, solid sodium chloride, or solid silver nitrate. In all cases, the final color of the system indicates the shift in the equilibrium. A sample of the equilibrium mixture, with the same volume as the other samples, is used as a control in all tests.

Materials

lab apron
safety glasses
cobalt(II) chloride equilibrium mixture in ethanol
$NaCl_{(s)}$
$AgNO_{3(s)}$
distilled water
crushed ice
100 mL beaker
(2) 400 mL beakers
(2) small test tubes with stoppers

- Problem
- Prediction
- Design
- Materials
- Procedure
- ✔ **Evidence**
- ✔ **Analysis**
- Evaluation
- Synthesis

INVESTIGATION

14.4 pH of Common Substances

One reason for the wide acceptance of the pH scale is the availability of a convenient, rapid, and precise measuring instrument. The purpose of this investigation is to show the technological advantages of a pH meter.

Problem

What generalizations can be made about the pH of foods and cleaning agents?

Experimental Design

The pH of a variety of solutions is measured. An attempt is made to develop generalizations concerning the pH of foods and cleaning agents.

Materials

lab apron
safety glasses
pH meter and pH 7 buffer solution
wash bottle of distilled water
400 mL waste beaker
several 100 mL beakers
various cleaning agents, such as ammonia, drain cleaner, and shampoo
various food products, such as juices, pop, vinegar, and milk

Procedure

Substances must be dissloved in water before measuring the pH.

1. Rinse the electrode of the pH meter with distilled water.
2. Place the pH meter electrode in a standard buffer solution and calibrate the instrument by adjusting the meter to read the pH of the buffer.
3. Rinse the pH meter electrode with distilled water.
4. Place the electrode in a beaker containing a sample and record the pH reading.
5. Rinse the pH meter electrode with distilled water.
6. Repeat steps 4 and 5 with each sample provided.

CAUTION

Some of the materials being tested are very corrosive. Do not allow them to come into contact with eyes, skin, or clothing.

- ✔ Problem
- ✔ Prediction
- ✔ Design
- ✔ Materials
- ✔ Procedure
- ✔ Evidence
- ✔ Analysis
- ✔ Evaluation
- ▨ Synthesis

CAUTION

Iron(III) compounds are irritant. Potassium thiocyanate is toxic.

INVESTIGATION

14.5 Studying a Chemical Equilibrium System

The purpose of this investigation is to solve a problem concerning the effect of an energy change on the following equilibrium system.

$$Fe^{3+}_{(aq)} + SCN^-_{(aq)} \rightleftharpoons FeSCN^{2+}_{(aq)}$$

almost colorless   colorless   red

Write a problem statement and then design and carry out an investigation to determine the role of energy in this equilibrium system.

- Problem
- ✔ **Prediction**
- ✔ **Design**
- Materials
- ✔ **Procedure**
- ✔ **Evidence**
- ✔ **Analysis**
- ✔ **Evaluation**
- Synthesis

CAUTION

Chemicals used include toxic, corrosive, and irritant materials. Avoid eye and skin contact. If you spill any of the acid or base solutions on your skin, immediately wash the area with lots of cool water.

INVESTIGATION

15.1 Testing Arrhenius's Acid-Base Definitions

The purpose of this investigation is to test Arrhenius's definitions of acid and base. A number of common substances in solution are identified as acid, base, or neutral, using one or more diagnostic tests. In your experimental design, be sure to identify all variables, including any controls.

Problem

Which of the substances tested may be classified as acid, base, or neutral?

Prediction

Experimental Design

Materials

lab apron
safety glasses
aqueous 0.10 mol/L solutions of:
   hydrogen chloride
   sodium carbonate (soda ash)
   sodium hydrogen carbonate (baking soda)
   sodium hydrogen sulfate
   sodium hydroxide (lye)
   calcium hydroxide (saturated solution)
   sulfur dioxide
   magnesium oxide (saturated solution)
   ammonia
   hydrogen acetate (vinegar)
conductivity apparatus, blue litmus paper, red litmus paper, and
   any other materials necessary for diagnostic tests

- ▦ Problem
- ✔ **Prediction**
- ▦ Design
- ✔ **Materials**
- ✔ **Procedure**
- ✔ **Evidence**
- ✔ **Analysis**
- ✔ **Evaluation**
- ▦ Synthesis

INVESTIGATION

15.2 Testing Brønsted-Lowry Predictions

The purpose of this investigation is to test the Brønsted-Lowry concept of acids and bases and the five-step method for predicting acid-base reactions.

Problem

What reactions occur when the following substances are mixed? (Hints for diagnostic tests are in parentheses.)

1. ammonium chloride and sodium hydroxide solutions (odor)
2. hydrochloric acid and sodium acetate solutions (odor)
3. sodium benzoate and sodium hydrogen sulfate solutions (solubility)
4. hydrochloric acid and aqueous ammonium chloride (odor)
5. solid sodium chloride added to water (litmus)
6. solid aluminum sulfate added to water (litmus)
7. solid sodium phosphate added to water (litmus)
8. solid sodium hydrogen sulfate added to water (litmus)
9. solid sodium hydrogen carbonate added to hydrochloric acid (pH)
10. solid sodium hydrogen carbonate added to sodium hydroxide solution (pH)
11. solid sodium hydrogen carbonate added to sodium hydrogen sulfate solution (pH)

Prediction

(The prediction is presented on an attached sheet.)

Experimental Design

Each prediction of a reaction using the established procedure is accompanied by a diagnostic test of the prediction. The certainty of the evaluation is increased by performing as many diagnostic tests as possible, complete with controls.

CAUTION

Chemicals used include toxic, corrosive, and irritant materials. Avoid eye and skin contact. If you spill any of the solutions on your skin, immediately wash the area with lots of cool water. Remember to detect odors cautiously by wafting air toward your nose from the container.

▨ Problem
▨ Prediction
▨ Design
▨ Materials
▨ Procedure
☑ **Evidence**
☑ **Analysis**
▨ Evaluation
▨ Synthesis

◆ **CAUTION**

Acids and bases are corrosive. Avoid skin and eye contact. If you spill any of the acid or base solutions on your skin, immediately wash the area with lots of cool water.

INVESTIGATION

15.3 Demonstration of pH Curves

The purpose of this demonstration is to study pH curves and the function of an indicator in an acid-base reaction.

Problem

What are the shapes of the pH curves for the continuous addition of hydrochloric acid to a sample of a sodium hydroxide solution and to a sample of a sodium carbonate solution?

Experimental Design

Small volumes of hydrochloric acid are added continuously to a measured volume of a base. After each addition, the pH of the mixture is measured. The volume of hydrochloric acid is the manipulated variable and the pH of the mixture is the responding variable.

Materials

lab apron
safety glasses
0.10 mol/L $HCl_{(aq)}$
0.10 mol/L $NaOH_{(aq)}$
0.10 mol/L $Na_2CO_{3(aq)}$
bromothymol blue indicator
methyl orange indicator
pH 7 buffer solution for calibration of pH meter
distilled water
pH meter
50 mL buret and funnel
150 mL beaker
(2) 250 mL beakers
(2) 50 mL graduated cylinders

Procedure

1. Set the temperature on the pH meter and calibrate it by adjusting it to indicate the pH of the known pH 7 buffer solution.

2. Place 50 mL of sodium hydroxide in a 150 mL beaker and add a few drops of bromothymol blue indicator.

3. Measure and record the pH of the 0.10 mol/L sodium hydroxide solution.

4. Successively add small quantities of $HCl_{(aq)}$, measuring the pH and noting any color changes after each addition, until about 80 mL of acid has been added.

5. Repeat steps 1 to 4 for 50 mL of 0.10 mol/L sodium carbonate with hydrochloric acid, using a 250 mL beaker and methyl orange indicator. Continue until 130 mL of $HCl_{(aq)}$ has been added.

- ✔ **Problem**
- ✔ **Prediction**
- ✔ **Design**
- ✔ **Materials**
- ✔ **Procedure**
- ✔ **Evidence**
- ✔ **Analysis**
- ✔ **Evaluation**
- ▨ Synthesis

INVESTIGATION

15.4 Buffers

The purpose of this investigation is to prepare and test a common buffer, such as the dihydrogen phosphate ion-hydrogen phosphate ion pair. For convenience, prepare a 0.10 mol/L buffer solution.

 CAUTION

Acids and bases are corrosive. Avoid skin and eye contact. If you spill any of the solutions on your skin, immediately wash the area with lots of cool water.

- Problem
- Prediction
- ✔ **Design**
- ✔ **Materials**
- ✔ **Procedure**
- ✔ **Evidence**
- ✔ **Analysis**
- ✔ **Evaluation**
- Synthesis

CAUTION

Ammonia irritates skin and mucous membranes. Hydrochloric acid is a corrosive acid. Avoid eye and skin contact. If you spill any of the solutions on your skin, immediately wash the area with lots of cool water.

INVESTIGATION

15.5 Ammonia Analysis

Ammonia is most often used by consumers as a household cleaner. In its simplest form, ammonia is sold as an aqueous solution with dilution instructions for various cleaning applications, such as washing laundry, cleaning glass, and removing wax. Ammonia is a relatively weak base and requires a strong acid to meet the quantitative reaction requirements for a titration. The pH curve for ammonia titrated with hydrochloric acid is shown in Figure 15.29.

The purpose of this investigation is to determine the molar concentration of a household ammonia solution, using a titration with hydrochloric acid as the experimental design. Sodium carbonate may be used as a primary standard for standardizing the hydrochloric acid. (See the pH curve for this reaction in Figure 15.20, page 482.)

Problem

What is the molar concentration of the household ammonia sample provided?

Prediction

According to the literature, the concentration of a fresh household ammonia solution varies from 3.0% to 29% by mass. This concentration corresponds to a range of 1.8 mol/L to 17 mol/L.

25.0 mL of 0.45 mol/L $NH_{3(aq)}$ Titrated with 0.50 mol/L $HCl_{(aq)}$

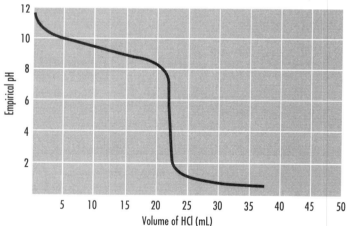

Figure 15.29
pH curve for the addition of
0.50 mol/L $HCl_{(aq)}$ to 25.0 mL of
0.45 mol/L $NH_{3(aq)}$.

ADDITIONAL INVESTIGATIONS

The additional investigations can be used as instruction (Gravimetric Stoichiometry Demonstration), as explorations of technology (Titration Analysis of Aspirin® Tablets), as motivation (Preparation of Nylon 6-10), and to assess performance (The Nature of the $Al_{(s)}$ / $CuCl_{2(aq)}$ Reaction System). These investigations are commonly used in school laboratory and work well. Some of the investigations lend themselves to a lab-test context while the demonstrations add concrete experience to otherwise abstract concepts.

The placement of the investigations may vary. For example, both titrations (in Investigations 7.A2 and 7.A3) involve organic reagents and could add some laboratory activities to *Nelson Chemistry* Chapter 9, Organic Chemistry. Investigation 15.A2 should be done at the end of the course after electrochemisty, acids and bases, and thermochemistry. Take advantage of this beginning by adding some of your favorite investigations to this section.

- The front of each sheet in this section is a blackline master for the additional investigation, which becomes the information and assignment for the student.

- The back of each sheet has suggested teaching tips and lists pertinent technical information for the instructor and/or lab technician.

- Each additional investigation has been reviewed for safety. Appropriate warnings concerning potential safety hazards are included where applicable.

MASTER LIST OF ADDITIONAL INVESTIGATIONS

TEXT CREDITS:

p. 348   Smarties® is a trademark of Sociétédes Produits Nestlé S.A. of Vevey Switzerland

p. 348   "M&M's"® Chocolate Candies is the registered trademark of Effem Foods Ltd.

Problem
✔ **Prediction**
Design
Materials
Procedure
✔ **Evidence**
✔ **Analysis**
✔ **Evaluation**
Synthesis

INVESTIGATION

7.A1 Gravimetric Stoichiometry (Demonstration)

The purpose of this demonstration is to test the stoichiometric method by comparing the mass of silver produced from the reaction of excess copper with a silver nitrate sample with the mass predicted by stoichiometric calculation.

Problem
What mass of silver is produced from the reaction of excess copper with 2.75 g of silver nitrate in solution?

Prediction

Experimental Design
A specific mass (2.75 g) of silver nitrate is measured and dissolved in water. Excess copper is added, and the reaction allowed to proceed until no further change is observed. The silver product is separated by filtration and dried, and the mass of silver measured. This experimental value is then compared with the predicted value.

Materials

lab apron	filtration apparatus
safety glasses	filter paper
centigram balance	250 mL beaker
$AgNO_{3(s)}$	400 mL beaker
heavy gauge (#18 – 22) Cu wire	laboratory scoop
wash bottle of pure water	stirring rod

Procedure – Day 1
1. Obtain a 2.75 g sample of $AgNO_{3(s)}$ in a 250 mL beaker.
2. Add about 100 mL of pure water and stir to dissolve.
3. Coil about 30 cm of $Cu_{(s)}$ wire and place in the solution so that most (but not all) of the wire is submerged. Record evidence of chemical reaction.

Procedure – Day 2
4. Remove the wire coil, shaking/rinsing off any silver that clings to it.
5. Measure and record the mass of a piece of filter paper.
6. Filter the mixture to separate the solid silver from the filtrate.
7. Place the filter paper and contents on a paper towel to dry overnight.

Procedure – Day 3
8. Measure and record the mass of dried silver and filter paper.

CAUTION

Silver nitrate is toxic and an irritant. Wash your hands thoroughly at the end of each activity.

D **DISPOSAL TIP**

Rinse all solutions down a drain with plenty of water.

Notes

This demonstration is an excellent way to introduce gravimetric stoichiometry, section 7.2, page 169. Students can refer to Figure 4.9 on page 99 to anticipate or confirm the visual changes. During each of the first two lab periods, about 15 – 20 minutes will be needed for the procedure, and several more minutes for a detailed and methodical instruction in the gravimetric stoichiometry required for the prediction calculation. Students can be assigned the problems on page 171 for practice during the remaining time. The comparison of experimental and predicted values as a percent difference is used to evaluate the stoichiometric method. As part of the demonstration, lead students in a discussion of the Evaluation of the investigation report.

Caution

Silver nitrate is toxic and an irritant.

Waste Disposal

The solutions in this investigation may be disposed of by rinsing down a drain with plenty of water, and the solids can be disposed of as regular (landfill) garbage.

Prediction (sample)

$$Cu_{(s)} + 2\ AgNO_{3(aq)} \rightarrow Cu(NO_3)_{2(aq)} + 2\ Ag_{(s)}$$

$$\begin{array}{ccc} & 2.75\ g & & m \\ & 169.88\ g/mol & & 107.87\ g/mol \end{array}$$

$$n_{AgNO_3} = 2.75\ g \times \frac{1\ mol}{169.88\ g} = 0.0162\ mol$$

$$n_{Ag} = 0.0162\ mol \times \frac{2}{2} = 0.0162\ mol$$

$$m_{Ag} = 0.0162\ mol \times \frac{107.87\ g}{1\ mol} = 1.75\ g$$

According to the stoichiometric method, the mass of silver produced by the reaction of excess copper with 2.75 g of silver nitrate in solution is 1.75 g.

Evidence (sample)

mass of filter paper = 0.97 g
mass of filter paper + silver = 2.69 g
• The solution turned blue and a silver-grey precipitate was formed.

Analysis (sample)

mass of silver produced = 2.69 g – 0.97 g = 1.72 g

According to the evidence gathered, the mass of silver produced by the reaction of 2.75 g of silver nitrate with excess copper is 1.72 g.

Problem
Prediction
Design
Materials
Procedure
✔ **Evidence**
✔ **Analysis**
✔ **Evaluation**
Synthesis

CAUTION

Phenolphthalein indicator solution (alcohol based) is flammable. Sodium hydro-xide and vinegar are corrosive.

INVESTIGATION

7.A2 Titration Analysis of Vinegar

The purpose of this investigation is to practice the technological skill of titration as part of a quantitative chemical analysis. The indicator used for this analysis, phenolphthalein, changes from colorless to pink (magenta) as the endpoint of the reaction. Each titration should continue until the addition of a single drop from the buret causes the endpoint color change to a permanent pale pink.

Problem

What is the molar concentration of acetic acid, $CH_3COOH_{(aq)}$, in a commercial "white" vinegar?

Experimental Design

A standard solution of sodium hydroxide is prepared. Measured samples of a commercial vinegar are titrated with this sodium hydroxide solution using phenolphthalein as the indicator until three reliable results, consistent to within 0.1 mL, are obtained.

Materials

lab apron
safety glasses
phenolphthalein indicator
$NaOH_{(s)}$
5% or 7% vinegar (see label)
medicine dropper
laboratory scoop
centigram balance
wash bottle of pure water
stirring rod

400 mL waste beaker
(2) 150 mL beakers
(2) 250 mL Erlenmeyer flasks
small funnel
100 mL volumetric flask with stopper
50 mL buret and clamp
10 mL pipet and pipet bulb
laboratory stand
meniscus finder

Procedure

1. Precisely measure and record the mass of a 2 – 3 g sample of $NaOH_{(s)}$ in a clean, dry 150 mL beaker.
2. Prepare 100.0 mL solution of $NaOH_{(aq)}$, and transfer it to a clean, dry, labelled 150 mL beaker.
3. Place 70 mL to 80 mL of commercial vinegar in a clean, dry, labelled 150 mL beaker.
4. Set up the buret to contain the standard $NaOH_{(aq)}$.
5. Pipet a 10.00 mL sample of vinegar into a clean Erlenmeyer flask and add 1 to 3 drops of phenolphthalein indicator.
6. Record the initial buret reading (to 0.1 mL).
7. Titrate the vinegar sample with $NaOH_{(aq)}$ until a single drop produces a permanent change in color from colorless to pink.
8. Record the final buret reading (to 0.1 mL).
9. Repeat steps 4 to 7 until three consistent results are obtained.
10. Dispose of all solutions into the sink.
11. Wash your hands thoroughly before leaving the laboratory.

Notes

- This investigation is similar to Lab Exercise 7H on page 191.
- $NaOH_{(s)}$ is not a (highly precise) primary standard compound, but for students working in Chapter 7 this is not yet a point of concern.
- Commercial 5% white vinegar is about 0.87 mol/L concentration, so each titration trial will use about 11 mL to 17 mL of $NaOH_{(aq)}$.
- Conversion
  5% (by volume) of one litre is 50 mL.
  Density of $CH_3COOH_{(l)}$ is 1.05 g/mL.
  In one litre of vinegar, $m_{CH_3COOH} = 50 \text{ mL} \times 1.05 \text{ g/mL} = 52.5 \text{ g}$
  $M_{CH_3COOH} = 60.06 \text{ g/mol}$

  $$n_{CH_3COOH} = \frac{52.5 \text{ g}}{60.06 \text{ g/mol}} = 0.874 \text{ mol}$$

Minilab Version

Chemical consumption can be reduced by dilution (see alternatives below), or by having pairs or triads of students alternate titrations and share evidence. However. the chemicals used are inexpensive, readily available, and non-toxic when diluted, making this a good choice for a realistic titration "wet" analysis experience.

Alternative (Enrichment)

You can have the students dilute the commercial vinegar in a 5:1 volume ratio to make an approximate 1% solution, and use a prepared 0.100 mol/L solution of $NaOH_{(aq)}$. (This reduces the consumption of chemicals while providing practice in diluting stock solutions at the same time.)

Several brands of vinegar can be analyzed and the results compared for consistency. Each brand's result of analysis should be compared with its label information — a percent difference calculation is useful here.

Alternative (as a Chapter 15 Investigation)

The purpose of the investigation can be changed to a test of the manufacturer's label. You can have students titrate diluted vinegar with a standard solution of sodium hydroxide, $NaOH_{(aq)}$, prepared to be approximately 0.1 mol/L. Have students standardize the $NaOH_{(aq)}$ first by titrating with 0.100 mol/L potassium hydrogen phthalate, $KHC_8H_4O_{4(aq)}$ (a primary standard) using phenolphthalein as the indicator.

- Preparation of solutions used
0.500 mol/L $NaOH_{(aq)}$	20.00 g of NaOH / L
0.100 mol/L $NaOH_{(aq)}$	4.00 g of NaOH /
0.100 mol/L $KHC_8H_4O_{4(aq)}$	20.42 g of $KHC_8H_4O_4$ / L

Caution

Sodium hydroxide is very caustic. Neither the solution nor solid should come in contact with skin. Rinse very thoroughly with water if this happens. Phenolphthalein solution (alcohol based) is flammable.

Waste Disposal

All of the chemicals in this investigation may be disposed of by flushing down a drain with plenty of water.

Problem
Prediction
Design
Materials
Procedure
✔ **Evidence**
✔ **Analysis**
✔ **Evaluation**
Synthesis

CAUTION

Phenolphthalein indicator solution is flammable. Sodium hydroxide is corrosive. Methanol is toxic and flammable.

INVESTIGATION

7.A3 Titration Analysis of an Aspirin® Tablet

Acetylsalicylic acid, ASA, is the most commonly used drug — with over 10 000 t manufactured in North America every year. This compound is an "organic" acid and reacts with strong bases such as sodium hydroxide according to the following reaction equation.

$$C_8H_7O_2COOH_{(alc)} + NaOH_{(aq)} \rightarrow NaC_8H_7O_2COO_{(aq)} + H_2O_{(l)}$$

The purpose of this investigation is to test the manufacturer's claim for the mass of ASA in a commercial tablet.

Problem

What is the quantity of "active ingredient" in a commercial ASA tablet ?

Experimental Design

Commercial "headache" tablets are dissolved in methanol and titrated with 0.150 mol/L sodium hydroxide solution using phenolphthalein as the indicator.

Materials

lab apron	400 mL waste beaker
safety glasses	(2) 150 mL beakers
phenolphthalein indicator	(2) 250 mL Erlenmeyer flasks
0.150 mol/L NaOH$_{(aq)}$	50 mL graduated cylinder
ASA tablets	small funnel
methanol (methyl alcohol)	50 mL buret and clamp
wash bottle of pure water	meniscus finder
stirring rod with rubber policeman	laboratory stand

Procedure

1. Obtain 60 mL to 70 mL of 0.150 mol/L NaOH$_{(aq)}$ in a clean, dry, labelled 150 mL beaker.

2. Obtain several commercial ASA tablets and record the label information about the mass of these tablets.

3. Set up the buret to contain the standard NaOH$_{(aq)}$.

4. Add 20 mL to 40 mL of methanol to one ASA tablet in a clean Erlenmeyer flask, and stir. (These tablets usually have "inert" ingredients, so the tablet may not dissolve completely.)

5. Record the initial buret reading (to 0.1 mL).

6. Add 1 to 3 drops of phenolphthalein indicator to the flask, and slowly titrate the tablet sample with NaOH$_{(aq)}$.

7. Record the final buret reading (to 0.1 mL).

8. Repeat steps 4 to 7 until three consistent results are obtained.

9. Dispose of all chemicals by flushing them down the drain with plenty of water.

10. Wash your hands thoroughly before leaving the laboratory.

Notes

Since the solid compound is more soluble in alcohol than in water, methanol is used to dissolve the active ingredient in the tablet. The phenolphthalein indicator used for this analysis changes color from colorless to (magenta) pink as the endpoint of the reaction. Each titration should continue until the addition of a single drop from the buret causes the endpoint color change to a permanent pale pink (a pink color that does not disappear quickly as the flask contents are swirled).

- The endpoint color change can be slow. Tell students to swirl the flask constantly to aid the reaction. It is useful to compare brand and generic tablets — different student groups can titrate different brands for comparison. **Do not** use "buffered" headache tablets — they will react unpredictably.

- A variety of brands, including some extra strength 500 mg tablets, could be used as part of a performance assessment. Place four tablets from the same bottle into numbered envelopes. Each student gets a different envelope and must report the ASA content.

- A 325 mg tablet will react with approximately 12 mL of 0.150 mol/L $NaOH_{(aq)}$; about three titrations are possible without refilling the buret.

- Some students could use ethanol as a solvent to see if this change affects the results.

- Remind students that quantities listed on commercial labels are guaranteed minimums, not measurements.

- Preparation of solution used

  0.150 mol/L $NaOH_{(aq)}$ 6.00 g of NaOH/L

$NaOH_{(s)}$ is hygroscopic, absorbing water from the air, and is not a primary standard compound. If your stock bottle is kept closed, and you obtain the required mass fairly quickly, this will not be a problem.

Caution

Methanol is flammable and toxic — avoid breathing vapors and do not use around open flames. Quantity used should be 40 mL $\times$ 15 sets, i.e. 600 mL per class. It could be a problem if methanol collects in the drain or oxidizers are flushed down the drain soon after the methanol.

Sodium hydroxide is very caustic. Neither the solution nor solid should come in contact with skin. Rinse very thoroughly with water if this happens.

Waste Disposal

All of the chemicals in this investigation may be disposed of by flushing down a drain with plenty of water.

- ▨ Problem
- ▨ Prediction
- ▨ Design
- ▨ Materials
- ▨ Procedure
- ✔ **Evidence**
- ✔ **Analysis**
- ✔ **Evaluation**
- ▨ Synthesis

CAUTION

Methanol is flammable.
Vinegar is corrosive.

INVESTIGATION

8.A1 Paper Chromatography — Molecular Separation

In this kind of chromatography molecules (of visible substances) of different sizes and polarities are carried along a paper surface by a moving colorless solvent. Differences in attractions among the molecules of the substances, solvent and surface material cause the visible substances to move at different rates — separating them into groups according to molecular structure. Many different types of chromatography are widely used for identifying and analyzing molecular substances — from hydrocarbon gas analysis to specific protein identifications.

Problem

What is the solvent and type of ink used to make a given (unknown) chromatogram?

Experimental Design

Various inks from different sources are tested for visible molecular component mobility along porous paper using several "carrier" molecular solvents of varying molecular polarity.

Materials

lab apron
safety glasses
methanol
vinegar
distilled water

(3) 25×200 mm test tubes
(3) stoppers to fit test tubes
porous paper strip
test tube stand
various marker pens

Procedure

1. Cut strips of paper to a length slightly longer than a test tube.
2. Add about 1 cm of solvent to a test tube; stopper and let stand.
3. Imagine a line across the width of the paper strip where the surface of the solvent will touch the strip when one end of the strip is dropped to the bottom of the test tube. Place a small dot of the ink to be investigated about 1 cm above this line.
4. Open the test tube and set the paper in place — using the stopper to hold the paper in position when the test tube is resealed. The end of the strip must be in the liquid solvent, and the ink dot must be above the liquid solvent.
5. Observe and record the separation of colors as molecules of different substances are carried up the paper strip by the solvent.
6. Repeat steps 2 to 5 with the other two solvents.
7. Repeat steps 2 to 6 with the other samples of ink.
8. Compare your chromatograms with the given unknown to identify the ink and solvent used to make it.
9. Dispose of all chemicals as directed by your teacher.
10. Wash your hands thoroughly before leaving the laboratory.

Notes

Try out this investigation before you have students do it. The only sure way to get materials that give good results is trial and error. Try a variety of paper including filter paper, coffee filters, etc. Black marker pens and overhead transparency markers in dark colors usually work well. Color coatings in candies such as Smarties® and M & M's® chocolate candies also work well in this laboratory activity.

Prepare a sample unknown (or more than one) in advance. This investigation is fun for students and provides a great lead-in to discussions of intermolecular bonding and substance identification. For enrichment, have students research other types of chromatography and applications of these techniques.

Commercial methyl hydrate (fondue fuel or gas-line antifreeze) is fine for this lab. 7% white (pickling or extra strength) vinegar (acetic acid) works well.

A variation for this investigation is to use disposable glass dropping pipets and powdered alumina. Simple, inexpensive column chromatography can be done this way.

Caution

Methanol and methyl hydrate are flammable and toxic — avoid breathing vapors and do not use around open flames. Vinegar is corrosive.

Waste Disposal

All of the chemicals in this investigation may be disposed of by flushing down the drain with plenty of water. Avoid pouring (large quantities of) methanol down the drain because of its possible reactions with oxidizers flushed down the drain by other classes or cleaners.

- Problem
- ✔ **Prediction**
- Design
- Materials
- Procedure
- Evidence
- Analysis
- Evaluation
- Synthesis

CAUTION

Trichlorotrifluoroethane, sebacyl chloride, and 1,6-hexanediamine are corrosive, toxic, and irritant. Use the fume hood for this demonstration.

INVESTIGATION

9.A1 The Preparation of Nylon 6-10 (Demonstration)

The purpose of this demonstration is to illustrate the formation of a polymer.

Problem

How does the combination of sebacyl (sebacoyl) chloride and 1,6-hexanediamine form the polymer known as Nylon 6-10?

Prediction

(*Write a structural equation showing the condensation reaction between a molecule of sebacyl chloride, $COCl(CH_2)_8COCl_{(l)}$, and a molecule of 1,6-hexanediamine, $NH_2(CH_2)_6NH_{2(l)}$. Join the two structures by "condensing" out a molecule of HCl. Indicate how this process can be expected to form a continuous polymer.*)

Materials

lab apron
safety glasses
100 mL or 150 mL beaker
50 mL beaker
forceps (tweezers)
laboratory stand with
   iron ring
rinsing tray or sink

disposable latex or polyethylene gloves
(2) 10 mL graduated cylinders
sebacyl chloride, $COCl(CH_2)_8COCl_{(l)}$
1,6-hexanediamine, $NH_2(CH_2)_6NH_{2(l)}$
cardboard tubes (paper towel rolls)
wash bottle of distilled water
trichlorotrifluoroethane, $C_2Cl_3F_{3(l)}$
paper towels

Procedure

1. Perform this experiment in a fume hood.
2. Dissolve 1mL of sebacyl chloride in about 50 mL of trichlorotrifluoroethane in a 100 mL beaker.
3. Dissolve 1 mL of 1,6-hexanediamine in about 25 mL of water in a 50 mL beaker.
4. Carefully pour the 1,6-hexanediamine solution into the 100 mL beaker along the side so that it forms a layer on top of the sebacyl chloride solution.
5. With the forceps, carefully take hold of the film that forms at the interface of the two solutions and slowly pull it straight up, forming a continuous filament of Nylon 6-10.
6. Wash the nylon strand with water and place the reaction mixture into an organic waste container inside the fume hood.
7. Wash your hands thoroughly before leaving the laboratory.

Notes

The nylon strand can be pulled until the 1,6-hexanediamine is used up. One easy way to handle the strand is to roll it around a large graduated cylinder. With two people and two cylinders a long strand can be strung over a considerable distance. The plastic should be washed with water before handling — better yet, handle only with disposable gloves on (see **Caution** below).

Caution

The chemicals used in this demonstration are corrosive, toxic, and irritant, and may react with heat and oxidizers. Therefore, precautions should be taken for the handling, storage, and disposal of the chemicals, and the demonstration should be done in a fume hood. Disposable gloves should be worn and skin contact avoided. **Wash** the nylon strand with water before allowing it to be handled.

Waste Disposal

The chemical mix should be disposed of into an organic waste container inside a fume hood, and stored for later toxic waste disposal.

Problem
Prediction
Design
Materials
Procedure
✔ **Evidence**
✔ **Analysis**
✔ **Evaluation**
Synthesis

CAUTION

2-methyl-2-propanol and pentane are flammable, toxic, and irritant. Avoid skin and eye contact.

INVESTIGATION

9.A2 Fractional Distillation (Demonstration)

Fractional distillation is the process used commercially to separate molecular components of petroleum by size, using boiling point differences. The same technique can be used to separate components of most liquid mixtures, provided a significant boiling point difference exists. Alcohol-water mixtures can be separated this way because alcohol has a much lower boiling point than water. In this demonstration, the vapors of the boiling mixture rise into a fractionating column, where initially they all condense and fall back into a flask. When the temperature in the column rises sufficiently, vapors of the component with the lower boiling point pass through the column and enter a condenser which cools and condenses the component back to a liquid for collection. The component with the higher boiling point still condenses in the fractionating column, which effectively separates the two liquids. When nearly all of the first "fraction" has gone, the temperature in the column will rise, and the second component will begin to pass through to the condenser.

Problem

What is the evidence for the separation of pentane from 2-methyl-2-propanol?

Experimental Design

A mixture of the two liquids is heated in a fractional distillation apparatus while the temperature is measured at regular intervals in a fume hood.

Materials

lab apron
safety glasses
round bottom distillation flask
electric heating mantle
(2) small collecting flasks
timer (clock or stopwatch)
fume hood
thermometer to fit column stopper

pentane, $C_5H_{12(l)}$
2-methyl-2-propanol,
    $CH_3C(CH_3)OHCH_{3(l)}$
fractionating column to fit flask
large (400 mL – 600 mL) beaker
condenser with tubing and
    fittings

Procedure

1. Add 25 mL of each of the two liquids to the distillation flask.
2. Heat the flask slowly, taking the vapor temperature at regular 30 s intervals.
3. After most of the pentane has boiled off, when the column temperature rises noticeably, change collection flasks at the outflow of the condenser.
4. As soon as most of the alcohol has boiled off, remove the heating mantle.
5. Dispose of the organic chemicals in an organic waste container.
6. Wash your hands thoroughly before leaving the laboratory.

Notes

- Heat the mixture slowly at first to avoid boiling off the pentane too quickly. Pentane boils at 36°C while 2-methyl-2-propanol (*t*-butanol boils) at 82°C. The whole sample should distill over in 10 – 15 minutes. If you collect the fractions in graduated cylinders you can record the volumes after distillation.

- Have students complete a vapor temperature versus time graph as you collect the evidence. Label boiling point plateaus when the graph is finished.

- An alternative problem is to separate pentane, 2-methyl-2-propanol, and pentacosane (paraffin wax). Prepare this mixture before class using 25 mL of each liquid and 1 – 2 g of paraffin wax. After the pentane and alcohol have been distilled, the molten wax (b.p. > 450°C) remains in the flask where it will solidify. If you are careful in handling the hot flask, you can pour the molten wax onto the wall of a beaker at room temperature where the wax will solidify instantly.

Caution

The chemicals used are highly flammable and irritant — dangerous fire and explosion risk. Keep away from any flame or spark.

The demonstration should be done in a fume hood to avoid breathing the vapors as they are toxic and an irritant to the eyes and skin.

Waste Disposal

The chemicals should be disposed of in an approved Organic Waste container.

Problem
✔ **Prediction**
Design
Materials
Procedure
✔ **Evidence**
✔ **Analysis**
✔ **Evaluation**
Synthesis

CAUTION

Ammonia and sulfuric acid are both corrosive. Use a face shield near the end of the evaporating process because of the possible splattering of the dried solid crystals.

INVESTIGATION

15.A1 Preparation of Ammonium Sulfate

Ammonium sulfate is a very common commercial chemical used in fertilizers, in water treatment, in fireproofing, and even as a food additive. The commercial preparation is essentially the same process as this investigation — a neutralization followed by crystallization of the product.

Problem

What is the mass and percent yield of ammonium sulfate prepared by reacting 50.0 mL of 0.200 mol/L sulfuric acid with excess aqueous ammonia?

Prediction

Experimental Design

A solution of aqueous ammonia is added to a sulfuric acid sample until the ammonia is in excess, and the product separated by heating and crystallization.

Materials

lab apron	stirring rod
safety glasses	50 mL graduated cylinder
1.0 mol/L $NH_{3(aq)}$	100 mL beaker
0.200 mol/L $H_2SO_{4(aq)}$	250 mL beaker
centigram balance	hotplate

Procedure

1. Measure and record the mass of a clean, dry 250 mL beaker.

2. Use a graduated cylinder to transfer 50.0 mL of 0.200 mol/L $H_2SO_{4(aq)}$ to the 250 mL beaker.

3. Obtain about 50 mL of 1.0 mol/L $NH_{3(aq)}$ in a 100 mL beaker.

4. **Slowly**, with stirring, in a fume hood or well ventilated area, add the ammonia solution to the sulfuric acid.

5. On a hotplate set initially to high heat and in a fume hood, evaporate the reaction mixture until the liquid volume becomes quite small.

6. When the volume of liquid becomes small, reduce the heat to low until the liquid is all evaporated and only dry crystals remain. (The solid compound will decompose at temperatures over 235°C, so it is important to watch the evaporation process and reduce the heat at the end!)

7. After cooling, measure the mass of the beaker and contents.

8. If time permits, reheat and then reweigh the beaker and contents.

9. Wash your hands thoroughly before leaving the laboratory.

Notes

This reaction allows students to use crystallization as a separation process. The chemicals are very common and inexpensive, and the connection with society is strong — the Sherritt® plant at Fort Saskatchewan, Alberta uses this method to manufacture ammonium sulfate by the kilotonne for use in fertilizers.

According to the Brønsted-Lowry concept, five-step method, and stoichiometry, a mass of 1.32 g of ammonium sulfate should be produced. This corresponds to a 100% yield based on the assumption of a quantitative reaction represented by the following equation.

$$H_2SO_{4(aq)} + 2\,NH_{3(aq)} \rightarrow (NH_4)_2SO_{4(aq)}$$

Discuss why the ammonia is used as the excess reagent in the design of this experiment. (It readily evaporates. Concentrated sulfuric acid, on the other hand, takes a very high temperature and a great deal of time to evaporate.)

- Prepare the reacting solutions as follows from standard laboratory concentrated (28% or 14.8 mol/L) aqueous ammonia and standard laboratory concentrated (95% or 17.8 mol/L) sulfuric acid.

1.0 mol/L $NH_{3(aq)}$	68 mL of concentrated base / L
0.200 mol/L $H_2SO_{4(aq)}$	11.2 mL of concentrated acid / L

Caution

When preparing the acid solution, add the concentrated $H_2SO_{4(aq)}$ to the water, with stirring, to disperse released heat.

Ammonia and sulfuric acid are both corrosive, skin contact should be avoided.

Ammonia vapors are irritating and toxic in high concentration — use a fume hood.

Waste Disposal

All chemicals used may be safely rinsed in to a sink with plenty of water.

© NELSON CANADA,
A DIVISION OF THOMSON CANADA LIMITED, 1994

INVESTIGATION

- ✔ **Problem**
- ✔ **Prediction**
- ✔ **Design**
- ✔ **Materials**
- ✔ **Procedure**
- ✔ **Evidence**
- ✔ **Analysis**
- ✔ **Evaluation**
-   Synthesis

15.A.2 The Nature of the $Al_{(s)}$ / $CuCl_{2(aq)}$ Reaction System

Record your qualitative observations of the reaction between some $Al_{(s)}$ wire and 0.5 mol/L $CuCl_{2(aq)}$. Then propose some problems that can be investigated with the chemistry knowledge and laboratory materials available to you.

Notes

This investigation is designed to be as "open-ended" as possible. The intent is to ask students to use almost everything they have learned throughout the course to structure an investigation completely, from problem to evaluation. The reaction selected uses inexpensive and easily handled materials which react in a way not predictable by simple theory. Students will tend to predict a simple single replacement reaction, but the evolution of heat and hydrogen gas soon convince them that this is unacceptable. The purpose of the investigation is the process — there are no simple answers to the questions they will propose. As in most "real" science, the questions just lead to further questions.

An easy way to set up the initial observation is to use 13×100 mm test tubes containing 0.5 mol/L $CuCl_{2(aq)}$ to a depth of 2 cm. Have students place a 12-cm piece of #8 gauge $Al_{(s)}$ wire into the test tube and observe the subsequent reaction.

- Radio Shack® sells 12 ft of #8 gauge $Al_{(s)}$ antenna ground wire for about $5.00 — the catalogue number is 15–035.

After observing the reaction, class discussion should be used to generate a list of problems to be investigated. This can be as directed as you like, but of course, the actual investigations will depend on the facilities available. It is effective to split the class into 5 to 6 groups, each dealing with a different problem, and then have them report results later, compiling a combined report. The following list of typical problems can be used.

1. What is the percent yield of the gas-producing reaction?
2. What is the percent yield of the solid-producing reaction?
3. What is the enthalpy of reaction of the system?
4. How do the acid-base characteristics change during the reaction?
5. What is the cell potential of this reaction?
6. What is the percent yield of the soluble product?

There is really no limit to this investigation or the discussion it can generate — set your own limits according to the abilities and interests of your class.

- Preparation of solution used
  0.500 mol/L $CuCl_{2(aq)}$ 116 g of $CuCl_2 \cdot 2H_2O$ / L

Caution

The reaction rate varies dramatically with solution concentration and solid surface area. **Try out** the demonstration reaction first, and dilute the CuCl2(aq) if the test tube gets too hot or if the reaction threatens to bubble out of the tube. You may substitute $Al_{(s)}$ strip, but do not use foil — you need to have a solid piece which can be pulled out of the tube if necessary, and foil reacts very quickly. Remember to wear a lab apron and safety glasses, and wash your hands thoroughly before leaving the laboratory.

Waste Disposal

The chemicals in this investigation pose no special hazard — solids can go in the (landfill) garbage, and solutions should be rinsed down a sink with plenty of water.

PERFORMANCE ASSESSMENT INSTRUMENTS

This section of the *Nelson Chemistry Teacher's Resource Masters* provides some basic instruments for monitoring and/or assessing the performance of students in homework and laboratory work. The criteria used in creating and evaluating the use of the instruments are efficiency, practicality, and logical consistency. There are many methods of monitoring and assessing students' work — this section suggests a basic set of checklists that should facilitate a wide range of classroom and laboratory approaches. For further discussion of performance assessment and portfolios, see *Nelson Chemistry Teacher's Edition* pages 43 to 48.

Monitoring Homework

A homework monitoring and assessment checklist is provided for all the Exercises, Lab Exercises, and Investigations in *Nelson Chemistry* Chapters 1 to 9, and another one for Chapters 10 to 15. This instrument can be printed back-to-back with the Performance Assessment sheet and used in different ways. For example, on the first day of class, the students may each be given a Homework/Performance Assessment sheet on which they write their name, teacher's name, course, and period. A student number is usually provided by numbering the students in order of the seating plan. At the beginning of any class period when homework is to be checked, student 1 is given the set of sheets to be distributed to the class. Sheets for absent students are put at the bottom of the pile as the sheets are passed around. Students who receive their sheet should orient it for the teacher to record a $\sqrt{}$, $\frac{1}{2}$, **X**, or **a** (for absent). These symbols/values are used because they are difficult to change. Red ink is recommended. At the end of the monitoring/assessment time, have students hand in their sheets in numerical order. A blank column is also provided for homework assessment by peers or by the teacher.

Monitoring homework and laboratory work has many benefits.

- Students who do their work are rewarded for their efforts — if only by recognition.
- Students who do not normally do their work would be more motivated to increase their effort.
- The quality of the work generally increases — as do the students' marks.
- The amount of time spent marking decreases, both because monitoring replaces some of the time needed for assessing, and because the quality of the work increases — it is faster to mark higher quality work.

There are nearly as many methods for monitoring and assessing homework as there are teachers. The above example is just one of the methods that have worked well. Adapt it as you see fit.

Monitoring Performance Assessment

Performance assessment is used herein to refer to laboratory or laboratory-like work. Problem solving, processes, and process skills are monitored and/or assessed using the instruments provided. In *Nelson Chemistry*, the headings for different sections of the investigation report are used to help list the processes involved in scientific problem solving. As you can see from the checklists, some extra categories of assessment are provided, e.g., clean-up and safety. Also note the list of Investigations and Lab Exercises at the end of this section.

Teachers using the performance assessment checklists in *Nelson Chemistry* have found that the quality of student work increased significantly. Each student is provided with an individual performance assessment sheet that is distributed and collected as described above for monitoring homework. In the classroom, the performance assessment sheet is used to check Lab Exercises and pre- and post-lab work. Each student starts with 10 out of 10 in each category and may gradually lose marks over the course of the year. The deduction method is much more efficient than providing an affirmative mark for every student for each process and skill. Experience indicates that this system works well — students respond well to it.

In the laboratory, the sheets can be displayed near the edge of each work station — facing in such a direction that teachers can easily place a $\sqrt{}$, $\frac{1}{2}$, **X**, or a under any process heading. For example, it is easy to check whether all students are wearing their safety glasses and laboratory aprons, and to place an **X** on the performance sheet of an offender. At the end of the lab period, students leave their sheet at their stations and you can circulate to see that everything is clean and organized.

Laboratory Report Assessment

Assessment of laboratory reports is only one part of performance assessment. Taking in and marking reports is not the only method to provide assessment on laboratory performance. Monitoring and/or assessing lab exercises as homework assignments checked in class is one method (see checklists). Peer marking of lab reports by using the sample reports in *Nelson Chemistry Solutions Manual* is another method.

A sample marking scheme for laboratory report assessment is provided herein. Each student starts off with 100%, then loses marks for specified errors and/or omissions and gains marks for exceptional work. You may duplicate these two pages and provide them to all students. Laboratory reports improve just by being specific about your expectations.

Homework Monitoring/
Homework Assessment
Chapters 1–9

Student Profile for _____ Number _____

Course _____ Teacher _____ Period _____

Exercise	$\sqrt{}, \frac{1}{2}, X, a$	Assessment	Lab Ex.	$\sqrt{}, \frac{1}{2}, X, a$	Assessment	Invest.	$\sqrt{}, \frac{1}{2}, X, a$	Assessment
C1. 1–28			1A			1.1		
C1. 29–31			2A			1.2		
C1. OV			2B			4.1		
C2. 1–16			3A			5.1		
C2. OV			3B			5.2		
C3. 1–7			3C			5.3		
C3. OV			4A			5.4		
C4. 1–8			4B			5.5		
C4. OV			5A			6.1		
C5. 1–6			5B			7.1		
C5. 7–11			5C			7.2		
C5. 12–20			5D			7.3		
C5. 21–28			5E			7.4		
C5. 29–32			5F			7.5		
C5. 33–35			6A			7.6		
C5. OV			6B			7.7		
C6. 1–30			6C			8.1		
C6. OV			7A			8.2		
C7. 1–14			7B			8.3		
C7. 15–17			7C			8.4		
C7. 18–22			7D			8.5		
C7. 23–25			7E			9.1		
C7. 26–43			7F			9.2		
C7. 44–45			7G			9.3		
C7. OV			7H			9.4		
C8. 1–32			7I			9.5		
C8. 33–37			7J					
C8. 38–40			7K					
C8. OV			7L					
C9. 1–4			8A					
C9. 5–7			8B					
C9. 8–13			8C					
C9. 14–19			8D					
C9. 20–23			8E					
C9. 24–30			9A					
C9. OV			9B					
			9C					
			9D					

Homework Monitoring/
Homework Assessment
Chapters 10–15

Student Profile for _____ Number _____

Course _____ Teacher _____ Period _____

Exercise	$\sqrt{}$, $\frac{1}{2}$, X, a	Assessment	Lab Ex.	$\sqrt{}$, $\frac{1}{2}$, X, a	Assessment	Invest.	$\sqrt{}$, $\frac{1}{2}$, X, a	Assessment
C10. 1–13			10A & 10B			10.1		
C10. 14–26			10C & 10D			10.2		
C10. 27–31			10E			10.3		
C10. 32			11A & 11B			10.4		
C10. 33–35			11C & 11D			11.1		
C10. OV			11E			12.1		
C11. 1–10			11F			12.2		
C11. 11–14			11G			12.3		
C11. 15–24			11H			12.4		
C11. OV			12A & 12B			13.1		
C12. 1–5			12C			13.2		
C12. 6–13			12D			13.3		
C12. 14			12E			13.4		
C12. 15–18			12F			13.5		
C12. 19–32			12G			13.6		
C12. 33–38			12H			13.7		
C12. 39–47			12I			14.1		
C12. 48			12J			14.2		
C12. OV			12K			14.3		
C13. 1–5			12L			14.4		
C13. 6–13			13A			14.5		
C13. 14–27			13B & 13C			15.1		
C13. 28–29			13D			15.2		
C13. 30–33			13E			15.3		
C13. 34–37			13F			15.4		
C13. OV			14A			15.5		
C14. 1–8			14B					
C14. 9–11			14C					
C14. 12–22			14D					
C14. 23–25			14E					
C14. 26–33			15A & 15B					
C14. OV			15C					
C15. 1–15			15D					
C15. 16–23			15E					
C15. 24–29			15F & 15G					
C15. 30–47			15H					
C15. OV								

NELSON CHEMISTRY
TEACHER'S RESOURCE MASTERS

Performance Monitoring/
Performance Assessment

Student Profile for _____ Number _____

Course _____ Teacher _____ Period _____

Activity	PS	Level (X—subtract 1, $\frac{1}{2}$—subtract $\frac{1}{2}$)						Summary	
		10/9	8/7	6/5	4/3	2/1	Bonus Initial	Mark	Comment
Purpose	A								
Problem	A								
Prediction	A								
Experimental Design	A								
Materials	B								
Procedure	B								
Manipulation	B								
Evidence	B								
Analysis	C, D								
Evaluation	E, F								
Synthesis	E								
Problem-Solving	A–F								
Clean-up	B								
Safety	B								
Listening Skills, etc.	A								
Scientific Attitudes	A–E								

Final Mark _____

Investigations: 1.1, 1.2, 4.1, 5.1, 5.2, 5.3, 5.4, 5.5, 6.1, 7.1, 7.2, 7.3, 7.4, 7.5, 7.6, 7.7, 8.1, 8.2, 8.3, 8.4, 8.5, 9.1, 9.2, 9.3, 9.4, 9.5, 10.1, 10.2, 10.3, 10.4, 11.1, 12.1, 12.2, 12.3, 12.4, 13.1, 13.2, 13.3, 13.4, 13.5, 13.6, 13.7, 14.1, 14.2, 14.3, 14.4, 14.5, 15.1, 15.2, 15.3, 15.4 and 15.5. (*Circle all investigations that are monitored/assessed.*)

Lab Exercises: 1A, 2A, 2B, 3A, 3B, 3C, 4A, 4B, 5A, 5B, 5C, 5D, 5E, 5F, 6A, 6B, 6C, 7A, 7B, 7C, 7D, 7E, 7F, 7G, 7H, 7I, 7J, 7K, 7L, 8A, 8B, 8C, 8D, 8E, 9A, 9B, 9C, 9D, 10A, 10B, 10C, 10D, 10E, 11A, 11B, 11C, 11D, 11E, 11F, 11G, 11H, 12A, 12B, 12C, 12D, 12E, 12F, 12G, 12H, 12I, 12J, 12K, 12L, 13A, 13B, 13C, 13D, 13E, 13F, 14A, 14B, 14C, 14D, 14E, 15A, 15B, 15C, 15D, 15E, 15F, 15G and 15H. (*Circle all lab exercises that are monitored/assessed.*)

Problem Solving Categories: **A**: Initiating and planning, **B**: Collecting and recording, **C**: Organizing and Communicating, **D**: Analyzing, **E**: Connecting, synthesizing, and integrating, **F**: Evaluating the process or outcomes

Performance Monitoring/
Performance Assessment

Student Profile for _____ Number _____

IB Course _____ Teacher _____ Period _____

| Activity | PS | Level (X—subtract 1, $\frac{1}{2}$—subtract $\frac{1}{2}$) | | | | | | Summary | |
		10/9	8/7	6/5	4/3	2/1	Bonus Initial	Mark	Comment
Purpose	1								
Problem	1								
Prediction	5, 1								
Experimental Design	5, 1								
Materials	5, 1								
Procedure	5, 1								
Manipulation	6								
Evidence	4, 1								
Analysis	3, 1								
Evaluation	3, 1								
Synthesis	3, 1								
Problem-Solving	3, 1								
Clean-up	2								
Safety	2								
Listening Skills, etc.	2								
Scientific Attitudes	2								

International Baccalaureate: **1**: Communication   **2**: Attitude   **3**: Interpretation
4: Observation   **5**: Planning   **6**: Manipulation

Investigations: 1.1, 1.2, 4.1, 5.1, 5.2, 5.3, 5.4, 5.5, 6.1, 7.1, 7.2, 7.3, 7.4, 7.5, 7.6, 7.7, 8.1, 8.2, 8.3, 8.4, 8.5, 9.1, 9.2, 9.3, 9.4, 9.5, 10.1, 10.2, 10.3, 10.4, 11.1, 12.1, 12.2, 12.3, 12.4, 13.1, 13.2, 13.3, 13.4, 13.5, 13.6, 13.7, 14.1, 14.2, 14.3, 14.4, 14.5, 15.1, 15.2, 15.3, 15.4 and 15.5. *(Circle all investigations that are monitored/assessed.)*

Lab Exercises: 1A, 2A, 2B, 3A, 3B, 3C, 4A, 4B, 5A, 5B, 5C, 5D, 5E, 5F, 6A, 6B, 6C, 7A, 7B, 7C, 7D, 7E, 7F, 7G, 7H, 7I, 7J, 7K, 7L, 8A, 8B, 8C, 8D, 8E, 9A, 9B, 9C, 9D, 10A, 10B, 10C, 10D, 10E, 11A, 11B, 11C, 11D, 11E, 11F, 11G, 11H, 12A, 12B, 12C, 12D, 12E, 12F, 12G, 12H, 12I, 12J, 12K, 12L, 13A, 13B, 13C, 13D, 13E, 13F, 14A, 14B, 14C, 14D, 14E, 15A, 15B, 15C, 15D, 15E, 15F, 15G and 15H. *(Circle all lab exercises that are monitored/assessed.)*

Performance Assessment   **Filtration**

Student	Mark	1	2	3	4	5	Checkpoints
							1. assembles apparatus
							2. prepares filter paper
							3. decants solution
							4. transfers and rinses solid
							5. dries and measures solid
							Use the general assessment instrument (*Teacher's Edition*) to determine the rating (0–5) and the specific checkpoints to note flaws or assistance given.
							or
							Assess each specific checkpoint (√ if OK) and determine the rating based on the number of checks.
							Investigations
							5.3
							7.1
							7.3
							14.1

Performance Assessment   Pipetting

Student	Mark	1	2	3	4	5	Checkpoints
							1. initial rinses
							2. pipet hand
							3. bulb hand
							4. set level of solution
							5. transfer of solution
							Use the general assessment instrument (*Teacher's Edition*) to determine the rating (0–5) and the specific checkpoints to note flaws or assistance given.
							or
							Assess each specific checkpoint (√ if OK) and determine the rating based on the number of checks.
							Investigations
							55
							7.7
							12.4
							15.5

Performance Assessment   **Titration**

Student	Mark	1	2	3	4	5	Checkpoints
							1. assembles, rinses and disassembles apparatus
							2. hand positions
							3. positions flask
							4. uses light background for flask
							5. obtains consistent results
							Use the general assessment instrument (*Teacher's Edition*) to determine the rating (0–5) and the specific checkpoints to note flaws or assistance given.
							or
							Assess each specific checkpoint (√ if adequate) and determine the rating based on the number of checks.
							Investigations
							7.7
							12.4
							15.5

Performance Assessment　　　　　**Procedure/Skill** _____

Student	Mark	1	2	3	4	5	Checkpoints	
							Investigations	

Nelson Chemistry
TEACHER'S RESOURCE MASTERS

LABORATORY REPORT ASSESSMENT

Investigation # and title:
Your name:
Station: Partner:
Block: Date of Inv.:
Purpose:

Assessment
–10% for *omitting* this section
    from the laboratory report
–5% for being *incomplete*
–5/10/15/20%+ for *late* reports

Problem
Write problem statement as a question.
Use a question mark.
Use variables i/a.

Assessment
–10% for *omitting* the Problem
–5% for a *poorly worded* problem statement

Prediction
Predict the answer to the problem as, "According to a (scientific concept or another authority), ..."
Provide the reasoning behind the prediction.

Assessment
–10/15% for *omitting* the Prediction
–5/10% for omitting the statement or the "*According to*" phrase
–5% for omitted or wrong *reasoning*
–5% for wrong *certainty* i/a

Experimental Design
Write a one to three sentence paragraph as an overview.
List assumptions being made in the experimental design and diagnostic tests being used.
Identify the variables i/a.

Assessment
–10% for *omitting* the Design
–5% for being too *short* or too *specific* or not being in paragraph form
–5% for omitting list of *variables* i/a.
+5% for *exceptional* design

Materials
List all chemicals and equipment in two columns, complete with quantities and sizes. Draw a diagram i/a.

Assessment
–10% for *omitting* the Materials
–5% for an *incomplete list*
+5% for listing materials *often missed* by others

Procedure
Write about six steps.
Number the steps.
Use capitals and periods in each statement.
Add any safety or waste disposal cautions and procedures i/a.
Add any special notes on control of variables i/a.

Assessment
–10% for *omitting* the Procedure
–5% for the procedure being too *long* or too *short* or for *missing steps*
–5% for not *numbering* the steps
–5% for not using *capitals* and *periods* in the procedural statements
–5% for not including *safety* and *waste* notes i/a
+5% for *exceptional* procedure

Key: i/a means "if appropriate".
         c/w means "complete with".

Evidence

List all qualitative and quantitative evidence required to answer the problem.

Prepare a well labelled table i/a.

Add any additional observations that collaborate the evidence or experimental design.

Assessment

–10% for *omitting* the Evidence

–5% for omitting some *evidence* or for poor organization of evidence

–5% for unacceptable *labelling* of tables or wrong precision

+5% for *exceptional* evidence or tables

Analysis

Complete any calculations.

Draw any graphs or complete any tables to re-present the evidence. Answer the Problem question.

Use "According to the evidence..." to answer the Problem question.

Assessment

–15% for *omitting* the Analysis

–5/10% for not showing *calculations* or for incorrect calculations

–5% for wrong *certainty*

–5% for unacceptable or omitted *labelling* of tables and graphs

–10% for not *answering* the question

–5% for not using the "*According to the evidence*" phrase

+5% for *exceptional* presentation

Evaluation

Evaluate the experimental design. Evaluate the procedure (c/w reason).

Evaluate the skills of the experimenter (c/w reason). Indicate alternative designs, procedures, and skills, and indicate the certainty of the results.

Calculate the percent difference (i/a).

Use the evidence or percent difference to evaluate the prediction (c/w reason).

Use the evaluation of the prediction to evaluate the authority used for the prediction (c/w reason).

Indicate the confidence you have in the your evaluation of the results.

Assessment

–20% for *omitting* the Evaluation

–5/10% for errors in the evaluation of the *design* or in the reasoning

–5/10% for errors in the evaluation of the *procedure* or in the reasoning

–5/10% for errors in the evaluation of the technological *skills* or in the reasoning

–5/10% for an error in or omitting the *percent difference* calculation (i/a)

–5/10% for errors in the evaluation of the *prediction* or in the reasoning

–5/10% for errors in the evaluation of the *authority* or in the reasoning

+5% for exceptional scientific *language*, e.g. expression of certainty

+5% for suggesting *exceptional* alternative experimental designs

General

+5/10% bonus for showing exceptional understanding of the purpose of scientific or technological research even though there may be several technical errors in the report. The relationship or connections among Problem, Prediction, Experimental Design, Evidence, Analysis, and Evaluation must be apparent. The report must typify logical reasoning and be well communicated.

General Assessment

–5% for wrong *certainty* (i.e. wrong significant digits)

–5% for not following SI or IUPAC rules of communication

–5/10% for not following the rules for communicating in English

–5–20% for a messy and disorganized report that continually requires extra time by the reader to interpret it

Listing of Investigations and Lab Exercises

X — Section to be completed in the investigation report

Activity	Problem	Prediction	Exp. Design	Materials	Procedure	Evidence	Analysis	Evaluation	Synthesis
Invest. 1.1						X	X		
Invest. 1.2						X	X	X	
Lab Ex. 1A							X	X	
Lab Ex. 2A							X		
Lab Ex. 2B		X						X	
Lab Ex. 3A							X		
Lab Ex. 3B		X					X	X	
Lab Ex. 3C							X		
Lab Ex. 4A							X		
Invest. 4.1		X				X	X	X	
Lab Ex. 4B		X	X						
Lab Ex. 5A							X		
Invest. 5.1			X	X	X	X	X		
Lab Ex. 5B							X	X	
Invest. 5.2			X			X	X		
Invest. 5.3				X	X	X	X		
Lab Ex. 5C							X		
Invest. 5.4					X				
Lab Ex. 5D								X	
Invest. 5.5		X				X	X	X	
Lab Ex. 5E		X					X	X	
Lab Ex. 5F							X	X	

Activity	Problem	Prediction	Exp. Design	Materials	Procedure	Evidence	Analysis	Evaluation	Synthesis
Lab Ex. 6A							X		
Lab Ex. 6B							X		
Invest. 6.1		X				X	X	X	
Lab Ex. 6C							X		
Lab Ex. 7A							X		
Invest. 7.1				X	X	X	X	X	
Invest. 7.2		X			X	X	X	X	
Lab Ex. 7B		X					X	X	
Lab Ex. 7C		X	X	X	X		X	X	
Invest. 7.3		X	X	X	X	X	X	X	
Lab Ex. 7D							X	X	
Invest. 7.4		X				X	X	X	
Lab Ex. 7E		X					X	X	
Lab Ex. 7F							X		
Lab Ex. 7G		X							
Invest. 7.5					X				
Invest. 7.6					X				
Lab Ex. 7H							X		
Invest. 7.7						X	X	X	
Lab Ex. 7I				X			X	X	
Lab Ex. 7J				X			X	X	
Lab Ex. 7K				X			X	X	
Lab Ex. 7L							X	X	

Activity	Problem	Prediction	Exp. Design	Materials	Procedure	Evidence	Analysis	Evaluation	Synthesis
Invest. 8.1			X	X	X	X	X	X	
Lab Ex. 8A							X		
Lab Ex. 8B							X		
Invest. 8.2						X	X	X	
Invest. 8.3						X	X		
Lab Ex. 8C		X					X	X	
Invest. 8.4						X	X		
Lab Ex. 8D		X					X	X	X
Invest. 8.5		X			X	X	X	X	
Lab Ex. 8E		X				X	X	X	
Invest. 9.1						X	X		
Invest. 9.2						X	X		
Invest. 9.3						X	X		
Lab Ex. 9A							X		
Invest. 9.4						X	X		
Lab Ex. 9B							X	X	
Invest. 9.5						X	X	X	
Lab Ex. 9C							X		
Lab Ex. 9D							X		

Activity	Problem	Prediction	Exp. Design	Materials	Procedure	Evidence	Analysis	Evaluation	Synthesis
Invest. 10.1		X		X		X	X	X	
Invest. 10.2				X		X	X		
Lab Ex. 10A		X							
Lab Ex. 10B		X					X	X	
Invest. 10.3			X	X	X	X	X	X	
Invest. 10.4		X		X	X	X	X	X	
Lab Ex. 10C							X		
Lab Ex. 10D							X	X	
Lab Ex. 10E							X	X	
Lab Ex. 11A							X		
Lab Ex. 11B		X					X	X	
Invest. 11.1					X	X	X	X	
Lab Ex. 11C		X					X	X	
Lab Ex. 11D							X		
Lab Ex. 11E			X						
Lab Ex. 11F							X		
Lab Ex. 11G		X					X		
Lab Ex. 11H		X					X	X	

Activity	Problem	Prediction	Exp. Design	Materials	Procedure	Evidence	Analysis	Evaluation	Synthesis
Invest. 12.1		X	X			X	X	X	
Invest. 12.2		X	X		X	X	X	X	
Lab Ex. 12A							X	X	
Lab Ex. 12B							X		X
Lab Ex. 12C							X		
Lab Ex. 12D		X	X						
Invest. 12.3		X	X			X	X	X	
Lab Ex. 12E								X	
Lab Ex. 12F							X		
Lab Ex. 12G							X		
Lab Ex. 12H							X	X	
Lab Ex. 12I							X		
Invest. 12.4		X				X	X	X	
Lab Ex. 12J							X	X	
Lab Ex. 12K							X		
Lab Ex. 12L							X		
Invest. 13.1		X				X	X	X	
Invest. 13.2		X			X	X	X	X	
Lab Ex. 13A						X	X		
Invest. 13.3		X				X	X		
Lab Ex. 13B							X		
Lab Ex. 13C		X							
Invest. 13.4		X	X		X	X	X	X	
Invest. 13.5						X	X		
Invest. 13.6		X			X	X	X	X	
Invest. 13.7		X		X		X	X	X	
Lab Ex. 13D		X					X	X	
Lab Ex. 13E		X	X	X					
Lab Ex. 13F		X							

Activity	Problem	Prediction	Exp. Design	Materials	Procedure	Evidence	Analysis	Evaluation	Synthesis
Invest. 14.1		X			X	X	X	X	
Lab Ex. 14A							X		
Lab Ex. 14B							X		
Invest. 14.2		X				X	X	X	
Invest. 14.3		X			X	X	X	X	
Lab Ex. 14C		X	X						
Invest. 14.4						X	X		
Lab Ex. 14D			X				X		
Lab Ex. 14E							X		
Invest. 14.5	X	X	X	X	X	X	X	X	
Invest. 15.1		X	X		X	X	X	X	
Lab Ex. 15A							X		
Lab Ex. 15B			X						
Lab Ex. 15C							X		
Lab Ex. 15D		X					X	X	
Invest. 15.2		X		X	X	X	X	X	
Lab Ex. 15E							X		
Invest. 15.3						X	X		
Invest. 15.4	X	X	X	X	X	X	X	X	
Lab Ex. 15F							X		
Lab Ex. 15G			X				X		
Invest. 15.5			X	X	X	X	X	X	
Lab Ex. 15H		X					X	X	X
Totals:	2	49	22	19	25	50	119	65	3

Investigations: 52
Lab Exercises: 82
Total: 134